An Introductio...

BOUNDARY
ELEMENT
METHODS

Prem K. Kythe

University of New Orleans

CRC Press
Taylor & Francis Group
Boca Raton London New York

CRC Press is an imprint of the
Taylor & Francis Group, an **informa** business

This book was typeset with $\TeX$, $\mathcal{A}_{\mathcal{M}}\mathcal{S}$-$\TeX$, and Mathematica.

$\TeX$ and $\mathcal{A}_{\mathcal{M}}\mathcal{S}$-$\TeX$ are registered trademarks of the American Mathematical Society. Mathematica is a registered trademark of Wolfram Research, Inc. Turbo C is a registered trademark of Borland International, Inc.; Quick C is a trademark of Microsoft Corporation; UNIX C is a registered trademark of AT&T.

Published 1995 by CRC Press
Taylor & Francis Group
6000 Broken Sound Parkway NW, Suite 300
Boca Raton, FL 33487-2742

© 1995 by Taylor & Francis Group, LLC
CRC Press is an imprint of Taylor & Francis Group, an Informa business

First issued in paperback 2019

No claim to original U.S. Government works

ISBN-13: 978-0-367-44914-8 (pbk)
ISBN-13: 978-0-8493-7377-0 (hbk)

Visit the Taylor & Francis Web site at
http://www.taylorandfrancis.com

and the CRC Press Web site at
http://www.crcpress.com

Library of Congress Cataloging-in-Publication Data

Catalog record is available from the Library of Congress

Errata

Prem K. Kythe:
An Introduction to Boundary Element Methods

- While running the programs, when asked, type the title with no spaces.

- Page 96, line 4, Equation 5.2 should read:
$$\nabla^2 u^* = -\delta(i\,), \; u^* = u^*(\mathbf{x},\mathbf{x}'), \; \mathbf{x}' = x_i \equiv i.$$

- Page 143, line 19, Example 5.14, and page 144, Figure 5.26:
  Replace $T_a = 100$ by $T_a = 50$.

- Page 145
  Line 3: node 13 should be 11
  Line 9: k should be β.

To Kiran, Dave, Jay, and Bharti:
With deep affection

The Author

Prem Kishore Kythe (formerly Kulshrestha), *b.* India, 29 January 1930; U. S. citizen; *Educ.:* Ph. D. (Mathematics), Aligarh Muslim University, India, 1961; *Prof. Exp.:* Faculty member at Aligarh Muslim University 1958–60, at Indian Institute of Technology, Bombay, 1960–67; at University of New Orleans (UNO) 1967–; Professor of Mathematics at UNO since 1974; Invited speaker at the NATO Advanced Institute for Automatic Translation (from Russian) at Venice, Italy, July 1962; UNESCO Fellow in Linguistic Data Processing (Machine Translation) 1963; Consultant, Institute of Human Learning, University of California at Berkeley 1964; Guest participant in the Summer Linguistics Institute, University of Washington, Seattle, WA, June-Aug 1963; Participant in Summer School on Complex Function Theory, University of Cork, Ireland, 1971; Visitor at Mathematics Department, Imperial College, London, Fall 1973; Visiting Professor, Department of Computer Science, University of Illinois at Urbana-Champaign, Spring 1986; Reviewer for the Bulletin, Institute of Mathematics, Academia Sinica; Reviewer for the JEMT ASME; Reviewer for NSF for a research grant; Reviewer for Zentralblatt für Mathematik, and for Applied Mechanics Reviews; Over 15 books/monographs and some research papers translated from Russian into English; Over 40 research publications in the areas of univalent functions, boundary value problems in continuum mechanics, differential equations, Laplace transform, wave theory, wave structure in rotating flows; Numerous citations in research articles, and graduate text/monographs; Listed in American Men and Women of Science; has been teaching finite element analysis and boundary element methods at UNO.

Contents

Preface

The finite element and the boundary element methods are the two most important developments in numerical mathematics to occur in this century. These techniques have evolved, within only a few decades, into a widely used and richly varied computational approach for various scientific and technological fields. There is a continuing rapid expansion in the number of users, range of applications and job market for these methods which cover the areas of solid mechanics, fluid mechanics, geomechanics, aeromechanics, biomechanics, coupled systems, chemical reactions, neutron flux, plasmas, acoustics, electric and magnetic fields, and many other related specialized fields.

Due to the growing need of introduction of numerical methods in engineering to solve boundary value problems, many engineering and mathematics graduate curricula now include courses in both finite and boundary element methods. Traditionally numerical methods have always been mathematically oriented. But it is equally important to merge the understanding of the basic methodology to actual physical, engineering and industrial problems. Recent advances in computational facilities, interactive computer usage and graphics output through sophisticated software have made it imperative to produce textbooks which encompass not only the numerical but the computational aspects. Both theory and applications are needed in applied courses. This is, in particular, true for a textbook in the boundary element methods.

This textbook has evolved out of lecture notes developed while teaching the course to graduate students in mathematics, mechanical, civil and electrical engineering, physics, and geophysics at the University of New Orleans since the Spring Semester, 1992. Much effort has gone into the organization of the subject matter in order to make the course attractive to students and the textbook easy to read. Although there are quite a number of published monographs and books now available, there has been a need for a textbook in this subject. This publication, based on the classroom experience, fulfils such a need. The mathematical contents of the book are simple enough for the average student to understand the methodology and the fine points of the boundary element methods. The computational part has been presented in detail to bring out the salient features of the methods. The chapters present a balanced course material, starting with the first chapter which reviews the important results from calculus, linear algebra and analysis. Chapter 2 is devoted to the weak variational formulation and the weighted residual methods, especially the Galerkin or Rayleigh–Ritz method, which eventually becomes the mathematical basis for the development of boundary element integral equations. The material of this chapter is supplemented by a short appendix on variational calculus in an effort to present the basic concepts for students to understand. Although one–dimensional problems are hardly ever treated through the boundary element methods, they are presented in Chapter 3 to impart the very basic ideas at the simplest setting. Chapter 4 is devoted to derivation of some fundamental solutions which in themselves are very helpful and valuable to lay the foundation for the theory of boundary element methods. Another appendix provides relevant material on potential theory.

Two–dimensional potential problems, involving the Laplace and the Poisson equations, are developed in Chapter 5. This chapter builds the foundation for the boundary element methods. Although the 'direct' method has been followed throughout the textbook, a section on the 'indirect' method is provided to show the difficulty in that approach. The problems associated with the domain integrals are raised in this chapter, but the dual reciprocity method (DRM), Fourier series method (FSM), and multiple reciprocity method (MRM), to transform the domain integrals into boundary integrals, are dealt with in Chapter 9.

Linear elasticity, related two– and three–dimensional boundary element methods, torsion problem, and axisymmetric problems are studied in Chapter 6. The Navier–Stokes equation and its development for the ocean acoustics and related Helmholtz equation are presented in Chapter 7. This chapter also includes underwater scattering and the associated CHIEF algorithm. Chapter 8 deals with the problems of groundwater flows and the D'Arcy equation; flows with singularities are analyzed. As mentioned earlier, Chapter 9 describes the DRM, FSM, and MRM for transforming domain integrals to the boundary. The last Chapter 10 investigates the transient problems, and presents the time marching process and the Laplace transform BEM for the Poisson equation.

This chapter ends with the transient DRM and MRM for the Fourier–Kirchhoff and the Helmholtz equation by using the θ–scheme. An appendix on the Gauss quadrature formulas is included, and computer programs in C are provided on a DOS diskette to complete the computational part of the textbook. These programs, tested on some benchmark problems, produce more accurate and better numerical results.

Almost every chapter has a set of problems toward the end. These have been selected from related literature; some of them may be challenging. Their role is to extend the theory to applications. Answers or hints to all those problems which do not have exact solutions are provided in as much detail as possible. The last section in each chapter provides references and bibliography for additional reading. Every effort has been made to make this list up to date. Sections, equations, and problems are numbered in decimal form. Thus, Example 5.8 is the eighth example in Chapter 5.

Although it is not possible to include every topic connected with the boundary element methods without making the text unwieldy, the topics included in this textbook are carefully selected and meticulously presented. The scope of the book can be determined from the table of contents.

This book has more than average flexibility for classroom use. For a one semester course, Chapters 1 through 5 and Chapter 9 form the basic core which should be supplemented by one of the Chapters 6, 7, 8, or 10, depending on the student's background and interest and the discretion of the instructor. Chapters 6 through 10 are substantially independent of each other, and any one of them may be chosen as a special final project which the student completes in as much detail as possible by using the references and related bibliography provided at the end of each chapter.

There are a few different notational styles available in literature on the boundary element methods. But experience has taught me that the one adopted by Professor C. A. Brebbia is best suited for textbooks. It is the virtue of simplicity that I have adopted the Brebbia notation throughout this textbook.

Acknowledgements

A number of people have provided me help, advice, constructive comments and suggestions, and pure encouragement during the period of preparation of this book. I wish to recognize and thank the following persons: Ms. Nancy Radonovich (UNO Library) for logistics support; Mr. John R. Stelly, and Dr. Lew Lefton for some TEXnical support; Mr. Jay K. Kythe for artistic support; and my friend and colleague Dr. Pratap Puri for valuable advice and constructive commentary, and Dr. Joseph E. Murphy for some helpful suggestions. Thanks are also due to my students in Math 4240 classes, especially to Mr. Praveen Ghantasala, who helped in programming and debugging. The staff at the CRC Press, and the editors deserve my sincere gratitude for support, encouragement, and valuable contributions. Special thanks are due to Mr. M. R. Schäferkotter of TechType Works for complete technical support and assistance in producing the camera–ready copy of the manuscript.

I would like to take this opportunity to thank my wife who has been a source of continuous support and motivation over the years, including the period of preparation of this book.

Notation

$B(x, y)$	beta function
b_i	body forces
B_i	domain integral
c	sound speed
C	boundary of a two–dimensional region R $(C = C_1 \cup C_2)$
$\tilde{C}$	discretized boundary C
$\tilde{C}_j$	j–th boundary element
$C^n(R)$	class of continuous functions with n-th derivative continuous in R
$D_{\mathbf{u}}f$	directional derivative of f along a unit vector $\hat{\mathbf{u}}$
e_{jk}	permutation symbol
$\mathrm{erf}(x)$	error function
$\mathrm{erfc}(x)$	complementary error function
E	Young's modulus
$E(m, \pi/2)$	complete elliptic integral of second kind
$E_n(z)$	exponential integral function of order n
f	approximating function (DRM)
$\bar{f}$	Laplace transform of f
$\tilde{F}$	Fourier transform
$F(\mathbf{x}, \theta)$	rotation transformation about z–axis
g	acceleration due to gravity
G	shear modulus $(= \mu)$
$G(\mathbf{x}, \mathbf{x}')$	Green's function
$H(x - x')$	Heaviside unit step function
H_n^1, H_n^2	Hankel functions
I_n	$n \times n$ identity matrix

J	Jacobian
J_0, J_1	Bessel functions
k	conductivity
k_n	eigenfrequencies
K_n	modified Bessel functions
$K(m, \pi/2)$	complete elliptic integral of first kind
L	differential operator; length
L_i	length of the boundary element $\tilde{C}_i$
$\hat{\mathbf{i}}, \hat{\mathbf{j}}, \hat{\mathbf{k}}$	unit vectors along the x, y, z axes
$m(x)$	moment (Chapter 3)
$m^*(x, x')$	virtual moment (Chapter 3)
M	number of interior cells
M, M_∞	Mach numbers (Chapter 8)
$\mathbf{n}$	outward normal vector
$\hat{\mathbf{n}}$	unit normal vector
N	number of nodes
p, q	$p = \partial u/\partial x, q = \partial u/\partial y$, in variational calculus
p	acoustic pressure (Chapter 7)
p^*	fundamental solution for $(\nabla^2 + k^2)p = 0$
$\mathbf{p}$	traction vector
p_i	tractions
$P^*_{rr}, \ldots$	axisymmetric virtual tractions (Chapter 6)
$P_n(\zeta)$	Legendre polynomials
q	flux $(= \partial u/\partial n)$
q^*	virtual flux $= \partial u^*/\partial n$
r	radial distance $(= |\mathbf{x} - \mathbf{x}'|)$
r	$= u_{xx}$, in variational calculus
R	a two–dimensional region
$\tilde{R}$	discretized two–dimensional (polygonal) region
s	variable of the Laplace transform
s	storativity (Chapter 8)
s	$= u_{xy}$, in variational calculus
$s(x)$	shear (Chapter 3)
$s^*(x, x')$	virtual shear (Chapter 3)
S	boundary surface of a three–dimension region V
$\tilde{S}_j$	(surface) boundary elements
t, t'	time
t	$= u_{yy}$, in variational calculus
T	torque (Chapter 6); time interval $[0, t_0]$ (Chapter 10)

T	transmissivity (Chapter 8); temperature
u	displacement
u^*	fundamental solution
$u^*(x, x')$	virtual displacement (Chapter 3)
$\mathbf{u}$	displacement vector
$\tilde{u}, \tilde{u}_n$	approximate values of u
u_m	eigenvectors (displacements)
$\bar{u}(\mathbf{x}, s)$	Laplace transform of u
$U^*_{rr}, \dots$	axisymmetric virtual displacements (Chapter 6)
$\mathbf{v}$	velocity vector $(= (u, v, w))$
V	a three–dimensional region
V_{ext}	exterior region
V_{int}	interior region
w_i	weights in logarithmic Gauss quadrature
W_i	weights in Gauss quadrature
$\mathbf{x}$	field point
$\mathbf{x}'$	source point
Y_0, Y_1	Bessel functions
α	variable of the Fourier transform
Γ	circulation
δ_{ij}	Kronecker delta
$\delta(\mathbf{x}, \mathbf{x}')$	or $\delta(x, x')$ Dirac delta function
δu	variation of u
Δt	time step
$\varepsilon_{\theta\theta}$	Hoop strain
ε_{ii}	normal strain $(= u_{i,i}, \; i = 1, 2, 3)$
ε_{ij}	shearing strain $(= (u_{i,j} + u_{j,i})/2\,)$
ϕ	velocity potential (piezometric head in Chapter 8)
ϕ_i	interpolation functions
$\phi(x, y)$	stress function (Prandtl theory of torsion)
ψ	stream function
ψ_α	shape functions, $\alpha = 1, 2, \dots$
κ_i	material constants (anisotropic D'Arcy's equation)
κ	permeability tensor
λ	scalar eigenvalue
λ, μ	Lame's constants
ν	Poisson's ratio
$\theta(x)$	rotation (Chapter 3)
$\theta^*(x, x')$	virtual rotation (Chapter 3)

$\theta_{\mathrm{u}}, \theta_{\mathrm{q}}$	weights in the θ–scheme
θ_{nm}	normal modes
ρ	density
σ	source density
$\sigma_x, \sigma_y, \sigma_z$	normal stresses
σ_{ij}	normal stresses if $i = j$; shearing stresses if $i \neq j$
$\sigma_{\theta\theta}$	Hoop stress
$\tau_{xy}, \ldots$	shearing stresses
ξ	boundary point on the scatterer S
ξ_1, ξ_2, ξ_3	coordinates of a triangle
ζ_i	Gauss points
ω	radian frequency
$\boldsymbol{\omega}$	vorticity vector
ω_{nm}	eigenfrequencies
$\boldsymbol{\Omega}$	rotation vector
∇	$= \mathrm{grad} = \hat{\mathbf{i}}\dfrac{\partial}{\partial x} + \hat{\mathbf{j}}\dfrac{\partial}{\partial y} + \hat{\mathbf{k}}\dfrac{\partial}{\partial z}$
∇^2	Laplacian $= \dfrac{\partial^2}{\partial x^2} + \dfrac{\partial^2}{\partial y^2} + \dfrac{\partial^2}{\partial z^2}$
$\mathcal{L}\{u(\mathbf{x}, t)\}$	Laplace transform of $u(\mathbf{x}, t)$
r, θ, z	polar cylindrical coordinates ($r > 0$, $0 \leq \theta \leq 2\pi$, $-\infty < z < \infty$)
r, θ, ϕ	spherical coordinates ($r > 0$, $0 \leq \theta \leq \pi$, $0 \leq \phi \leq 2\pi$)
x, y, z	or x_1, x_2, x_3, cartesian coordinates
$<, >$	inner product

Acronyms :

BE	boundary element
BEM	boundary element method
BIEM	boundary integral equation method
DRM	dual reciprocity method
FSM	Fourier series expansion method
MRM	multiple reciprocity method

Abbreviations :

BE Eq	boundary element equation
BI Eq	boundary integral equation
dof	degree(s) of freedom
Eq	equation when followed by its reference number
iff	if and only if

Introduction

1. Historical Background

Boundary element methods (BEM's) constitute a recent development in computational mathematics for the solution of boundary value problems in various branches of science and technology. These methods evolved from integral equation methods which are known as boundary integral equation methods (BIEM's). There were, in fact, two types of BIEM's which, though seemingly alike, had totally different approaches of formulation. One of them is the so called 'indirect' method, which relies on the physical aspect of the problem and involves the transformation of the boundary surface (or curve) by a surface (or curve) of sources/sinks of adjustable strengths. This approach was mostly used in the early research in fluid mechanics and electromagnetic fields. The other type of boundary integral equation formulation, now known as the 'direct' method, is based on the mathematical aspect of finding Green's function solutions of partial differential equations. The methodology is derived from the fact that once the Green's function of a given equation, together with the prescribed boundary conditions on a geometrically well–defined boundary, is known, then the solution of such a boundary value problem is also known in the form of an integral equation and can be numerically computed. More than the first half of this century has been devoted to research in obtaining the Green's functions. The breakthrough in this approach came when it was found that the use of the Green's function in the free space, the fundamental solution,

— the one which satisfies the governing equation at a point source without satisfying any prescribed boundary conditions — would reduce the dimension of the problem by unity in the homogeneous case, whereby the volume integrals reduce to surface integrals, the surface integrals to line integrals, and the governing differential equations to integral equations. The BIEM thereafter became very popular during the period from about 1960 through 1975. The work of M. A. Jawson (1963), G. T. Symm (1963), and T. A. Cruise and F. J. Rizzo (1968), using the BIEM, are the forerunners of what was to be known as the BEM in 1978 at the Second Conference at Southampton, UK, and on the publication of the first monograph–style text on the BEM by C. A. Brebbia (1978). Since then the research groups headed by C. A. Brebbia, P. K. Banerjee, R. P. Shaw, J. A. Liggett, H. Antes and others, have advanced the subject to a high degree of adaptability in numerical solution of boundary value problems in most of physical and technological fields, to the extent that this subject has entered into graduate curricula in many universities.

2. What is BEM?

The BEM is based on integral equation formulation of boundary value problems and requires discretization of only the boundary (surface or curve) and not the interior of the region under consideration. Unlike the 'domain type' methods, e.g., the finite difference or the finite element methods (FEM's), the order of dimensionality reduces by unity in boundary element formulation, thus simplifying the analysis and the computer code to a large extent by solving a small system of algebraic equations. This method is suitable for problems with complicated boundaries and unbounded regions. It has been used in different types of problems in physics and engineering, such as potential theory, elastostatics, elastodynamics, viscoelasticity, plasticity, thermoelasticity, fracture analysis, heat transfer, fluid flows, water waves, acoustics, groundwater flows, soil–structure and fluid–structure interfaces, and many other areas.

The 'direct' BEM which is presented and pursued in this textbook is based on the Galerkin–type weak variational formulation where the discretized boundary element equations are formulated with the help of the fundamental solutions of the related field equations. The prescribed boundary conditions are used to connect the unknown boundary values to the known values.

Because of the simplicity in computer code and data structure, the BEM is developing very rapidly in solving boundary value problems. However, like other numerical methods, it does not solve each and every problem; it even fails to tackle some of those problems which can be solved with ease by the finite difference or finite element methods. At present the limitations of the BEM are rooted in the complexity of nasty, nonlinear problems, and those where turbulent flows are involved. In future as research in this area develops, many unsolved problems will be solved by advanced boundary element techniques in conjunction with other numerical methods. There are hybrid methods available to solve some difficult problems by combining the BEM and the FEM. Any advancement in research in the numerical solutions of integral equations will help advance the scope of the BEM.

A practical comparison of the BEM with the FEM can be made on the basis of numerical and computational criteria. While the FEM uses very sophisticated techniques and the computational power of modern computers equipped with automatic grid generation and other computational advantages, the BEM begins with elementary solutions and relies on the use of computers only at the last stage. There are a lot of problems on both analysis and numerical methods that need solutions before the BEM can become a powerful tool. The current state in BEM research can be compared to the early days of the FEM when it was also thought to be very limited in scope in solving boundary value problems.

The simplicity in computer code in the development of the BEM is sometimes prone to errors when appropriate numerical techniques are not used. This in turn relies on the kind of mathematical basis that justifies the numerical technique. This textbook contains computer software for some very basic problems and it is hoped that the knowledge and the training gained through this book will prepare the readers for advanced, practical, industrial, and engineering problems.

3. Boundary Elements

Although we shall study boundary elements in detail in Chapters 5 and 6, it is appropriate to point out that, unlike the FEM where we find linear, quadratic and higher order elements, there are constant, linear, quadratic and higher order boundary elements in the BEM as a result of the reduction in dimensions. We shall represent in the figure given below different types of boundary elements with appropriate number of nodes. Thus, a constant element has only one node, a linear two nodes in both continuous and discontinuous types, and so on. In fact, there are infinitely many types of boundary elements. For example, we can choose a four–node element (§1.2), or a cubic spline element for a cubic geometry. The strength of the BEM lies in an unlimited choice of boundary elements.

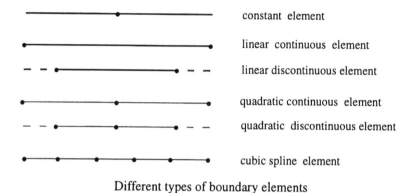

Different types of boundary elements

References and Bibliography for Additional Reading

H. Antes, *Anwendungen der Methode der Randelemente in der Elastodynamik und der Fluiddynamik*, B. G. Teubner, Stuttgart, 1988.

D. E. Beskos, *Introduction to the Boundary Element Methods*, Ch. 1 in Boundary Element Methods in Mechanics (D. E. Beskos, Ed.), North–Holland, Amsterdam, 1987.

C. A. Brebbia, *The Boundary Element Method for Engineers*, Pentech Press, London, 1978.

———— and J. Dominguez, *Boundary Elements*, Computational Mechanics Publications, Southampton and McGraw–Hill, New York, 1992.

————, J. C. F. Telles and L. C. Wrobel, *Boundary Element Techniques*, Springer–Verlag, Berlin, 1984.

T. A. Cruise and F. J. Rizzo, *A direct formulation and numerical solution of the general transient elasto–dynamic problem, I*, J. Math. Anal. Appl. **22** (1968), 244–259.

M. A. Jawson, *Integral equation methods in potential theory, I*, Proc. Roy. Soc., Ser. A **275** (1963), 23–32.

J. A. Liggett, *Fluid Mechanics*, Ch. 8 in Boundary Element Methods in Mechanics (D. E. Beskos, Ed.), North–Holland, Amsterdam, 1987.

G. T. Symm, *Integral equation methods in potential theory, II*, Proc. Roy. Soc., Ser. A **275** (1963), 33–46.

1

Mathematical Preliminaries

We will first cite some useful formulas and results, and discuss the notation used in this book. Proofs for these results can be found in textbooks on calculus, linear algebra, differential equations, and analysis. This chapter should, therefore, be either glanced through quickly or studied carefully depending on the reader's background.

1.1. Results from Calculus

Integration. The formula

$$\int_0^1 x^{\mu-1}\left(1-x^\lambda\right)^{\nu-1} dx = \frac{1}{\lambda}B(\mu/\lambda, \nu), \qquad (1.1a)$$

where $\Re\mu > 0$, $\Re\nu > 0$, $\lambda > 0$, and $B(x, y)$ is the beta function, yields, e.g., for positive integers m and n

$$\int_0^1 x^m(1-x)^n dx = \frac{m!\,n!}{(m+n+1)!}. \qquad (1.1b)$$

Integration by Parts. Let f, $g \in C^2(a, b)$, and $h(x) \in C(a, b)$. Then using integration by parts we have

$$\int_a^b g\frac{df}{dx} dx = -\int_a^b f\frac{dg}{dx} dx + fg\Big|_a^b, \qquad (1.2a)$$

$$\int_a^b g \frac{d}{dx}\left[\frac{df}{dx}\right] dx = \int_a^b g \frac{d^2 f}{dx^2} dx = -\int_a^b \frac{df}{dx}\frac{dg}{dx} dx + g\frac{df}{dx}\Big|_a^b, \quad (1.2b)$$

$$\int_a^b g \frac{d^2}{dx^2}\left[h(x)\frac{d^2 f}{dx^2}\right] dx = \int_a^b h(x)\frac{d^2 f}{dx^2}\frac{d^2 g}{dx^2} dx+$$

$$+ g\left[\frac{d}{dx}\left[h(x)\frac{d^2 f}{dx^2}\right] - h(x)\frac{d^2 f}{dx^2}\frac{dg}{dx}\right]_a^b.$$
$$(1.2c)$$

Differentiation under the integral sign. Let $f(x, \alpha)$ be a continuous function of two variables x and α where x varies from X_1 to X_2 and α varies between certain fixed limits α_1 and α_2. Let

$$F(x, \alpha) = \int_{X_1}^{X_2} f(x, \alpha)\, dx.$$

Then

$$\frac{dF}{d\alpha} = \int_{X_1}^{X_2} \frac{\partial f}{\partial \alpha}(x, \alpha)\, dx + \frac{dX_2}{d\alpha} f(X_2, \alpha) - \frac{dX_1}{d\alpha} f(X_1, \alpha). \quad (1.3)$$

Gradient Theorems. In two–dimensional problems, integration by parts is carried out through the gradient and divergence theorems. Note that

$$\nabla \equiv \hat{\mathbf{i}}\frac{\partial}{\partial x} + \hat{\mathbf{j}}\frac{\partial}{\partial y} + \hat{\mathbf{k}}\frac{\partial}{\partial z}, \qquad \nabla^2 \equiv \frac{\partial^2}{\partial x^2} + \frac{\partial^2}{\partial y^2} + \frac{\partial^2}{\partial z^2}.$$

Let V denote a body with the bounding surface S. Then

$$\iiint_V \nabla F\, dV = \iint_S \hat{\mathbf{n}} F\, dS, \qquad (1.4)$$

or

$$\iiint_V \left(\hat{\mathbf{i}}\frac{\partial F}{\partial x} + \hat{\mathbf{j}}\frac{\partial F}{\partial y} + \hat{\mathbf{k}}\frac{\partial F}{\partial z}\right) dV = \iint_S \left(\hat{\mathbf{i}}n_x + \hat{\mathbf{j}}n_y + \hat{\mathbf{k}}n_z\right) F\, dS.$$
$$(1.4a)$$

Divergence Theorem.

$$\iiint_V \nabla \cdot \mathbf{G}\, dV = \iint_S \mathbf{G} \cdot \hat{\mathbf{n}}\, dS, \qquad (1.5)$$

or

$$\iiint_V \left(\frac{\partial G_x}{\partial x} + \frac{\partial G_y}{\partial y} + \frac{\partial G_z}{\partial z} \right) dV = \iint_S (n_x G_x + n_y G_y + n_z G_z) \, dS. \tag{1.5a}$$

Also

$$\iiint_V \nabla F \cdot \mathbf{G} \, dV = - \iiint_V F \nabla \cdot \mathbf{G} \, dV + \iint_S F\mathbf{G} \cdot \hat{n} \, dA, \tag{1.5b}$$

or

$$\iiint_V \left(G_x \frac{\partial F}{\partial x} + G_y \frac{\partial F}{\partial y} + G_z \frac{\partial F}{\partial z} \right) dV =$$
$$- \iiint_V F \left(\frac{\partial G_x}{\partial x} + \frac{\partial G_y}{\partial y} + \frac{\partial G_z}{\partial z} \right) dV + \iint_S F (G_x n_x + G_y n_y + G_z n_z) \, dS. \tag{1.5c}$$

Note that $\hat{n}$ is the outward unit vector normal to the surface S of the region V; $n_{x_i} = \cos(x_i, \hat{n})$ are the rectangular components of $\hat{n}$, and $\cos(x_i, \hat{n})$ is the cosine of the angle between the positive $x_i \equiv (x, y, z)$ direction and the unit vector $\hat{n}$.

Green's Theorems. Different forms of the Green's theorem are as follows:

First Form:

$$\iiint_V (u \nabla^2 v + \nabla u \cdot \nabla v) \, dV = \iint_S u \nabla v \cdot \hat{n} \, dS. \tag{1.6a}$$

Second Form:

$$\iiint_V (u \nabla^2 v - v \nabla^2 u) \, dV = \iint_S \left(u \frac{\partial v}{\partial n} - v \frac{\partial u}{\partial n} \right) dS. \tag{1.6b}$$

Vector Form: In a two–dimensional space, we consider a region R with boundary C. Then

$$\iint_R \left(\nabla \times \mathbf{F} \cdot \hat{k} \right) dA = \int_C \mathbf{F} \cdot d\mathbf{r}, \tag{1.6c}$$

which has the scalar form

$$\iint_R \left(\frac{\partial N}{\partial x} - \frac{\partial M}{\partial y} \right) dA = \int_C M \, dx + N \, dy, \tag{1.6d}$$

where $dA = dx\, dy = dy\, dx$.

Green's first identity.

$$\iiint_V u\nabla^2 v\, dV + \iiint_V \nabla u \cdot \nabla v\, dV = \iint_R u\frac{\partial v}{\partial n}\, dS. \qquad (1.7)$$

Green's second identity. Because of symmetry in the middle term in (1.7), if we interchange u and v and subtract the corresponding terms, we get

$$\iiint_V \left(u\nabla^2 v - v\nabla^2 u\right)\, dV = \iint_R \left(u\frac{\partial v}{\partial n} - v\frac{\partial u}{\partial n}\right)\, dS. \qquad (1.8)$$

This is also known as Green's reciprocity theorem. The formula (1.7) can be easily derived by taking $\mathbf{G} = \nabla u$ in (1.5). In two–dimensional case these formulas become

$$\iint_R u\nabla^2 v\, dx\, dy + \iint_R \nabla u \cdot \nabla v\, dx\, dy = \int_C u\frac{\partial v}{\partial n}\, ds. \qquad (1.7a)$$

$$\iint_R \left(u\nabla^2 v - v\nabla^2 u\right)\, dx\, dy = \int_C \left(u\frac{\partial v}{\partial n} - v\frac{\partial u}{\partial n}\right)\, ds. \qquad (1.8a)$$

Directional Derivative. The directional derivative $D_{\mathbf{u}}f(x, y, z)$ of a function $f(x, y, z)$ in the direction of a unit vector $\hat{\mathbf{u}}$ is defined by

$$D_{\mathbf{u}}f(x, y, z) = \nabla f(x, y, z) \cdot \hat{\mathbf{u}}. \qquad (1.9)$$

Example 1.1:
By using the Green's formula (1.6d), we have

$$\iint_R \left[\frac{\partial u}{\partial x}\frac{\partial w}{\partial x} + \frac{\partial u}{\partial y}\frac{\partial w}{\partial y}\right] dx\, dy =$$

$$= \iint_R \left[\frac{\partial}{\partial x}\left(w\frac{\partial u}{\partial x}\right) + \frac{\partial}{\partial y}\left(w\frac{\partial u}{\partial y}\right)\right] dx\, dy -$$

$$- \iint_R \left(\frac{\partial^2 u}{\partial x^2} + \frac{\partial^2 u}{\partial y^2}\right) w\, dx\, dy$$

$$= \int_C \left(\frac{\partial u}{\partial x}w\, dy - \frac{\partial u}{\partial y}w\, dx\right) - \iint_R w\, \nabla^2 u\, dx\, dy$$

$$= -\iint_R w\, \nabla^2 u\, dx\, dy + \int_C \frac{\partial u}{\partial n}w\, ds. \qquad (1.10)$$

Here C denotes the boundary of a region R in the (x, y)–plane. Note that the last step in (1.10) follows from

$$
\begin{aligned}
\frac{\partial u}{\partial x} \, dy - \frac{\partial u}{\partial y} \, dx &= \left(\frac{\partial u}{\partial x} \frac{dy}{ds} - \frac{\partial u}{\partial y} \frac{dx}{ds} \right) ds \\
&= \left[\cos(s, y) \frac{\partial u}{\partial x} - \cos(s, x) \frac{\partial u}{\partial y} \right] ds \\
&= \left[\cos(n, x) \frac{\partial u}{\partial x} + \cos(n, y) \frac{\partial u}{\partial y} \right] ds.
\end{aligned}
$$

Indicial Notation. Let $i, j, k = 1, 2, 3$. Then for outer products, there are 9 combinations of u_i and v_j components, thus

$$
u_i v_j = v_j u_i = w_{ij}, \tag{1.11}
$$

where the symbols u_i and v_j can be in any order. Also,

$$
\sigma_{ij} n_k = s_{ijk}, \qquad \sigma_{ij} \varepsilon_{kl} = E_{ijkl}.
$$

For the inner products, we have for the scalar product:

$$
\phi = \mathbf{u} \cdot \mathbf{v} = u_1 v_1 + u_2 v_2 + u_3 v_3 = \sum_{i=1}^{3} u_i v_i. \tag{1.12}
$$

For the strain energy:

$$
\begin{aligned}
U &= \frac{1}{2} \left(\sigma_{11} \varepsilon_{11} + \sigma_{12} \varepsilon_{12} + \sigma_{13} \varepsilon_{13} + \sigma_{21} \varepsilon_{21} + \cdots + \sigma_{33} \varepsilon_{33} \right) \\
&= \sum_{i=1}^{3} \sum_{j=1}^{3} \sigma_{ij} \varepsilon_{ij}. \tag{1.13}
\end{aligned}
$$

For the matrix product for an $n \times n$ system:

$$
\sum_{j=1}^{n} A_{ij} x_j = y_i. \tag{1.14}
$$

We will, therefore, stipulate that a repeated suffix implies summation on it over its full range, provided that no suffix appears more than twice in an expression. Then the above three examples can be written simply as

$$
\phi = u_i v_i, \qquad U = \sigma_{ij} \varepsilon_{ij}, \qquad A_{ij} x_j = y_i.
$$

The suffixes on which summation is carried out are called dummy indices and the symbols used for them are interchangeable. Thus,

$$A_{ij}u_j + B_{ik}u_k = A_{il}u_l + B_{il}u_l. \qquad (1.15)$$

Note that the free indices must match on both sides of an equation, as seen in (1.15). In view of this indicial notation, e.g., the stress–strain relation becomes

$$\sigma_{ij} = C_{ijkl}\varepsilon_{kl}, \qquad (1.16)$$

and the Laplacian can be written as

$$\nabla^2 u = \frac{\partial^2 u}{\partial x_1^2} + \frac{\partial^2 u}{\partial x_2^2} + \frac{\partial^2 u}{\partial x_3^2} = \frac{\partial^2 u}{\partial x_i \partial x_i} = u_{,ii}, \qquad (1.17)$$

where a comma denotes partial differentiation (thus, $u_{,i} = \partial u/\partial x_i$, $u_{i,j} = \partial u_i/\partial x_j$, and so on). Note that in (1.15) the sequence of writing A, B is not important, thus $A_{ij}B_{kl} = B_{kl}S_{ij}$; however, $A_{ij} \neq A_{ji}$ unless A is symmetric in both i and j.

Kronecker delta (or substitution tensor).

$$\delta_{ij} = \begin{cases} 1, & \text{if } i = j \\ 0 & \text{if } i \neq j, \end{cases} \qquad (1.18)$$

such that

$$u_i\delta_{ij} = u_j, \quad u_iv_j\delta_{ik} = u_kv_j,$$

$$u_iv_j\delta_{ij} = u_iv_i, \quad \sigma_{ij}\varepsilon_{kl}\delta_{ik}\delta_{lj} = \sigma_{ij}\varepsilon_{ij}. \qquad (1.19)$$

The expressions in (1.19) are said to be contracted after multiplication by δ_{ij}, i.e., they are reduced in rank by 2 after each such multiplication. Note that the scalars are called quantities of rank 0, vectors (u_i) of rank 1, and so on, and C_{ijkl} of rank 4. Also note that $\delta_{ii} = 3$.

Permutation Symbol. It is defined as

$$e_{ijk} = \begin{cases} 0, & \text{if any two suffixes are equal} \\ 1, & \text{for cyclic suffix order 123 231 312} \\ -1, & \text{for anticyclic suffix order 321} \dots. \end{cases}$$

This symbol arises in the evaluation of determinants, thus

$$\det \|J_{ij}\| = e_{rst}J_{1r}J_{2s}J_{3t}, \qquad (1.20)$$

or as components in vector cross–products $\mathbf{W} = \mathbf{U} \times \mathbf{V}$, i.e.,

$$w_i = e_{ijk} u_j v_k. \tag{1.21}$$

Note that $e_{ijk} e_{jki} = 6$.

Transformation Rules. In the rectangular cartesian coordinate system, let a set of axes y_i be translated by h_i and rotated with respect to x_i. Let

$$c_{ij} = \frac{\partial y_i}{\partial x_j} = \frac{\partial x_j}{\partial y_i}$$

denote the direction cosine of rotation, i.e., $c_{12} = \cos y_1 x_2, \ldots$. Then we have the transformation

$$y_i = c_{ij}(x_j - h_j), \tag{1.22}$$

or

$$dy_i = c_{ij} dx_j = \frac{\partial y_i}{\partial x_j} dx_j = \frac{\partial x_j}{\partial y_i} dx_j. \tag{1.23}$$

Thus,

$$\begin{aligned}
ds^2 = dy_i dy_i &= \left(\frac{\partial y_i}{\partial x_j} dx_j \right) \left(\frac{\partial x_k}{\partial y_i} dx_k \right) \\
&= \left(\frac{\partial y_i}{\partial x_j} \frac{\partial x_k}{\partial y_i} \right) dx_j \, dx_k = \delta_{jk} \, dx_j dx_k = dx_j dx_j.
\end{aligned} \tag{1.24}$$

Cartesian tensors transform according to the following rule (here the primed quantities refer to the tensor components in y; unprimed in x):

Rank 0 (scalars):
$$\phi'(y) = \phi(x).$$

Rank 1 (vectors like force, displacements, gradients):
$$v_i' = c_{ij} v_j = \frac{\partial y_i}{\partial x_j} v_j = \frac{\partial x_j}{\partial y_i} v_j.$$

Rank 2 (stress, strain, conductivity):
$$\sigma_{ij}' = c_{ik} c_{jl} \sigma_{kl}.$$

Rank 3 (elastic compliances):
$$c_{ijkl}' = c_{im} c_{jn} c_{kp} c_{lq} C_{mnpq}. \tag{1.25}$$

Note that the coordinates x_i are, in general, not tensors, although dx_i is rank 1; also, c_{ij} is a rank 2 tensor. It is again worth remembering that although we can write vectors as column matrices and rank 2 tensors as matrix arrays, the converse is in general not true.

Example 1.2:
Invariance of strain energy under transformation:

$$2U = \sigma'_{ij}\varepsilon'_{ij}$$
$$= c_{ik}c_{jl}\sigma_{kl\_}imc_{jn}\varepsilon_{nn}$$
$$= \sigma_{mn}\varepsilon_{mn} = \sigma_{ij}\varepsilon_{ij}.$$

Note that $||c_{ij}|| = I = 1.$ ∎

1.2. Interpolation Functions

The transformation from the global coordinate system x to a local system ξ, when the origin is at the center of an element and scaled such that $\xi = -1$ is at the left–end node and $\xi = 1$ at the right–end node, is given by (see Fig. 1.1 for notation)

$$\xi = \frac{2x - (x_i + x_{i+1})}{h_i}. \qquad (1.26)$$

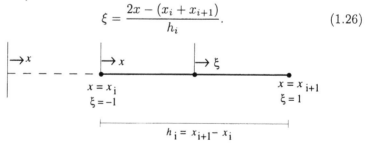

Fig. 1.1. Line element.

The ξ–coordinates are called normal (or natural) coordinates since these coordinates are normalized (non–dimensional) with values between -1 and 1. The formula (1.26) establishes the transformation between points x ($x_i \leq x \leq x_{i+1}$) and the points ξ ($-1 \leq x \leq 1$).

The (Lagrange) interpolation functions for a linear element in normal coordinate system are:

$$\phi_1 = \frac{1}{2}(1-\xi)$$

$$\phi_2 = \frac{1}{2}(1+\xi). \tag{1.27}$$

2 - node element
(linear)

$$\phi_1 = -\frac{1}{2}\xi(1-\xi)$$

$$\phi_2 = (1-\xi^2) \tag{1.28}$$

$$\phi_3 = \frac{1}{2}\xi(1+\xi).$$

3 - node element
(quadratic)

$$\phi_1 = -\frac{9}{16}(1-\xi)(1/9-\xi^2)$$

$$\phi_2 = \frac{27}{16}(1-\xi^2)(1/3-\xi)$$

$$\phi_3 = \frac{27}{16}(1-\xi^2)(1/3+\xi) \tag{1.29}$$

$$\phi_4 = -\frac{9}{16}(1+\xi)(1/9-\xi^2).$$

4 - node element
(cubic)

Note that the interpolation functions $\phi_i(\xi_j)$ are chosen such that

$$\phi_i(\xi_j) = \delta_{ij} = \begin{cases} 1, & \text{if } i = j \\ 0, & \text{if } i \neq j, \end{cases}$$

where ξ_j denotes the ξ–coordinate of the j-th node of the element, ϕ_i ($i = 1, 2, \cdots, n$) are polynomials of degree $n-1$ (n being the number of nodes in the element), and δ_{ij} is the Kronecker delta.

Another method of constructing the functions $\phi_i(\xi_j)$ is as follows: Form the product of $n-1$ linear functions $\xi - \xi_j$ ($j = 1, \cdots, i-1, i+1, \cdots, n; j \neq i$), i.e.,

$$\phi_i = c_i(\xi - \xi_1)(\xi - \xi_2) \cdots (\xi - \xi_{i-1})(\xi - \xi_{i+1}) \cdots (\xi - \xi_n).$$

Here $\phi_i = 0$ at all nodes except the node i. Now, determine c_i such that $\phi_i = 1$ at $\xi = \xi_i$, i.e.,

$$c_i = \left[(\xi_i - \xi_1)(\xi_i - \xi_2) \cdots (\xi_i - \xi_{i-1})(\xi_i - \xi_{i+1}) \cdots (\xi_i - \xi_n) \right]^{-1}.$$

This gives the required interpolation functions ϕ_i associated with the node i as

$$\phi_i(\xi) = \frac{(\xi - \xi_1)(\xi - \xi_2) \cdots (\xi - \xi_{i-1})(\xi - \xi_{i+1}) \cdots (\xi - \xi_n)}{(\xi_i - \xi_1)(\xi_i - \xi_2) \cdots (\xi_i - \xi_{i-1})(\xi_i - \xi_{i+1}) \cdots (\xi_i - \xi_n)}. \tag{1.30}$$

1.3. Distributions

Inner Product. In a one–dimensional euclidean space, the *inner product* $\langle f_1, f_2 \rangle$ of two functions f_1, f_2 is defined as

$$\langle f_1, f_2 \rangle = \int_a^b f_1(x) f_2(x) \, dx \tag{1.31}$$

for $x \in (a, b)$. In a two–dimensional space, the definition becomes

$$\langle f_1, f_2 \rangle = \iint_R f_1(x, y) f_2(x, y) \, dx \, dy \tag{1.32}$$

for $(x, y) \in R$. The properties of the inner product are:

$$\langle f_1, f_2 \rangle = \langle f_2, f_1 \rangle, \tag{1.33}$$

$$c \langle f_1, f_2 \rangle = \langle c f_1, f_2 \rangle, \tag{1.34}$$

$$\langle f_1, f_2 + f_3 \rangle = \langle f_1, f_2 \rangle + \langle f_1, f_3 \rangle, \tag{1.35}$$

the norm:

$$\|f\| = \sqrt{\langle f, f \rangle}, \tag{1.36}$$

the translation property:

$$\langle f_1(x - a), f_2(x) \rangle = \langle f_1(x), f_2(x + a) \rangle, \tag{1.37}$$

and the differentiation property (which is another way of writing integration by parts formula (1.2a)):

$$\left\langle \frac{df_1}{dx}, f_2 \right\rangle = -\left\langle f_1, \frac{df_2}{dx} \right\rangle + f_1 f_2 \Big|_a^b. \tag{1.38}$$

Self–adjoint Operators. An *operator* L, when applied to a function u, produces another function f:

$$L u = f. \tag{1.39}$$

An operator is said to be *linear* if

$$L(\alpha u_1 + \beta u_2) = \alpha L u_1 + \beta L u_2. \tag{1.40}$$

In a two–dimensional space, an operator L^* is said to be *adjoint* to the operator L iff

$$\iint_R w \, Lu \, dx \, dy - \iint_R u \, L^* w \, dx \, dy \tag{1.41}$$

is a function of u, w and their derivatives evaluated on the boundary.

The definition (1.41) is written as

$$\langle Lu, \, w \rangle = \langle u, \, L^* w \rangle + \int_C \{F(w)G(u) - F(u)G^*(w)\} \, ds. \tag{1.42}$$

In a one–dimension space, (1.42) becomes

$$\langle Lu, \, w \rangle = \langle u, \, L^* w \rangle + F(w)G(u) - F(u)G^*(w). \tag{1.43}$$

If the operator L^* is *adjoint* to the operator L, i.e., if $L^* = L$, then L is said to be *self–adjoint*. In this case $G^* = G$ also. Thus, selfadjointness of an operator is similar to the symmetry of a matrix.

The formulas (1.42) and (1.43) represent the variational formulation for the equation (1.39). Here $F(u)$ generates *essential boundary conditions*, whereas $G(u)$ generates *nonessential boundary conditions*, also called the natural boundary conditions, which must be enforced at some point so as to have a unique solution. See §2.5 for a detailed account of the procedure and specification of the essential and natural boundary conditions.

In a two–dimensional space, the positive–definite property of a self–adjoint operator L is defined by

$$\iint_R u \, Lu \, dx \, dy > 0, \tag{1.44a}$$

or

$$\langle u, Lu \rangle > 0, \tag{1.44b}$$

for all nontrivial u.

Analogous properties of symmetry and positive definiteness for a matrix are defined as follows: Let $A = [a_{ij}]$ be a square $(n \times n)$ matrix $(i, j = 1, \cdots, n)$. Then A is symmetric when $A^T = A$, where $A^T = [a_{ji}]$ is the transpose of A.

We shall denote a vector X and its transpose X^T by

$$X = \left\{ \begin{array}{c} x_1 \\ \vdots \\ x_n \end{array} \right\}, \qquad X^T = [x_1 \cdots x_n]. \tag{1.45}$$

Note that the vector X is an $n \times 1$ matrix (also called a column vector), and X^T is a $1 \times n$ matrix (also called a row vector). The inner (scalar) product of two vectors X and Y is defined by

$$\langle X, Y \rangle = X^T Y = x_1 y_1 + \cdots + x_n y_n.$$

Symmetry of a matrix A can also be defined by requiring that

$$\langle Y, AX \rangle = \langle X, AY \rangle^T$$

for arbitrary vectors X, Y. Since $(BC)^T = C^T B^T$, we have

$$\langle Y, AX \rangle = Y^T AX, \qquad \langle X, AY \rangle = X^T AY$$

$$\Rightarrow \langle X, AY \rangle^T = Y^T A^T X.$$

Hence $A^T = A$.

A matrix A is positive definite if $\langle X, AX \rangle > 0$ for all vectors X. The positive definiteness is an important property for constructing variational expressions and solution schemes in BEM.

Functionals. A functional is an expression of the form

$$I(u) = \int_a^b F(x, u, u') \, dx, \tag{1.46}$$

where $F(x, u, u')$ is a known function of x, u and $u' = du/dx$. Although the value $I(u)$ depends on u, yet for a given u, the value of $I(u)$ is a scalar quantity. The functional $I(u)$ represents a function defined by integrals whose arguments are themselves functions. In fact, this functional is an operator I

which maps u into a scalar value $I(u)$. Its domain is the set of all functions $u(x)$, whereas its range which is a subset of the real field is the set of images of all functions u under the map I.

We will define a linear functional by $l(u)$ iff

$$l(\alpha u + \beta v) = \alpha l(u) + \beta l(v). \tag{1.47}$$

A functional $b(u, v)$ is said to be bilinear if it is linear in both u and v. See §2.5 for linear and bilinear functionals in the weak variational formulation of boundary value problems.

1.4. Boundary Conditions

We shall consider a partial differential equation

$$L u = f, \tag{1.48}$$

where the operator L is, in general, of order 2:

$$L \equiv a\frac{\partial^2}{\partial x^2} + b\frac{\partial^2}{\partial x \partial y} + c\frac{\partial^2}{\partial y^2} + F\left(x, y, u, \frac{\partial}{\partial x}, \frac{\partial}{\partial y}\right), \tag{1.49}$$

f is a given function of x, y, and a, b, c are in general functions of x, y. (If a, b, c also depend on u, then the operator L is non–linear.)

Let u be the solution of Eq (1.48) in a two–dimensional region R, and satisfy the following homogeneous conditions on the boundary C of R:

Dirichlet boundary condition : $u = 0$, $\qquad$ (1.50a)

Neumann boundary condition : $\dfrac{\partial u}{\partial n} = 0$, $\qquad$ (1.50b)

Mixed boundary condition : $\dfrac{\partial u}{\partial n} + k(s)\, u = 0$,

$\qquad$ (1.50c)

where s is the arc–length along C, measured from some fixed point on C, and $\partial u/\partial n$ represents differentiation along the outward normal $\mathbf{n}$ to C.

Example 1.3 (One–dimensional):
Let

$$L\,u = \frac{d^2u}{dx^2}, \ 0 < x < 1.$$

Then, using the formula (1.2b) for integration by parts, we have

$$\langle w, L\,u \rangle = \int_0^1 wL\,u\,dx$$

$$= \left[w\frac{du}{dx} \right]_0^1 - \int_0^1 \frac{dw}{dx}\frac{du}{dx}\,dx$$

$$= \int_0^1 u\frac{d^2w}{dx^2}\,dx + \left[w\frac{du}{dx} - u\frac{dw}{dx} \right]_0^1$$

$$\equiv \langle u, L^*\,w \rangle + [F(w)G(u) - F(u)G^*(w)]_0^1,$$

$$L^* = L, \ \text{(operator is self–adjoint)}$$

$$F(w) = w, \ G(u) = du/dx \ \text{(natural boundary conditions)},$$

$$F(u) = u, \ G^*(w) = dw/dx \ \text{(essential boundary conditions)}. \ \blacksquare$$

Example 1.4 (One–dimensional):
Let

$$L\,u = \frac{d^2u}{dx^2} + \frac{du}{dx} + u.$$

Then

$$\langle w, L\,u \rangle =$$

$$= \int_0^1 w\left[\frac{d^2u}{dx^2} + \frac{du}{dx} \right] dx + \int_0^1 uw\,dx$$

$$= w\left[\frac{du}{dx} + u \right]_0^1 - \int_0^1 \frac{dw}{dx}\left(\frac{du}{dx} + u \right) dx + \int_0^1 uw\,dx$$

$$= w\left[\frac{du}{dx} + u \right]_0^1 - \int_0^1 \frac{dw}{dx}\frac{du}{dx}\,dx - \int_0^1 \frac{dw}{dx}u\,dx + \int_0^1 uw\,dx$$

$$= \left[w\frac{du}{dx} + uw - \frac{dw}{dx}u \right]_0^1 + \int_0^1 \frac{d^2w}{dx^2}u\,dx - \int_0^1 \frac{dw}{dx}u\,dx +$$

$$+ \int_0^1 uw\,dx$$

$$= \int_0^1 \left[\frac{d^2w}{dx^2} - \frac{dw}{dx} + w \right] u\,dx + \left[w\left(\frac{du}{dx} + u\right) - u\frac{dw}{dx} \right]_0^1$$

$$\equiv \langle u, L^* w \rangle + [F(w)G(u) - F(u)G^*(w)]_0^1 .$$

Here $L \neq L^*$, and

$$F(u) = u \text{ (essential boundary condition, prescribed)},$$
$$G(u) = du/dx \text{ (natural boundary condition, prescribed)}.$$

$G(u)$ and $G^*(w)$ are different since $L \neq L^*$. ■

Example 1.5 (Two–dimensional) :
Let $L = -\nabla^2$, and consider the Poisson equation

$$Lu \equiv -\nabla^2 u \equiv -\frac{\partial^2 u}{\partial x^2} - \frac{\partial^2 u}{\partial y^2} = f.$$

Now

$$\iint_R (wL u - uL^* w) \, dA = \iint_R (u\nabla^2 w - w\nabla^2 u) \, dA$$
$$= \int_C \left(u\frac{\partial w}{\partial n} - w\frac{\partial u}{\partial n} \right) ds,$$

by using the second form of the Green's Theorem (1.8a). Thus, the operator L is self–adjoint.

Again,

$$\iint_R u \left(-\nabla^2 u\right) \, dA = -\iint_R \nabla \cdot (u\nabla u) \, dA + \iint_R |\nabla u|^2 \, dA$$
$$= -\int_C u\frac{\partial u}{\partial n} \, ds + \iint_R |\nabla u|^2 \, dA, \tag{1.51}$$

by divergence theorem (1.5) and Problem 1.1. If u satisfies the boundary conditions (1.50a) or (1.50b), then (1.51) gives

$$\iint_R u \left(-\nabla^2 u\right) \, dA = \iint_R |\nabla u|^2 \, dA > 0,$$

which implies that the operator L is positive definite. If u satisfies the mixed boundary conditions (1.50c), then (1.51) gives

$$\iint_R u \left(-\nabla^2 u\right) \, dA = \int_C k(s)u^2 \, ds + \iint_R |\nabla u|^2 \, dA > 0,$$

provided that $k(s) > 0$. This implies that the operator L is positive definite. Note that

$$|\nabla u|^2 = \nabla u \cdot \nabla u = \left(\frac{\partial u}{\partial x}\right)^2 + \left(\frac{\partial u}{\partial y}\right)^2.$$

In Examples 1.3–1.5 we have also derived the variational formulation for the equations under consideration with the assumption that the appropriate essential and natural boundary conditions are prescribed. Since this technique is at the very basis for the BEM, we shall discuss the technique for the variational formulation in the next chapter.

1.5. Dirac Delta Function

For an impulsive force acting at a point x', the Dirac delta function is defined as

$$\delta(x, x') \stackrel{\text{def}}{=} \begin{cases} 0, & \text{if } x \neq x' \\ \infty, & \text{if } x = x'. \end{cases} \tag{1.52}$$

Note that $\delta(x, x')$ is also denoted as $\delta(x - x')$.) Mathematically, this function is a generalized function defined by

$$\delta(x, x') = \frac{1}{\pi} \lim_{n \to \infty} \frac{\sin(n(x - x'))}{x - x'}, \tag{1.53}$$

which, in the limit, is zero at every point $x \neq x'$ and infinite at $x = x'$. Thus, it represents a point singularity at the source point x', i.e.,

$$\delta(x, x') \begin{cases} \to \infty & \text{as } x \to x' \\ = 0, & \text{otherwise.} \end{cases} \tag{1.54}$$

This function has been used in defining a concentrated force in solid and fluid mechanics, a point mass in the theory of gravitational potential, a point charge in electronics, an impulsive force in acoustics, and other similar situations in physics and mechanics.

A basic property of this function is:

$$\iiint_V \delta(\mathbf{x}, \mathbf{x}')\phi(\mathbf{x}) \, dV = \begin{cases} \phi(\mathbf{x}'), & \text{if } \mathbf{x}' \in V \\ 0, & \text{if } \mathbf{x}' \notin V, \end{cases} \tag{1.55}$$

where V is a three–dimensional region in the neighborhood of a point $\mathbf{x}' = (x', y', z')$; $\mathbf{x} = (x, y, z)$ is some other point in V, and $\phi \in C(V)$. If $\phi(\mathbf{x}) = 1$ in (1.55), we get

$$\iiint_V \delta(\mathbf{x}, \mathbf{x}')\, dV = \begin{cases} 1, & \text{if } \mathbf{x}' \in V \\ 0, & \text{if } \mathbf{x}' \notin V. \end{cases} \tag{1.56}$$

Also,

$$\iiint_V \delta\left(k(\mathbf{x}, \mathbf{x}')\right) d\mathbf{x} = \iiint_V \frac{1}{k} \delta(\mathbf{x}, \mathbf{x}')\, d\mathbf{x}. \tag{1.57}$$

In one–dimensional case the basic property (1.55) becomes

$$\int_a^b \delta(x, x')\phi(x)\, dx = \phi(x'), \tag{1.58}$$

where a, b can be $-\infty$ or $+\infty$ for unbounded intervals. Thus,

$$\delta(-x) = \delta(x), \quad x\delta(x) = 0, \quad \delta(ax) = a^{-1}\delta(x)\ a > 0, \tag{1.59}$$

and

$$\frac{dH(x - x')}{dx} = \delta(x, x'), \quad \text{where } H(x - x') = \begin{cases} 0, & \text{if } 0 < x \leq x' \\ 1, & \text{if } x > x', \end{cases} \tag{1.60}$$

is the Heaviside unit step function. Note that $\delta(x, x')$ has the units $[(\text{ft})^{-1}]$ or $[(\text{m})^{-1}]$; $\delta(t, t')$ has the units $[(\text{sec})^{-1}]$ if t, t' denote time, and, in general, $\delta(\mathbf{x}, \mathbf{x}')$ has the units such that $\iiint_V \delta(\mathbf{x}, \mathbf{x}')\, dV = 1$. Also, we sometimes write $\delta(x)$ for $\delta(x, 0)$.

1.6. Fourier Series

Definition. A function $f(x)$ is said to be a periodic function of period $2a$ if

$$f(x + 2a) = f(x), \quad a > 0. \tag{1.61}$$

Let $f(x)$ be a periodic and integrable function of period 2π, and let it be represented in the interval $[-\pi, \pi]$ by the infinite trigonometric series

$$f(x) = a_0 + \sum_{n=1}^{\infty} \left(a_n^1 \cos nx + a_n^2 \sin nx\right), \tag{1.62}$$

where a_0, a_n^1, a_n^2 are called the Fourier coefficients for $n = 1, 2, \ldots$ and are determined by

$$a_0 = \frac{1}{2\pi} \int_{-\pi}^{\pi} f(x)\, dx,$$

$$a_n^1 = \frac{1}{\pi} \int_{-\pi}^{\pi} f(x) \sin nx\, dx, \qquad (1.63)$$

$$a_n^2 = \frac{1}{\pi} \int_{-\pi}^{\pi} f(x) \cos nx\, dx,$$

The following orthogonality relations hold:

$$\int_{-\pi}^{\pi} \sin nx\, dx = 0, \qquad \int_{-\pi}^{\pi} \cos nx\, dx = 0,$$

$$\int_{-\pi}^{\pi} \sin nx \sin mx\, dx = \pi \delta_{nm}, \qquad \int_{-\pi}^{\pi} \cos nx \cos mx\, dx = \pi \delta_{nm}, \qquad (1.64)$$

$$\int_{-\pi}^{\pi} \sin nx \cos mx\, dx = 0.$$

If $f(x)$ is a function of period $2a$ in the interval $[-\pi, \pi]$, then $f(y)$ will have period 2π if

$$y = \frac{\pi x}{a}. \qquad (1.65)$$

Substituting (1.65) in (1.62) we get

$$f(x) = a_0 + \sum_{n=1}^{\infty} a_n^1 \cos \frac{n\pi x}{a} + a_n^2 \sin \frac{n\pi x}{a}, \qquad (1.66)$$

where

$$a_0 = \frac{1}{2a} \int_{-a}^{a} f(x)\, dx,$$

$$a_n^1 = \frac{1}{a} \int_{-a}^{a} f(x) \cos \frac{n\pi x}{a}\, dx, \qquad (1.67)$$

$$a_n^2 = \frac{1}{a} \int_{-a}^{a} f(x) \sin \frac{n\pi x}{a}\, dx.$$

If $f(x)$ is an even function, then

$$a_0 = \frac{1}{a} \int_{0}^{a} f(x)\, dx, \quad a_n^1 = \frac{2}{a} \int_{0}^{a} f(x) \cos \frac{n\pi x}{a}\, dx, \quad a_n^2 = 0. \qquad (1.68)$$

If $f(x)$ is an odd function, then

$$a_0 = a_n^1 = 0, \quad a_n^2 = \frac{2}{a} \int_{0}^{a} f(x) \sin \frac{n\pi x}{a}\, dx. \qquad (1.69)$$

Fourier Theorem I. If $f(x)$ is a bounded periodic function of period 2π, which is piecewise continuous on the interval $(-\pi, \pi)$, then its Fourier series (1.66) converges to $f(x)$ at all points where $f(x)$ is continuous and converges to the average value

$$\frac{1}{2}\left[f(x+) + f(x-)\right] \tag{1.70}$$

at each point x $(-\infty < x < \infty)$, where the one–sided derivatives $f'(x+)$ and $f'(x-)$ both exist.

Definition. A function f is said to be of bounded variation on an interval $a \le x \le b$ if it is the sum of two monotone functions g and h:

$$f(x) = g(x) + h(x), \quad a \le x \le b, \tag{1.71}$$

such that g is nondecreasing and h is nonincreasing. Each such function f has the following properties:(i) one–sided limits $f(x+)$ and $f(x-)$ from the interior of the interval exist at each point; (ii) f has at most countably many discontinuities in the interval, and (iii) f is bounded and integrable over the interval.

Fourier Theorem II. Let $f(x)$ denote a periodic function of period 2π whose integral from $-\pi$ to π exists. If that integral is improper, let it be absolutely convergent. Then at each point x which is interior to an interval on which f is of bounded variation, the Fourier series for the function f converges to the average value (1.70).

The asymptotic behavior of the Fourier coefficients of a periodic function $f(x)$ is given by the following theorem:

Theorem. As $n \to \infty$, the Fourier coefficients a_n^1 and a_n^2 always approach zero at least as rapidly as c/n where c is a constant independent of n. If the function $f(x)$ is piecewise continuous, then either a_n^1 or a_n^2, and in general both, decrease no faster than c/n. In general, if $f(x)$ and its first $k-1$ derivatives satisfy the conditions of the Fourier theorems, then the Fourier coefficients a_n^1 and a_n^2 approach zero as $n \to \infty$ at least as rapidly as c/n^{k+1}. Moreover, if $f^{(k)}(x)$ is not everywhere continuous, then either a_n^1 or a_n^2, and in general both, approach zero no faster than c/n^{k+1}.

This theorem implies that the smoother the function f is the faster its Fourier series converges. It should be noted that the Fourier series for two–

and three–dimensional functions are similar to the above analysis for one–dimensional functions.

1.7. Problems

1.1. Prove that $u\nabla \cdot (h\,\nabla v) = \nabla \cdot (u\,h\,\nabla v) - \nabla u \cdot h\,\nabla v.$

1.2. Show that $u\nabla^2 v = \nabla \cdot (u\,\nabla v) - \nabla u \cdot \nabla v.$

1.3. Use Green's formulas to show that

$$\iint_R \nabla^2 u\,\nabla^2 w\,dx\,dy = \iint_R \nabla^4 u\,w\,dx\,dy + \int_C \nabla^2 u\frac{\partial w}{\partial n}\,ds-$$

$$- \int_C \frac{\partial}{\partial n}\nabla^2 u\,w\,ds.$$

1.4. Let $Lu = \nabla \cdot (h\nabla u)$, where h is a function of x, $y \in R$. Show that the operator L is self–adjoint and positive definite.

1.5. Let $L \equiv - (\nabla^2 + k^2)$ (Helmholtz operator), where k is real. Show that the operator L is self–adjoint.

1.6. Let $Lu = a_1\dfrac{d^4 u}{dx^4} + a_2\dfrac{d^2 u}{dx^2} + a_3 u$, where a_1, a_2, and a_3 are real. Show that the operator L is self–adjoint and positive definite.

1.7. For the transient heat conduction operator $L \equiv \dfrac{\partial}{\partial t} - \nabla^2$, show that the operator $L^* \equiv -\frac{\partial}{\partial t} - \nabla^2$ is self–adjoint, and that

$$\langle Lu, w \rangle - \langle u, L^*w \rangle = \iint_S \hat{\mathbf{n}} \cdot [uw\hat{\mathbf{e}}_t + u\nabla w - w\nabla u]\,dS,$$

where $\hat{\mathbf{e}}_t$ is a unit vector in the time direction.

1.8. If f is infinitely differentiable, show that

(a) $\langle f(x)\delta(x, x'), \phi \rangle = f(x')\phi(x').$

(b) $\langle f(x)\dfrac{\partial}{\partial x}\delta(x, x'), \phi \rangle = -f(x')\dfrac{\partial \phi}{\partial x}(x') - \dfrac{\partial f}{\partial x}(x')\phi(x').$

(c) $\langle \nabla^2 u, f \rangle = \langle u, \nabla^2 f \rangle.$

(d) $\dfrac{\partial}{\partial x'}\delta(x, x') = -\dfrac{\partial}{\partial x}\delta(x, x').$

1.9. Let D be a domain in R^n with boundary ∂D, and let

$$u_D = \begin{cases} u, & \mathbf{x} \text{ in } D \\ 0, & \text{otherwise,} \end{cases}$$

where u is twice differentiable in V. Show that

$$\langle \nabla^2 u_D, f \rangle = \iiint_D f \nabla^2 u \, dx + \iint_S \left(u \frac{\partial f}{\partial n} - f \frac{\partial u}{\partial n} \right) dS.$$

1.10. It is known that $\int_0^\infty e^{-ar^2} \, dr = \frac{1}{2}\sqrt{\pi/2}$, and $\int_0^\infty r e^{-ar^2} \, dr = 1/(2a)$, where $a > 0$ and $r = \sqrt{x_1^2 + \cdots + x_n^2}$. By repeated differentiation with respect to a, show that

(a) $\displaystyle\int_0^\infty r^{2n} e^{-ar^2} \, dr = \frac{(2n-1)!}{(n-1)!} \frac{1}{a^n 2^{2n}} \sqrt{\pi/2}.$

(b) $\displaystyle\int_0^\infty r^{2n+1} e^{-ar^2} \, dr = \frac{n!}{2a^{n+1}}.$

(c) $\lim_{a \to 0} e^{r^2/a^2} = (\pi)^{n/2} \delta(x_1, \cdots, x_n)$ for any positive integer n.

1.11. Find the Fourier series for the function $f(x)$ which is assumed to have the period 2π:

(a) $f(x) = \begin{cases} 1 & \text{if } -\pi/2 < x < \pi/2 \\ -1 & \text{if } \pi/2 < x < 3\pi/2. \end{cases}$

(b) $f(x) = x, \quad -\pi < x < \pi.$

(c) $f(x) = x^2, \quad -\pi < x < \pi.$

1.12. Find the Fourier series of the period function $f(x)$ of period T:

(a) $f(x) = \begin{cases} -1 & \text{if } -1 < x < 0 \\ 1 & \text{if } 0 < x < 1, \quad T = 2. \end{cases}$

(b) $f(x) = 1 - x^2, \quad -1 < x < 1, \quad T = 2.$

References and Bibliography for Additional Reading

M. Abramowitz and I. A. Stegun, *Handbook of Mathematical Functions*, Dover, New York, 1965.

R. Bellman, *Introduction to Matrix Analysis*, 2nd ed., McGraw–Hill, New York, 1970.

R. V. Churchill and J. W. Brown, *Fourier Series and Boundary Value Problems*, 3rd ed., McGraw–Hill, New York, 1978.

R. Courant, *Differential and Integral Calculus*, vol. 1, 2, Interscience, New York, 1964, 1965.

——— and D. Hilbert, *Methods of Mathematical Physics*, vol. 1, Interscience, New York, 1952.

W. F. Donoghue, Jr, *Distributions and Fourier Transforms*, Academic Press, New York, 1969.

P. R. Garabedian, *Partial Differential Equations*, Wiley, New York, 1964.

É. Goursat, *A Course in Mathematical Analysis*, Vol. I, Dover, New York, 1959.

Hildebrand F. B., *Methods of Applied Mathematics*, Prentice–Hall, Englewood Cliffs, NJ, 1965.

E. Kreyszig, *Advanced Engineering Analysis*, 5th ed., Wiley, New York, 1983.

P. Morse and H. Feshbach, *Methods of Theoretical Physics*, vol. I, II, McGraw–Hill, New York, 1953.

I. N. Sneddon, *Elements of Partial Differential Equations*, McGraw–Hill, New York, 1957.

J. J. Tuma, *Engineering Mathematics Handbook*, McGraw–Hill, New York, 1979.

A. G. Webster, *Partial Differential Equations of Mathematical Physics*, 2nd ed., Hafner, New York, 1947.

R. E. Williamson, R. H. Crowell and H. F. Trotter, *Calculus of Vector Functions*, 2nd ed., Prentice Hall, Englewood Cliffs, NJ, 1968.

C. R. Wylie and L. C. Barrett, *Advanced Engineering Analysis*, McGraw–Hill, New York, 1960.

2

Variation and Weighted Residual Methods

Variational formulation of boundary value problems originates from the fact that weighted variational methods provide approximate solutions of such problems.

The variational solution of a differential equation reduces the equation to an equivalent variational form. Then the approximate solution is taken as a linear combination $\sum c_i \phi_i$ of known approximate functions ϕ_i. The parameters a_i must then be determined from the variational form. This method has the disadvantage that the approximate functions ϕ_i for problems with arbitrary domains cannot be easily constructed.

2.1. Weak Variational Formulation

As in the direct finite element analysis, the weak variation formulation of boundary value problems is important to develop the direct BEM. This approach is derived from the fact that variational methods for finding approximate solutions of boundary value problems, viz., Galerkin, Rayleigh–Ritz, collocation, or other weighted residual methods, are based on the weak variational statements of the boundary value problems.

We shall consider a general form of a 2nd–order boundary value problem in a two–dimensional region, defined by (A.39) in Appendix A, i.e.,

$$\frac{\partial F}{\partial u} - \frac{\partial}{\partial x}\left(\frac{\partial F}{\partial p}\right) - \frac{\partial}{\partial y}\left(\frac{\partial F}{\partial q}\right) = 0 \quad \text{in} \quad R, \tag{2.1}$$

with

$$u = u_0 \quad \text{on } C_1 \text{ (prescribed essential boundary condition)}, \tag{2.1a}$$

$$\frac{\partial F}{\partial p}n_x + \frac{\partial F}{\partial q}n_y = q_0$$

$$\text{on } C_2 \text{ (prescribed natural boundary condition)}, \tag{2.1b}$$

where $F = F(x, y, u, p, q)$, $p = \partial u/\partial x$, $q = \partial u/\partial y$, and n_x, n_y are the direction cosines of the unit vector $\hat{n}$ normal to the boundary $C = C_1 \cup C_2$ of the region R. The prescribed natural boundary condition may define heat flux in a heat transfer problem.

For example, a special case of (2.1) is when F is defined as

$$F = \frac{1}{2}\left[k_1\left(\frac{\partial u}{\partial x}\right)^2 + k_2\left(\frac{\partial u}{\partial y}\right)^2\right] - f\,u.$$

This equation arises in heat conduction problems in a two–dimensional region with k_1, k_2 as thermal conductivities in the x, y directions, and f being the heat generation. Here

$$\frac{\partial F}{\partial p} = k_1\frac{\partial u}{\partial x}, \quad \frac{\partial F}{\partial q} = k_2\frac{\partial u}{\partial y}, \quad \frac{\partial F}{\partial u} = -f.$$

Then (2.1) becomes

$$-\frac{\partial}{\partial x}\left(k_1\frac{\partial u}{\partial x}\right) - \frac{\partial}{\partial y}\left(k_2\frac{\partial u}{\partial y}\right) = f \quad \text{in} \quad R, \tag{2.2}$$

$$u = u_0 \quad \text{on } C_1, \tag{2.2a}$$

$$\left(k_1\frac{\partial u}{\partial x}\right)n_x + \left(k_2\frac{\partial u}{\partial y}\right)n_y = q_0 \quad \text{on } C_2. \tag{2.2b}$$

If $k_1 = k_2 = 1$, then (2.2) yields $-\nabla^2 u = f$ with appropriate boundary conditions.

A one–dimensional example is given by

$$F = \frac{a}{2}\left(\frac{\partial u}{\partial x}\right)^2 + fu,$$

which leads to

$$-\frac{d}{dx}\left(a\frac{du}{dx}\right) = f, \quad 0 < x < l, \tag{2.3}$$

$$u(0) = u_0, \quad \left(a\frac{du}{dx}\right)(l) = q_0. \tag{2.3a,b}$$

Some examples of this boundary value problem are considered in Chapter 3.

Now, the weak variational formulation for Eq (2.1) can be obtained by the following three steps:

STEP 1: Multiply the Eq (2.1) by a test function w ($\equiv \delta u$) and integrate the product on R:

$$\iint_R \left[\frac{\partial F}{\partial u} - \frac{\partial}{\partial x}\left(\frac{\partial F}{\partial p}\right) - \frac{\partial}{\partial y}\left(\frac{\partial F}{\partial q}\right)\right] w \, dx \, dy = 0. \tag{2.4}$$

The test function w is arbitrary, but it must satisfy the homogeneous essential boundary conditions (2.1a) on u.

STEP 2: Use formula (1.5c) component–wise to the second and third terms in (2.4), in order to transfer the differentiation from the dependent variable u to the test function w, and identify the type of the boundary conditions admissible by the variational form:

$$\iint_R \left[w\frac{\partial F}{\partial u} + \frac{\partial w}{\partial x}\frac{\partial F}{\partial p} + \frac{\partial w}{\partial y}\frac{\partial F}{\partial q}\right] dx \, dy -$$
$$- \int_{C=C_1 \cup C_2} \left(\frac{\partial F}{\partial p}n_x + \frac{\partial F}{\partial q}n_y\right) w \, ds = 0. \tag{2.5}$$

Note that the formula (1.5c) does not apply to the first term in the integrand in (2.4).

Notice that this step also yields boundary terms which determine the nature of the essential and natural boundary conditions for the problem. The general rule to identify the essential and natural boundary conditions for (2.1) is as follows:

The essential boundary condition is specified by the dependent variable (u in this case) in the same form as w in the boundary integral in (2.5). Thus

$$u = u_0 \quad \text{on} \quad C_1$$

is the essential boundary condition for (2.1).

The natural boundary condition comes from specifying the coefficients of w and its derivatives in the boundary integral in (2.5). Thus,

$$\frac{\partial F}{\partial p} n_x + \frac{\partial F}{\partial q} n_y = q_0 \quad \text{on} \quad C_2$$

is the natural boundary condition in a Neumann boundary value problem.

In one–dimensional problems, use integration by parts formulas (1.2) instead of the divergence formula (1.5c).

Note that the purpose of the differentiation transfer in this step is to equalize the continuity requirements on u and w. It imparts weaker continuity requirements on the solution u in the variational problem than in the original equation.

In BEM we shall identify the variables involved in the essential boundary condition as primary variables and those in the natural boundary condition as secondary variables. In a problem the primary variables are always continuous, but the secondary ones need not be continuous.

STEP 3: Simplify the boundary terms by using the prescribed boundary conditions. This will affect the boundary integral in (2.5) which is split into two terms, one on C_1 and the other on C_2:

$$\iint_R \left[w \frac{\partial F}{\partial u} + \frac{\partial w}{\partial x} \frac{\partial F}{\partial p} + \frac{\partial w}{\partial y} \frac{\partial F}{\partial q} \right] dx\, dy -$$

$$- \int_{C_1 \cup C_2} \left(\frac{\partial F}{\partial p} n_x + \frac{\partial F}{\partial u_y} n_y \right) w\, ds = 0. \quad (2.6)$$

The integral on C_1 vanishes since $w = \delta u = 0$ on C_1. The natural boundary condition is substituted in the integral on C_2. Then (2.6) reduces to

$$\iint_R \left[w \frac{\partial F}{\partial u} + \frac{\partial w}{\partial x} \frac{\partial F}{\partial p} + \frac{\partial w}{\partial y} \frac{\partial F}{\partial q} \right] dx\, dy - \int_{C_2} w\, q_0\, ds = 0. \quad (2.7)$$

This is the weak variational form for the problem (2.1) (also called virtual work). We can write (2.7) in terms of the bilinear and linear differential forms as

$$b(w, u) = l(w),\qquad(2.8)$$

where

$$b(w, u) = \iint_R \left[\frac{\partial w}{\partial x}\frac{\partial F}{\partial p} + \frac{\partial w}{\partial y}\frac{\partial F}{\partial q}\right] dx\, dy,$$

$$l(w) = -\iint_R w\frac{\partial F}{\partial u}\, dx\, dy + \int_{C_2} w q_0\, ds.$$

$$(2.9)$$

The weak variational form (2.8), with a proper choice of w, is the basis for the direct BEM, as we shall see later. The quadratic functional associated with this variational form is given by

$$I(u) = \frac{1}{2}b(u, u) - l(u).\qquad(2.10)$$

Example 2.1:
We shall consider the one–dimensional second order equation (2.3), subject to the boundary conditions (2.3,a,b), where $a = a(x)$, $f = f(x)$, and q_0 are prescribed. This boundary value problem arises in many physical problems. Following the 3–step weak variational technique we have

$$0 = -\int_0^l \left[\frac{d}{dx}\left(a\frac{du}{dx}\right) - f\right] w\, dx$$

$$= \int_0^l \left(a\frac{du}{dx}\frac{dw}{dx} - fw\right) dx + w(0)\left(a\frac{du}{dx}\right)(0) - w(l)\left(a\frac{du}{dx}\right)(l),$$

$$(2.11)$$

which gives the quadratic functional as

$$I(u) = \frac{1}{2}\int_0^l \left\{a\left(\frac{du}{dx}\right)^2 - 2uf\right\} dx - q_0 u(l).\qquad(2.12)$$

Example 2.2:
We shall consider the 4–th order equation

$$\frac{d^2}{dx^2}\left(b\frac{d^2u}{dx^2}\right) + f = 0,\ 0 \le x \le l,\qquad(2.13)$$

where $b = b(x)$ and $f = f(x)$ are given functions, subject to two essential and two natural boundary conditions

$$u(0) = 0,\ \frac{du(0)}{dx} = 0,$$

$$\frac{d}{dx}\left(b\frac{d^2u}{dx^2}\right)\bigg|_{x=l} = F_l, \qquad b\frac{d^2u}{dx^2}\bigg|_{x=l} = M_l. \tag{2.14}$$

This boundary value problem represents the physical problem of bending of an elastic beam. The weak variational formulation is

$$
\begin{aligned}
0 &= \int_0^l \left[\frac{d^2}{dx^2}\left(b\frac{d^2u}{dx^2}\right) + f\right] w\, dx \\
&= \int_0^l \left[-\frac{d}{dx}\left(b\frac{d^2u}{dx^2}\right)\frac{dw}{dx} + fw\right] dx + \left[\frac{d}{dx}\left(b\frac{d^2u}{dx^2}\right)w\right]_{x=x_e}^{x=x_{e+1}} \\
&= \int_0^l \left(b\frac{d^2u}{dx^2}\frac{d^2w}{dx^2} + fw\right) dx + \left[\frac{d}{dx}\left(b\frac{d^2u}{dx^2}\right)w - b\frac{d^2u}{dx^2}\frac{dw}{dx}\right]_{x_e}^{x_{e+1}},
\end{aligned}
\tag{2.15}
$$

where $w (\equiv \delta u) \in C^2$ is the test function.

Example 2.3:
Consider the system of Navier–Stokes equations for a two–dimensional flow of a viscous, incompressible fluid (pressure – velocity fields):

$$
\begin{aligned}
u\frac{\partial u}{\partial x} + v\frac{\partial u}{\partial y} &= -\frac{1}{\rho}\frac{\partial p}{\partial x} + \nu\left(\frac{\partial^2 u}{\partial x^2} + \frac{\partial^2 u}{\partial y^2}\right), \\
u\frac{\partial v}{\partial x} + v\frac{\partial v}{\partial y} &= -\frac{1}{\rho}\frac{\partial p}{\partial y} + \nu\left(\frac{\partial^2 v}{\partial x^2} + \frac{\partial^2 v}{\partial y^2}\right), \\
\frac{\partial u}{\partial x} + \frac{\partial v}{\partial y} &= 0,
\end{aligned}
\tag{2.16}
$$

in R with boundary conditions $u = u_0$, $v = v_0$ on C_1, and

$$
\begin{aligned}
\nu\left(\frac{\partial u}{\partial x}n_x + \frac{\partial u}{\partial y}n_y\right) - \frac{1}{\rho}pn_x &= \hat{t}_x, \\
\nu\left(\frac{\partial v}{\partial x}n_x + \frac{\partial v}{\partial y}n_y\right) - \frac{1}{\rho}pn_y &= \hat{t}_y,
\end{aligned}
\tag{2.17}
$$

on C_2, where $\hat{t}_x, \hat{t}_y$ are the prescribed values of the secondary variables.

Let w_1, w_2, w_3 be the test functions, one for each equation, such that they satisfy the essential boundary conditions on u and v. Then

$$
0 = \iint_R \left[w_1\left(u\frac{\partial u}{\partial x} + v\frac{\partial u}{\partial y}\right) - \frac{1}{\rho}\frac{\partial w_1}{\partial x}p + \nu\left(\frac{\partial w_1}{\partial x}\frac{\partial u}{\partial x} + \frac{\partial w_1}{\partial y}\frac{\partial u}{\partial y}\right)\right] dx\,dy - \\
- \int_{C_2} w_1\hat{t}_x\, ds,
$$

$$0 = \iint_R \left[w_2 \left(u \frac{\partial v}{\partial x} + v \frac{\partial v}{\partial y} \right) - \frac{1}{\rho} \frac{\partial w_2}{\partial y} p + \nu \left(\frac{\partial w_2}{\partial x} \frac{\partial v}{\partial x} + \frac{\partial w_2}{\partial y} \frac{\partial v}{\partial y} \right) \right] dx\, dy -$$

$$- \int_{C_1} w_2 \hat{t}_y\, ds,$$

$$0 = \iint_R w_3 \left(\frac{\partial u}{\partial x} + \frac{\partial v}{\partial y} \right) dx\, dy.$$

Write $b((w_1, w_2, w_3), (u, v))$ and $l(w_1, w_2, w_3)$ from these equations, and complete the solution.

Some Weighted Residual Methods

2.2. Galerkin Method

Consider the boundary value problem

$$L\,u = f \qquad \text{in } R, \tag{2.18}$$

subject to the boundary conditions

$$u = g(s) \qquad \text{on } C_1, \tag{2.19}$$

$$\frac{\partial u}{\partial n} + k(s)\,u = h(s) \qquad \text{on } C_2, \tag{2.20}$$

where $C = C_1 \cup C_2$ is the boundary of the region R. Let us choose an approximation of the form

$$\tilde{u} = \sum_{i=1}^{n} c_i \phi_i. \tag{2.21}$$

An approximate solution does not, in general, satisfy the system (2.18)–(2.20). The residual (error) associated with an approximate solution is defined by

$$r(\tilde{u}) \equiv L\tilde{u} - f = L \left(\sum_{i=1}^{n} c_i \phi_i \right) - f. \tag{2.22}$$

Note that if u_0 is an exact solution of (2.18)–(2.20), then $r(u_0) = 0$. If the residual is orthogonalized with respect to the basis functions ϕ_i (which are also called the trial functions), then

$$\langle r, \phi_i \rangle = 0, \tag{2.23}$$

or

$$\int\int_R \{L(\tilde{u}) - f\}\phi_i \, dx \, dy = 0, \qquad i = 1, \cdots, n, \qquad (2.23a)$$

or

$$\sum_{j=1}^{n} c_j \int\int_R \phi_i \, L\phi_j \, dx \, dy = \int\int_R f\phi_i \, dx \, dy,$$

or

$$[A]\{c\} = \{b\}, \qquad (2.23b)$$

where

$$A_{ij} = \int\int_R \phi_i \, L\phi_j \, dx \, dy, \qquad b_i = \int\int_R f\phi_i \, dx \, dy. \qquad (2.23c)$$

Note from (2.23a) that in the Galerkin method the integral of the residual, weighted by the basis (trial) functions, is set equal to zero. The expression (2.23b) is the matrix form of (2.23a).

In the examples given below, we shall choose different values of n in (2.4) for the trial function $\tilde{u}$. There is some guidance from geometry for such choices; namely, they should satisfy the essential conditions and exhibit the nature of the approximation solutions vis-a-vis the exact solutions (see §2.4 for some choices). However, the larger the n, the better the approximation becomes.

Example 2.4:
Consider

$$u'' + u + x = 0, \qquad 0 < x < 1; \qquad u = 0 \text{ at } x = 0, 1.$$

Note that here $f = -x$. Choose a cubic trial function

$$\tilde{u}_1(x) = c_0 + c_1 x + c_2 x^2 + c_3 x^3.$$

Using the boundary conditions at $x = 0, 1$, we find that $c_0 = 0$, and $c_1 + c_2 + c_3 = 0$. We can taken $c_1 = a$, $c_3 = -b$; then $c_2 = b - a$, and the trial function becomes

$$\tilde{u}_1(x) = x(1 - x)(a + bx),$$

for which

$$r = x + a(-2 + x - x^2) + b(2 - 6x + x^2 - x^3).$$

Then (2.23c) gives

$$\int_0^1 r\,x(1-x)\,dx = 0, \qquad \int_0^1 r\,x^2(1-x)\,dx = 0.$$

Integrate and solve for a and b. This gives

$$\begin{bmatrix} 3/10 & 3/20 \\ 3/20 & 13/105 \end{bmatrix} \begin{Bmatrix} a \\ b \end{Bmatrix} = \begin{Bmatrix} 1/12 \\ 1/20 \end{Bmatrix},$$

which yields $a = 71/369$, $b = 7/41$. Thus, the approximate solution is

$$\tilde{u}_1 = x(1-x)\left(\frac{71}{369} + \frac{7}{41}x\right).$$

The exact solution of the problem is

$$u_0 = (\csc 1) \sin x - x.$$

These solutions are compared in the following table:

x	$\tilde{u}_1$	u_0
.25	.0440	.044014
.50	.0698	.069747
.75	.0600	.060056 ■

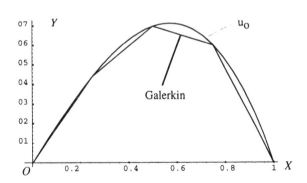

Example 2.5:
Consider the Poisson equation

$$-\nabla^2 u \equiv -\left(\frac{\partial^2 u}{\partial x^2} + \frac{\partial^2 u}{\partial y^2}\right) = c, \qquad u = 0 \text{ at } x = 0, a \text{ and } y = 0, b.$$

Choose the approximate solution as

$$\tilde{u}_1 = \alpha xy(x - a)(y - b).$$

Note this choice: It satisfies all four Dirichlet boundary conditions. The Galerkin equation (2.23a) gives

$$\int_0^b \int_0^a \left[-2\alpha(y^2 - by + x^2 - ax) - c \right] xy(x - a)(y - b)\, dx\, dy = 0$$

which simplifies to

$$\frac{\alpha}{90}\left[a^3 b^3 \left(a^2 + b^2 \right)\right] - \frac{a^3 b^3 c}{36} = 0.$$

Thus,

$$\alpha = \frac{5c}{2\left(a^2 + b^2\right)}.$$

Hence

$$\tilde{u}_1 = \frac{5c}{2\left(a^2 + b^2\right)}\, xy(x - a)(y - b).$$

We will solve this problem by choosing a different approximate function as

$$\tilde{u}_2 = \sum_j \sum_k \alpha_{jk} \sin \frac{j\pi x}{a} \sin \frac{k\pi y}{b},$$

which is a trigonometric series with a finite number of terms. Note that u_2 satisfies the boundary conditions. Also note the orthogonality condition

$$\int_0^a \sin \frac{m\pi x}{a} \sin \frac{n\pi x}{a}\, dx = \begin{cases} 0, & m \neq n \\ a/2, & m = n. \end{cases}$$

The Galerkin equation (2.23a) in this case gives

$$\int_0^b \int_0^a \left[\alpha_{jk}\left(\frac{j^2\pi^2}{a^2} + \frac{k^2\pi^2}{b^2} \right) \sin \frac{j\pi x}{a} \sin \frac{k\pi y}{b} - c \right] \times$$

$$\times \sin \frac{j\pi x}{a} \sin \frac{k\pi y}{b}\, dx\, dy = 0,$$

hence

$$\alpha_{jk}\frac{\pi^2}{4}\left(\frac{j^2}{a^2} + \frac{k^2}{b^2} \right) = \frac{c}{jk\pi^2}(\cos j\pi - 1)(\cos k\pi - 1),$$

or

$$\alpha_{jk} = \frac{4c(\cos j\pi - 1)(\cos k\pi - 1)a^2 b^2}{\pi^4 jk(b^2 j^2 + a^2 k^2)}.$$

Thus, this approximate solution is

$$\tilde{u}_2 = \sum_j \sum_k \frac{4a^2 b^2 c(1 - \cos j\pi)(1 - \cos k\pi)}{jk\pi^4(a^2 k^2 + b^2 j^2)} \sin \frac{j\pi x}{a} \sin \frac{k\pi y}{b}.$$

If the number of terms in each sum is infinite, then u_2 becomes the exact solution u_0.

At the center point (a/2,b/2), we have

$$\tilde{u}_{2,c} = \sum_j \sum_k \frac{4a^2 b^2 c(1 - \cos j\pi)(1 - \cos k\pi)}{jk\pi^4(a^2 k^2 + b^2 j^2)} \sin \frac{j\pi}{2} \sin \frac{k\pi}{2}.$$

If $a = b$, then at the center point $(a/2, a/2)$

$$\tilde{u}_{2,c} = \sum_j \sum_k \frac{4a^2 c(1 - \cos j\pi)(1 - \cos k\pi)}{jk\pi^4(j^2 + k^2)} \sin \frac{j\pi}{2} \sin \frac{k\pi}{2}$$

$$= \frac{a^2 c}{\pi^4}\left[8 + \frac{8}{15} + \frac{8}{15} + \frac{8}{81} + \cdots\right] \equiv u_0$$

$$\approx \frac{36.64}{\pi^4} c\left(\frac{a}{2}\right)^2. \blacksquare$$

2.3. Rayleigh–Ritz Method

Let us consider

$$-\nabla^2 u = f$$

with the homogeneous boundary conditions $u = 0$ on C_1 and $\partial u/\partial n = 0$ on C_2. Then, the weak variational formulation leads to

$$I(u) = \iint_R \left\{\frac{1}{2}|\nabla u|^2 - fu\right\} dx\, dy = 0. \qquad (2.24)$$

A generalization of the result in (2.24) for the case of the system $u\,Lu = f$ with the above homogeneous boundary conditions, where L is a linear self–adjoint and positive definite operator, leads to the functional

$$I(u) = \frac{1}{2} \iint_R \{uLu - 2fu\}\, dx\, dy. \qquad (2.25)$$

Theorem 2.1. If the operator L is self–adjoint and positive definite, then the unique solution of $Lu = f$ with homogeneous boundary conditions occurs at a minimum value of $I(u)$.

An application of Theorem 2.1 is the Rayleigh–Ritz method, where we find an approximate solution for the system $Lu = f$ with homogeneous boundary conditions as follows: Since u_0 is a function which gives a minimum value of $I(u)$, we can consider variations at u_0 given by the trial function

$$u = u_0 + \alpha\phi, \qquad (2.26)$$

where α is a (variable) parameter and ϕ is an arbitrary function. Then $I(u)$ attains its minimum when $\alpha = 0$, i.e.,

$$\left.\frac{dI}{d\alpha}\right|_{\alpha=0} = 0. \qquad (2.27)$$

In the Rayleigh–Ritz method we should choose a complete set of linearly independent basis functions ϕ_i, $i = 1, \cdots$. The exact solution u_0 is then approximated by a sequence of trial functions

$$\tilde{u} = \sum_{i=1}^{n} c_i\phi_i, \qquad (2.28)$$

where the constants c_i are chosen such that the functional $I(\tilde{u})$ is minimized at each stage. If $\tilde{u} \to u_0$ as $n \to \infty$, then the method yields a convergent solution. At each stage the method reduces the problem to that of solving a set of linear algebraic equations. The details for the boundary value problem $-\nabla^2 u = f$ with homogeneous boundary conditions are as follows: Using (2.28) in the functional (2.25) we get

$$I(\tilde{u}) = I(c_1, \cdots, c_n)$$
$$= \iint_R \left\{ \left(\frac{\partial \tilde{u}}{\partial x}\right)^2 + \left(\frac{\partial \tilde{u}}{\partial y}\right)^2 - 2\tilde{u}f \right\} dx\, dy$$
$$= \iint_R \left\{ \left(\sum c_i \frac{\partial \phi_i}{\partial x}\right)^2 + \left(\sum c_i \frac{\partial \phi_i}{\partial y}\right)^2 - 2f \sum c_i\phi_i \right\} dx\, dy,$$

thus

$$I(c_i) = c_i^2 \iint_R \left\{ \left(\frac{\partial \phi_i}{\partial x} \right)^2 + \left(\frac{\partial \phi_i}{\partial y} \right)^2 \right\} dx\,dy +$$

$$+ 2 \sum_{i \neq j} c_i c_j \iint_R \left(\frac{\partial \phi_i}{\partial x} \frac{\partial \phi_j}{\partial x} + \frac{\partial \phi_i}{\partial y} \frac{\partial \phi_j}{\partial y} \right) dx\,dy - 2c_i \iint_R \phi_i f \, dx\,dy.$$

Hence

$$\frac{\partial I}{\partial c_i} = 2A_{ii}c_i + 2 \sum_{i \neq j} A_{ij}c_j - 2h_i, \qquad (2.29)$$

and

$$A_{ij} = \iint_R \left(\frac{\partial \phi_i}{\partial x} \frac{\partial \phi_j}{\partial x} + \frac{\partial \phi_i}{\partial y} \frac{\partial \phi_j}{\partial y} \right) dx\,dy, \qquad (2.30)$$

$$h_i = \iint_R \phi_i f \, dx\,dy. \qquad (2.31)$$

Now, if we choose c_i such that $I(c_i)$ is a minimum (i.e., $\partial I / \partial c_i = 0$), then from (2.29) we get

$$\sum_{j=1}^n A_{ij}c_i = h_i, \quad i = 1, \cdots, n, \qquad (2.32)$$

which in the matrix notation is

$$[A]\{c\} = \{h\}, \qquad (2.33)$$

where the matrix $[A]$ has elements A_{ij} given by (2.30), $\{h\}$ has elements h_i given by (2.31), and $\{c\} = [c_1, \cdots, c_n]^T$. Note that (2.33) is a system of linear algebraic equations to be solved for the unknown parameter c_i, and $[A]$ is non–singular if L is positive definite.

Example 2.6:
Consider
$$-u'' = x^2; \quad u(0) = u(1) = 0, \ 0 \leq x \leq 1. \qquad (2.34)$$

Choose a cubic approximation, and let the trial function be

$$\tilde{u}_3 = c_0 + c_1 x + c_2 x^2 + c_3 x^3.$$

Using the Dirichlet boundary condition, we find that $c_0 = 0$, and by the boundary condition $\tilde{u}_3(1) = 0$ we get $c_1 + c_2 + c_3 = 0$, thus c_1, c_2, c_3 are linearly dependent. We take $c_1 = a$, $c_2 = -a + b$, $c_3 = -b$, thus

$$\tilde{u}_3 = ax(1 - x) + bx^2(1 - x),$$

where a, b are parameters, and $n = 2$ (2 dof). Here $\phi_1 = x - x^2$, $\phi_2 = x^2 - x^3$. To use (2.33) we need A_{11}, A_{22}, A_{12}, and h_1, h_2. Note that since the problem is one–dimensional, $\iint_R$ will become $\int_0^1$. Thus

$$A_{11} = A_{aa} = \int_0^1 \left(\frac{d\phi_1}{dx} \right)^2 dx = \int_0^1 (1 - 2x)^2 \, dx = \frac{1}{3},$$

$$A_{22} = A_{bb} = \int_0^1 \left(\frac{d\phi_2}{dx} \right)^2 dx = \int_0^1 (2x - 3x^2)^2 \, dx = \frac{2}{15},$$

$$A_{12} = A_{21} = A_{ab} = \int_0^1 \frac{d\phi_1}{dx} \frac{d\phi_2}{dx} \, dx = \int_0^1 (1 - 2x)(2x - 3x^2) \, dx = \frac{1}{6},$$

$$h_1 = \int_0^1 \phi_1 x^2 \, dx = \int_0^1 (x - x^2) x^2 \, dx = \frac{1}{20},$$

$$h_2 = \int_0^1 \phi_2 x^2 \, dx = \int_0^1 (x^2 - x^3) x^2 \, dx = \frac{1}{30},$$

which leads to

$$\begin{bmatrix} 1/3 & 1/6 \\ 1/6 & 2/15 \end{bmatrix} \begin{Bmatrix} a \\ b \end{Bmatrix} = \begin{Bmatrix} 1/20 \\ 1/30 \end{Bmatrix},$$

thus $a = 1/15$, $b = 1/6$, and the approximate solution is

$$\tilde{u}_3 = \frac{x}{30}(1 - x)(2 + 5x).$$

Note that the exact solution is

$$u_0 = \frac{1}{12} x (1 - x^3) = \frac{1}{12} x (1 - x)(1 + x + x^2). \ \blacksquare$$

Example 2.7:
Consider the same example as in Example 2.6, and take the approximate solution same as u_3 there. Then by the Galerkin method, we get

$$\int_0^1 r\phi_1 \, dx = 0, \quad \text{or} \quad \frac{1}{3} a + \frac{1}{6} b = \frac{1}{20},$$

$$\int_0^1 r\phi_2 \, dx = 0, \quad \text{or} \quad \frac{1}{6} a + \frac{2}{15} b = \frac{1}{30}.$$

This yields the same values of a and b as in Example 2.6. $\blacksquare$

Example 2.8:
Consider

$$Lu - f = \frac{d^2u}{dx^2} + u + x = 0; \quad u(0) = 0, \quad \frac{du}{dx}(1) = 0, \quad 0 < x < 1.$$

Take the first approximate solution as $\tilde{u}_1 = c_0 + c_1 x + c_2 x^2$, which in view of the boundary conditions and taking $c_1 = a$ gives

$$\tilde{u}_1 = ax\left(1 - \frac{x}{2}\right) = a\phi_1.$$

It satisfies both boundary conditions. Using the Galerkin method, we get

$$\int_0^1 r\left(x - \frac{x^2}{2}\right) dx = 0,$$

or

$$\int_0^1 \left[-a + a(x - \frac{x^2}{2}) + x\right]\left(x - \frac{x^2}{x}\right) dx = 0,$$

which gives $a = 25/24$.

In case we choose the second approximate solution as (from $\tilde{u}_2 = c_0 + c_1 x + c_2 x^2$, with $c_1 = a, c_2 = b$)

$$\tilde{u}_2 = ax + bx^2 = a\phi_1 + b\phi_2.$$

The above integral then gives

$$\int_0^1 r_2 \phi_1 \, dx = 0, \quad \text{and} \quad \int_0^1 r_2 \phi_2 \, dx = 0,$$

where the residual r_2 in $(0, 1)$ is

$$r_2 = \frac{d^2u_2}{dx^2} + u_2 + x = 2b + (a + 1)x + bx^2.$$

Thus,

$$\int_0^1 (2b + (a + 1)x + bx^2)x \, dx = 0, \quad \int_0^1 (2b + (a + 1)x + bx^2)x^2 \, dx = 0,$$

which yield

$$\frac{2}{3}a + \frac{3}{4}b = \frac{1}{3}, \quad \frac{3}{4}a + \frac{17}{15}b = \frac{1}{4},$$

or

$$\begin{bmatrix} 2/3 & 3/4 \\ 3/4 & 17/15 \end{bmatrix} \begin{Bmatrix} a \\ b \end{Bmatrix} = \begin{Bmatrix} 1/3 \\ 1/4 \end{Bmatrix},$$

thus, $a = 137/139$, $b = -60/139$ which gives $\dfrac{du(1)}{dx} = 17/139 \neq 1$. Notice that the last entry is not 0, but if we increased the order of the trial function, the solution would (almost) satisfy the boundary condition $du/dx = 0$ at $x = 1$. ∎

These examples show that if the boundary conditions are homogeneous, then both Galerkin and Rayleigh–Ritz methods give the same results.

The Rayleigh–Ritz method can be developed, alternately, by solving for u the equation (2.8), where we require that w satisfies the homogeneous essential conditions only. Then this problem is equivalent to minimizing the functional (2.10). In other words, we will find an approximate solution of (2.8) in the form

$$u_n = \sum_{j=1}^{n} c_j \phi_j + \phi_0, \qquad (2.35)$$

where the coefficients c_j are chosen such that Eq (2.8) is true for $w = \phi_i$, $i = 1, \cdots, n$, i.e.,

$$b(\phi_i, u_n) = l(\phi_i), \quad i = 1, \cdots, n,$$

or

$$b\left(\phi_i, \sum_{j=1}^{n} c_j \phi_j + \phi_0\right) = l(\phi_i),$$

thus,

$$\sum_{j=1}^{n} c_j b(\phi_i, \phi_j) = l(\phi_i) - b(\phi_i, \phi_0). \qquad (2.36)$$

This equation is a system of n linear algebraic equations in n unknowns c_j and has a unique solution if the coefficient matrix in (2.36) is nonsingular and thus has an inverse.

The functions ϕ_i must satisfy the following requirements: (i) ϕ_i should be well–defined such that $b(\phi_i, \phi_j) \neq 0$, (ii) ϕ_i must satisfy at least the essential homogeneous boundary condition, (iii) the set $\{\phi_i\}_{i=1}^{n}$ must be linearly independent, and (iv) the set $\{\phi_i\}_{i=1}^{n}$ must be complete.

Example 2.9:
Consider the boundary value problem

$$-u'' - u + x = 0, \quad 0 < x < 1; \quad u(0) = 0 = u(1).$$

The weak variational formulation gives

$$0 = \int_0^1 [-u'' - u + x] w \, dx$$

$$= \int_0^1 \left[\frac{dw}{dx} \frac{du}{dx} - uw + xw \right] dx.$$

Thus,

$$b(w, u) = \int_0^1 \left[\frac{dw}{dx} \frac{du}{dx} - uw \right] dx, \quad l(w) = - \int_0^1 xw \, dx.$$

The choice $\phi_i = x^i(1 - x)$ satisfies all four requirement mentioned above; thus, we take as an n–th approximation

$$\tilde{u}_n = \sum_{j=1}^n c_j x^j (1 - x).$$

Then substituting this $\tilde{u}_n$ and $w = \phi_i$ in $b(w, \tilde{u}_n) = l(w)$, we obtain

$$\int_0^1 \left[\frac{d\phi_i}{dx} \left(\sum_{j=1}^n c_j \frac{d\phi_j}{dx} \right) - \phi_i \left(\sum_{j=1}^n c_j \phi_j \right) \right] dx = - \int_0^1 \phi_i x \, dx$$

or

$$\sum_{j=1}^n c_j \left\{ \int_0^1 \left[i x^{i-1} - (i+1) x^i \right] \left[j x^{j-1} - (j+1) x^j \right] - x^{i+j} (1-x)^2 \right\} dx =$$

$$- \int_0^1 x^{i+1} (1 - x) \, dx$$

or, using (1.1b)

$$\sum_{j=1}^n c_j \left\{ \frac{2ij}{(i+j)[(i+j)^2 - 1]} - \frac{2}{(i+j+1)(i+j+2)(i+j+3)} \right\} =$$

$$- \frac{1}{(i+2)(i+3)}.$$

For $n = 2$, this equation leads to the system of equations

$$\begin{bmatrix} 3/10 & 3/20 \\ 3/20 & 13/105 \end{bmatrix} \begin{Bmatrix} c_1 \\ c_2 \end{Bmatrix} = \begin{Bmatrix} -1/12 \\ -1/20 \end{Bmatrix},$$

which yields $c_1 = -7/4, c_2 = -71/369$, thus

$$\tilde{u}_2 = -\frac{1}{369} x(1-x)(63+71x).$$

Taking $\tilde{u}_2 = a(x-x^2)+b(x^2-x^3)$ (i.e., $\phi_1 = x(1-x), \phi_2 = x^2(1-x)$), the Galerkin method for this problem leads to $r = a(2-x+x^2)-b(2-6x+x^2 = x^3) + x$, and then $\int_0^1 r\phi_1\, dx = 0$ and $\int_0^1 r\phi_2\, dx = 0$ give

$$\begin{bmatrix} 3/10 & 3/20 \\ 3/20 & 13/105 \end{bmatrix} \begin{Bmatrix} a \\ b \end{Bmatrix} = \begin{Bmatrix} -1/12 \\ -1/20 \end{Bmatrix},$$

which is the same as in the Rayleigh–Ritz method above. The exact solution is given by

$$u = x - \frac{\sin x}{\sin 1}. \quad \blacksquare$$

Example 2.10:
We will now consider the boundary value problem

$$-u'' - u + x^2 = 0, \quad 0 < x < 1; \quad u(0) = 0, u'(1) = 1.$$

Note that the essential and natural boundary conditions are of different type (one is Dirichlet and the other Neumann). The bilinear form $b(w,u)$ is the same as in Example 2.9, but the linear form is $l(w) = -\int_0^1 wx^2\, dx + w(1)$. In this case the proper choice is $\phi_i = x^i, i = 1, \cdots, n$ (this satisfies the essential boundary condition and the four requirements mentioned above). Then, $w(1) = \phi_i(1) = 1$, and assuming the approximate solution as $u_n = \sum_{j=1}^n c_j\phi_j$, we find from (2.8) that

$$\sum_{j=1}^n c_j \int_0^1 \left(ijx^{i+j-2} - x^{i+j} \right)\, dx = -\int_0^1 x^{i+2}\, dx + 1,$$

or

$$\sum_{j=1}^n c_j \left\{ \frac{ij}{i+j-1} - \frac{1}{i+j+1} \right\} = 1 - \frac{1}{i+3}.$$

For the case when $n = 2$, we get

$$\begin{bmatrix} 2/3 & 3/4 \\ 4/5 & 17/15 \end{bmatrix} \begin{Bmatrix} c_1 \\ c_2 \end{Bmatrix} = \begin{Bmatrix} 3/4 \\ 4/5 \end{Bmatrix},$$

which yields $c_1 = -3/7, c_2 = 29/21$.

With the choice of the approximate solution $u_2 = ax + bx^2$, the Galerkin method leads to

$$r = -u_2'' - u_2 + x^2 = -ax - b(2 + x^2) + x^2,$$

$$\int_0^1 r\phi_1 \, dx = 0 \Rightarrow \quad \frac{a}{3} + \frac{5}{4}b = \frac{1}{4},$$

$$\int_0^1 r\phi_2 \, dx = 0 \Rightarrow \quad \frac{a}{4} + \frac{13}{15}b = \frac{1}{5},$$

which gives $a = 24/17, b = -3/17$. Notice that the Rayleigh–Ritz and Galerkin methods do not produce the same results. Also, the solutions by these two methods do not satisfy the natural boundary condition. The exact solution is

$$u = \frac{2\cos(1-x) - \sin x}{\cos 1} + x^2 - 2. \blacksquare$$

Example 2.11:
Consider the boundary value problem

$$-u'' = x, \qquad 0 < x < 1; \ u(0) = 2, \ u'(1) = 3.$$

Choose a quadratic trial function $\tilde{u}_1 = c_0 + c_1 x + c_2 x^2$, which after using the boundary conditions gives

$$\tilde{u}_1 = 2 + ax + bx^2.$$

Note that the choice of the trial function must satisfy the essential (Dirichlet) boundary condition. This choice of $\tilde{u}_1$ is of the form (2.35) with $\phi_0 = 2$. Here $\phi_1 = x$, $\phi_2 = x^2$, $k = 0$, $h = 3$, $f = x$. Then $\phi_{1,2}(0) = 0$, $\phi_{1,2}(1) = 1$, and $d\tilde{u}_1(1)/dx = u'(1) = 3$. We will make a direct use of the weak variational formulation which gives

$$0 = \int_0^1 [-u'' - x] w \, dx$$

$$= \int_0^1 \left[\frac{dw}{dx} \frac{du}{dx} - xw \right] dx + w(0)u'(0) - w(1)u'(1).$$

Now, take $u = \tilde{u}_1$, and $w = \phi_1, \phi_2$ respectively. Then we get

$$\int_0^1 \left(\tilde{u}_1' \phi_1' - f\phi_1 \right) dx - h\phi_1 \bigg|_{x=1} = 0,$$

or

$$\int_0^1 \{(a+2bx) - x^2\}\, dx - 3x\Big|_{x=1} = 0,$$

thus,

$$a + b = 10/3.$$

Similarly,

$$\int_0^1 (\tilde{u}_1' \phi_2' - f\phi_2) - h\phi_2\Big|_{x=1} = 0,$$

or

$$\int_0^1 \{(a+2bx)2x - x^3\}\, dx - 3x^2\Big|_{x=1} = 0,$$

thus,

$$a + 4b/3 = 13/4.$$

Hence

$$\begin{bmatrix} 1 & 1 \\ 1 & 4/3 \end{bmatrix} \begin{Bmatrix} a \\ b \end{Bmatrix} = \begin{Bmatrix} 10/3 \\ 13/4 \end{Bmatrix} \qquad \text{i.e.,} a = \frac{43}{12},\ b = -\frac{1}{4},$$

which gives

$$\tilde{u}_1(x) = 2 + \frac{43}{12}x - \frac{1}{4}x^2.$$

The exact solution is $u_0 = 2 + 7x/2 - x^3/6$. ■ (Try the Galerkin method with the choice $\tilde{u}_1$ and see that it does not solve this problem.)

Example 2.12:
Consider the Bessel equation

$$x^2 u'' + xu' + (x^2 - 1)u = 0, \quad u(1) = 1, u(2) = 2.$$

Put $u = v + x$. Then the given equation and the boundary conditions become

$$x^2 v'' + xv' + (x^2 - 1)v + x^3 = 0, \quad v(1) = 0 = v(2).$$

In the self–adjoint form this equation is written as

$$xv'' + v' + \frac{x^2 - 1}{x}v + x^2 = 0.$$

For the 1st approximation, we take

$$v_1 = a_1 \phi_1 = a_1(x-1)(x-2).$$

Then using (2.23) we get $\int_1^2 (Lv_1 - f)\phi_1 \, dx = 0$, which gives

$$\int_1^2 \left[2a_1 x - (3 - 2x)a_1 + \frac{x^2 - 1}{x}(x-1)(x-2)a_1 + x^2 \right] (x-1)(x-2) \, dx = 0,$$

which, on integration, yields $a_1 = -0.811$, and thus,

$$u_1 = v_1 + x = -0.811(x - 1)(x - 2) + x.$$

The exact solution is $u = c_1 J_1(x) + c_2 Y_1(x)$, where $c_1 = 3.60756$, $c_2 = 0.75229$. A comparison with the exact solution in the following table shows that u_1 is a good approximation:

x	u_1	u_{exact}
1.3	1.4703	1.4706
1.5	1.7027	1.7026
1.8	1.9297	1.9294 ∎

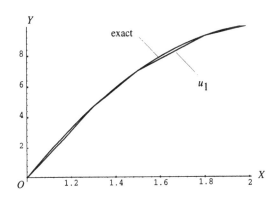

Example 2.13:
Consider the 4th–order equation

$$[(x + 2l)u'']'' + bu - kx = 0, \quad 0 < x < l,$$

with the boundary conditions: $u(l) = 0 = u'(l)$, $(x + 2l)u''(0) = 0$, $[(x + 2l)u'']'(0) = 0$. We choose the test functions

$$\phi_1(x) = (x - l)^2 (x^2 + 2lx + 3l^2)$$
$$\phi_2(x) = (x - l)^3 (3x^2 + 4lx + 3l^2).$$

For the 1st approximation, we have $u_1 = a_1 \phi_1(x)$. Then $\int_0^l (Lu_1 - f)\phi_1(x)\, dx = 0$, which gives

$$a_1(24l + 57.6 + 161bl^4/315 + 9ql^2/5) + kl/3 = 0.$$

If, e.g., we take $l = b = 1$, and $k = 3$, then $a_1 = 0.011917$, and

$$u_1 = 0.011917(x - 1)^2(x^2 + 2x + 3).$$

For the 2nd approximation, we take $u_2 = a_1 \phi_1(x) + a_2 \phi_2(x)$. Then

$$\int_0^l (Lu_2 - f)\phi_1\, dx = 0, \quad \text{and} \quad \int_0^l (Lu_2 - f)\phi_2(x)\, dx = 0$$

which, with $l = b = 1$, $k = 3$, yield

$$83.911a_1 - 67.313a_2 = 1$$
$$67.213a_1 - 91.882a_2 = 0.7143.$$

Thus $a_1 = 0.013743$, $a_2 = 0.002279$, and

$$u_2 = 0.013743(x - 1)^2(x^2 + 2x + 3) + 0.002279(x - 1)^3(3x^2 + 4x + 3).$$

Instead of determining the exact solution, we can compare u_1 and u_2. Thus, e.g., $u_1(0.5) = 0.012662$, and $u_2(0.5) = 0.012964$, which give good results. ∎

Example 2.14:
We shall do the same problem as in Example 2.4, and use the weak variational formulation and the fundamental solution to solve it. Although the fundamental solutions are studied in Chapter 4, they, and this example in its elementary setting, are very important in the development of the BEM. The weak variational formulation of the problem is given by

$$
\begin{aligned}
0 &= \int_0^1 \left(\frac{d^2 u}{dx^2} + u + x \right) w\, dx \\
&= \int_0^1 \left(\frac{d^2 w}{dx^2} + w \right) u\, dx + \int_0^1 xw\, dx + \left[\frac{du}{dx}w - u\frac{dw}{dx} \right]_0^1 .
\end{aligned}
\tag{2.37}
$$

If we replace w by the fundamental solution u^* which is such that

$$\frac{d^2 u^*}{dx^2} + u^* = \delta(x_0),
\tag{2.38}$$

where $\delta(x_0)$ is the Dirac delta function at a point $x_0 \in (0,1)$, such that $\delta(x) = 0$ for $x \ne x_0$, then Eq (2.37) becomes

$$0 = \int_0^1 \left(\frac{d^2 u^*}{dx^2} + u^* \right) u \, dx + \int_0^1 x u^* \, dx + [qu^* - uq^*]_0^1$$

$$= u(x_0) + \int_0^1 x u^* \, dx + [qu^*]_0^1,$$

where we have used the property (1.55) and the notation $q = du/dx$ and $q^* = du^*/dx$. The above equation can be rewritten as

$$u(x_0) = -\int_0^1 x u^* \, dx - [qu^*]_0^1$$

(2.39)

$$= -\int_0^{x_0} x u^* \, dx - \int_{x_0}^1 x u^* \, dx - [qu^*]_0^{x_0} - [qu^*]_{x_0}^1 .$$

Now, it can be verified (by substituting into (2.38)) that the fundamental solution is

$$u^* = \frac{1}{2} \sin r,$$ (2.40)

where

$$r = |x - x_0| = \begin{cases} x_0 - x, & \text{for } 0 < x < x_0, \\ x - x_0, & \text{for } x_0 < x < 1. \end{cases}$$

Let us denote $q_0 = q(0)$ and $q_1 = q(1)$. Then from (2.39) we find that

$$q_0 = \frac{1}{\sin 1}, \quad q_1 = \frac{\cos 1}{\sin 1} - 1.$$

Thus from (2.39), e.g.,

$$u(.25) = \frac{1}{2} \int_0^{.25} x \sin(x - .25) \, dx - \frac{1}{2} \int_{.25}^1 x \sin(x - .25) \, dx - $$
$$- q_1 \sin .25 + q_0 \sin .25$$
$$\approx 0.0440113654,$$

or

$$u(.5) = \frac{1}{2} \int_0^{.5} x \sin(x - .5) \, dx - \frac{1}{2} \int_{.5}^1 x \sin(x - .5) \, dx - $$
$$- q_1 \sin .5 + q_0 \sin .5$$
$$\approx 0.06974694. \blacksquare$$

2.4. Choice of Test Functions

Note that a suitable choice of the test functions $\phi_i(x, y)$ can be made by taking linear combinations of polynomials, or trigonometric functions, such that they satisfy the boundary conditions. For example, we can choose a system of functions

$$\phi_0 = g, \quad \phi_1 = gx, \quad \phi_2 = gy, \quad \phi_3 = gx^2, \quad \phi_4 = gxy, \ldots, \quad (2.41)$$

where $g = g(x, y)$. It can be shown that the system (2.41) is complete.

Some practical rules for constructing the functions $g(x, y)$ in (2.41) are as follows:
(i) For the rectangle $[-a, a; -b, b]$:

$$g(x, y) = (x^2 - a^2)(y^2 - b^2).$$

(ii) For a circle of radius r and center at origin:

$$g(x, y) = r^2 - x^2 - y^2.$$

(iii) If the boundary C of a region R is defined by $F(x, y) = 0$, where $F \in C^n$, then

$$g(x, y) = \pm F(x, y).$$

See (ii) above if C is a circle.
(iv) For the case of a convex polygon whose sides are defined by $a_1 x + b_1 y + c_1 = 0, \cdots, a_m x + b_m y + c_m = 0$, we have

$$g(x, y) = \pm (a_1 x + b_1 y + c_1) \cdots (a_m x + b_m y + c_m).$$

See (i) above for a rectangle.
(v) The choice in (iv) is also suitable in different types of regions bounded by curved lines; e.g., for a sector formed by the circles of radii r and $r/2$, as in Fig. 2.1, we have

$$g(x, y) = (r^2 - x^2 - y^2)(x^2 - rx + y^2).$$

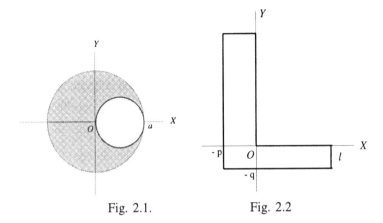

Fig. 2.1.   Fig. 2.2

(vi) For nonconvex polygons, the function $g(x, y)$ must be assigned piecewise in different parts of the region, and we must introduce moduli for any re–entrant angles. Thus, for the region in Fig. 2.2,

$$g(x, y) = \big(|x| + |y| - x - y\big)(x + p)(y + q)(l - x)(h - y)$$

$$= \begin{cases} -2y(x + p)(y + q)(l - x)(h - y), & \text{in } [0, l; -q, 0] \\ -2(x + y)(x + p)(y + q)(l - x)(h - y), & \text{in } [-p, 0; -q, 0] \\ -2x(x + p)(y + q)(l - x)(h - y) & \text{in } [-p, 0; 0, h]. \end{cases}$$

We can also take (2nd choice)

$$g(x, y) = \big(x^2 + y^2 - x|x| - y|y|\big)(x + p)(y + q)(l - x)(h - y).$$

In this case $g \in C^1$. A third choice is to assign the functions $u_n(x, y)$ separately in the three parts of the corner region of Fig. 2.2:

$$u_n(x, y) = \begin{cases} (x + p)x(h - y)(a_1 + a_2x + a_3y + \cdots + a_ny^m) \\ \quad \text{in } [-p, 0; 0, h] \\ (x + p)(y + q)(b_1 + b_2x + b_3y + \cdots + b_ny^m) \\ \quad \text{in } [-p, 0; -q, 0] \\ (y + q)(l - x)(c_1 + c_2x + c_3y + \cdots + c_ny^m) \\ \quad \text{in } [0, l; -q, 0], \end{cases}$$

where a_k, b_k, c_k $(k = 1, \cdots, n)$ are parameters which must be connected by a condition of conjugacy on the axes $x = 0$ and $y = 0$, viz,

$$(x + p)xh(a_1 + a_2x + a_4x^2 + \cdots) = (x + p)q(b_1 + b_2x + b_4x^2 + \cdots)$$
$$p(y + q)(b_1 + b_2y + \cdots + b_ny^m) = (y + q)yl(c_1 + c_2y + \cdots + c_ny^m).$$

In view of the above considerations, the test functions $\phi_i(x, y)$ are also called the shape functions for the region R.

Example 2.15:

Torsion of a prismatic rod of rectangular cross–section of length $2a$ and width $2b$ is defined by

$$\nabla^2 u = 2, \quad u = 0 \quad \text{on } C. \tag{2.42}$$

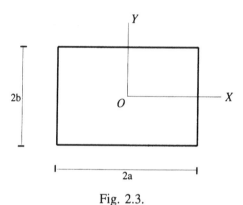

Fig. 2.3.

For the rectangle shown in Fig. 2.3, we choose $\phi(x, y) = (a^2 - x^2)(b^2 - y^2)$, and seek an approximate solution of the form

$$u_n(x, y) = (a^2 - x^2)(b^2 - y^2)\left(A_1 + A_2 x^2 + A_3 y^2 + \cdots + A_n x^{2i} y^{2j}\right).$$

First, for $n = 1$, we use (2.23a) with $f = 2$ and find that (with $\phi(x) = (a^2 - x^2)(b^2 - x^2)$)

$$0 = \iint_R \left(-\frac{\partial^2 u_1}{\partial x^2} - \frac{\partial^2 u_1}{\partial y^2} - 2\right)\phi \, dx \, dy$$

$$= 2\int_{-a}^{a}\int_{-b}^{b} \left[1 - A_1(a^2 - x^2) - A_1(b^2 - y^2)\right](a^2 - x^2)(b^2 - y^2) \, dy \, dx$$

$$= -\frac{128}{45}a^3 b^3 (a^2 + b^2)A_1 + \frac{32}{9}a^2 b^3$$

thus, $A_1 = 5/4(a^2 + b^2)$, and

$$u_1 = \frac{5}{4}\frac{(a^2 - x^2)(b^2 - y^2)}{a^2 + b^2}.$$

The torsional moment

$$M = 2G\theta \int_{-a}^{a} \int_{-b}^{b} u_1 \, dy \, dx = \frac{40}{9} \frac{G\theta a^3 b^3}{a^2 + b^2},$$

where G is the shear modulus, and θ is the angle of twist per unit length. The tangential stresses τ_{zx} and τ_{zy} are given by

$$\tau_{zx} = G\theta \frac{\partial u_1}{\partial y}, \quad \tau_{zy} = G\theta \frac{\partial u_1}{\partial x}.$$

For $a = b$, we find that $M = 20G\theta a^4/9 \approx 0.1388(2a)^4 G\theta$. The exact classical solution is given by

$$u = ax - x^2 - \frac{8a^2}{\pi^3} \sum_{n=1}^{\infty} \frac{\cosh \frac{(2n-1)\pi y}{2a}}{(2n-1)^3 \cosh \frac{(2n-1)\pi b}{a}} \sin \frac{(2n-1)\pi x}{a},$$

which gives

$$M = 2G\theta \left\{ \frac{a^3 b}{6} - \frac{32a^4}{\pi^5} \sum_{n=1}^{\infty} \frac{1}{(2n-1)^5} \tanh \frac{(2n-1)\pi b}{2a} \right\}.$$

For $a = b$, the exact value of M is $0.1406(2a)^4 G\theta$, which compares very well with the approximate value obtained above by the Galerkin method. ∎

Example 2.16:
Solve $\nabla^4 u = 0$ on the rectangle $[-a, a; -b, b]$ (Fig. 2.3) under the boundary conditions

$$\frac{\partial^2 u}{\partial x \partial y} = 0, \quad \frac{\partial^2 u}{\partial y^2} = c\left(1 - \frac{y^2}{b^2}\right) \quad \text{at } x = \pm a,$$

$$\frac{\partial^2 u}{\partial x \partial y} = 0, \quad \frac{\partial^2 u}{\partial x^2} = 0 \quad \text{at } y = \pm b,$$

where c is a constant. This problem pertains to the expansion of a rectangular plate under tensile forces.

First, we will reduce the above boundary conditions to homogeneous boundary conditions: The function

$$u_0 = \frac{c}{2} y^2 \left(1 - \frac{y^2}{6b^2}\right)$$

obviously satisfies the given boundary conditions (it follows by integrating each one of the above boundary conditions). Set $u = u_0 + \hat{u}$. Then $\nabla^4 \hat{u} = 2c/b^2$, and the boundary conditions become

$$\frac{\partial \hat{u}}{\partial x \partial y} = 0, \quad \frac{\partial \hat{u}}{\partial y^2} = 0 \quad \text{at } x = \pm a,$$

$$\frac{\partial^2 \hat{u}}{\partial x \partial y} = 0, \quad \frac{\partial^2 \hat{u}}{\partial x^2} = 0 \quad \text{at } y = \pm b.$$

These boundary conditions will be satisfied if the following conditions are met:

$$\hat{u} = 0, \quad \frac{\partial \hat{u}}{\partial x} = 0 \quad \text{at } x = \pm a,$$

$$\hat{u} = 0, \quad \frac{\partial \hat{u}}{\partial y} = 0 \quad \text{at } y = \pm b.$$

Thus, the given problem reduces to that of minimizing the integral

$$I(u) = \iint_R \left[(\nabla^2 \hat{u})^2 - \frac{4c}{b^2} \hat{u} \right] dx\, dy.$$

Then, by Rayleigh–Ritz (or Galerkin) method

$$\iint_R \left(\nabla^4 u_n - f \right) \phi_j\, dx\, dy = 0, \quad j = 1, \ldots, n, \qquad (2.43)$$

where u_n is the n–th approximate solution, which, in view of the geometric symmetry of the rectangle, is taken as

$$u_n = (x^2 - a^2)^2 (y^2 - b^2)^2 (a_1 + a_2 x^2 + a_3 y^2 + \cdots).$$

For $n = 1$, we find from (2.43) that

$$\int_{-a}^{a} \int_{-b}^{b} [24 a_1 (y^2 - b^2)^2 + 16 a_1 (3x^2 - a^2)(3y^2 - b^2) +$$

$$+ 24 a_1 (x^2 - a^2)^2 - \frac{2c}{b}](x^2 - a^2)^2 (y^2 - b^2)^2\, dy\, dx = 0,$$

or

$$\left(\frac{54}{7} + \frac{256}{49} \frac{b^2}{a^2} + \frac{64}{7} \frac{b^4}{a^4} \right) a_1 = \frac{c}{a^4 v^2},$$

which gives $a_1 = 0.043253 c/a^6$, and

$$u_1 = u_0 + \hat{u}_1 = \frac{c}{2} y^2 \left(1 - \frac{y^2}{6b^2}\right) + \frac{0.04253 c}{a^6}(x^2 - a^2)^2 (y^2 - b^2)^2. \ \blacksquare$$

2.5. Problems

Derive the variational formulation for the boundary value problems **2.1–2.10**
(Here a, b, f, g are functions of x; u_0, h_0, m_0, q_0, T_∞, u_∞ are constants):

2.1. $-\dfrac{d}{dx}\left(a\dfrac{du}{dx}\right) - f = 0$, $u(0) = u_0$, $a\dfrac{du}{dx}(l) = q_0$, $0 < x < l$ (one–
dimensional heat conduction).

2.2. $-\dfrac{d}{dx}\left(a\dfrac{du}{dx}\right) + f = 0$; $u(0) = 0$, with $\left[a\frac{du}{dx} + h_0(u - u_\infty)\right]_{x=1} = q_0$,
$0 < x < 1$ (one–dimensional heat conduction/convection).

2.3. $-\dfrac{d}{dx}\left(a\dfrac{du}{dx}\right) - cu + x^2 = 0$; $u(0) = 0$, $a\dfrac{du}{dx}(1) = 1$, $0 < x < 1$
(one–dimensional deformation of a bar).

2.4. $-\dfrac{d}{dx}\left(u\dfrac{du}{dx}\right) + f = 0$; $\dfrac{du}{dx}(0) = 0$, $u(1) = \sqrt{2}$, $0 < x < 1$. (one–
dimensional nonlinear equation).

2.5. $-\dfrac{d}{dx}\left(a\dfrac{du}{dx}\right) + \lambda u = 0$; $u(0) = 0$, with $\left[a\frac{du}{dx} + ku\right]_{x=l} = 0$, $0 < x < l$
(one–dimensional longitudinal deflection of a bar with one end spring).

2.6. $\dfrac{d^2}{dx^2}\left(a\dfrac{d^2u}{dx^2}\right) + f = 0$, $u(0) = \dfrac{du}{dx}(0) = 0$,

$a\dfrac{d^2u}{dx^2}\Big|_{x=l} = m_0$, $\dfrac{d}{dx}\left(a\dfrac{d^2u}{dx^2}\right)\Big|_{x=l} = f_0$ (one–dimensional heat conduc-
tion/convection).

2.7.

$$-\dfrac{\partial}{\partial x}\left(c_{11}\dfrac{\partial u}{\partial y} + c_{12}\dfrac{\partial u}{\partial x}\right) - \dfrac{\partial}{\partial y}\left(c_{21}\dfrac{\partial u}{\partial x} + c_{22}\dfrac{\partial u}{\partial y}\right) + f = 0$$

in R with boundary conditions $u = u_0$ on C_1, and $q_n \equiv \Big(c_{11}\dfrac{\partial u}{\partial x} +$
$c_{12}\dfrac{\partial u}{\partial y}\Big)n_x + \Big(c_{21}\dfrac{\partial u}{\partial x} + c_{22}\dfrac{\partial u}{\partial y}\Big)n_y = q_0$, on C_2, where c_{ij}, u_0, and q_0
are prescribed.

2.8. $-k\left(\dfrac{\partial^2 T}{\partial x^2} + \dfrac{\partial^2 T}{\partial y^2}\right) = f$ in the region R with boundary conditions
as shown in the Fig. 2.4 below. The following boundary conditions are
prescribed: $ku_x = q_0(y)$ on HA; $ku_x = -h(u - u_\infty)$ on BC; $u = u_0(x)$
on AB, and $\partial u/\partial n = q_0 = 0$ on $CDEFGH$ (insulated), where k is the
thermal conductivity of the material of the region R, h and u_∞ are ambient
quantities, and $\partial u/\partial n = -\partial u/\partial x = -u_x$ on HA (two–dimensional heat

conduction).

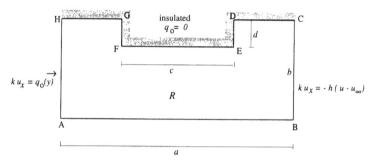

$k\,u_x = q_0(y)$

$k\,u_x = -h\,(u - u_\infty)$

Fig. 2.4.

2.9. $-\dfrac{d}{dx}\left\{a\left[\dfrac{du}{dx} + \dfrac{1}{2}\left(\dfrac{dv}{dx}\right)^2\right]\right\} + g = 0,$

$\dfrac{d^2}{dx^2}\left(b\dfrac{d^2v}{dx^2}\right) - \dfrac{d}{dx}\left\{a\dfrac{dv}{dx}\left[\dfrac{du}{dx} + \dfrac{1}{2}\left(\dfrac{dv}{dx}\right)^2\right]\right\} + f = 0;$

$u = v = 0$ at $x = 0, l$; $\left.\dfrac{dv}{dx}\right|_{x=0} = 0,$ $\left[b\dfrac{d^2v}{dx^2}\right]_{x=l} = m_0.$ (Large-deflection bending of a beam).

2.10. Find the functional $I(u)$ for the transverse deflection u of a membrane stretched across a frame, in the shape of a curve C, subjected to a pressure loading $f(x,y)$ per unit area. Assume that the tension T in the membrane is constant. Note that u satisfies the equation

$$-\nabla^2 u = \dfrac{f}{T}.$$

2.11. Consider the Poisson boundary value problem: $-\nabla^2 u = f$ in R, with the boundary conditions $u = 0$ on C_1 and $\partial u/\partial n = 0$ on C_2. Show that

$$I(u) = \dfrac{1}{2}\iint_R\left\{\left(\dfrac{\partial u}{\partial x}\right)^2 + \left(\dfrac{\partial u}{\partial y}\right)^2 - 2fu\right\}dx\,dy.$$

Use (a) Galerkin method, and (b) Rayleigh–Ritz method, to solve:

2.12. $-u'' = 2,\ \ 0 < x < 1;\ u(0) = u'(1) = 0.$

2.13. $-u'' = e^x,\ \ 0 < x < 1;\ u(0) = u(1) = 0.$

2.14. $-u'' = x^2,\ \ 0 < x < 1;\ u(0) = 1,\ u'(1) + 2u(1) = 1.$

2.15. $-(1+x)u'' - u' = x, 0 \le x \le 1;\ u(0) = u(1) = 0.$

2.16. $-u'' - u - x = 0, 0 < x < 1; u(0) = 0, u'(1) = 0.$

2.17. $\dfrac{d^2}{dx^2}\left(EI\dfrac{d^2u}{dx^2}\right) + f = 0, \quad 0 < x < l, \ EI > 0, \ f = \text{const}$, where EI is called the flexural rigidity of the beam, with

$$u(0) = 0 = \frac{du}{dx}(0), \quad EI\frac{d^2u}{dx^2}\Big|_{x=l} = M_0, \quad \frac{d}{dx}\left(EI\frac{d^2u}{dx^2}\right)\Big|_{x=l} = 0.$$

[Take $w = \phi_i = x^{i+1}$. The exact solution is

$$EI\,u = \frac{fl^4}{24} - \frac{fl^3}{6}x + \frac{M_0}{2}x^2 - \frac{f}{24}(l-x)^4.]$$

2.18. $-\nabla^2 u = 1$ in $R = \{(x,y) : 0 < x, y < 1\}$ such that

$$u(1,y) = 0 = u(x,1), \quad \frac{\partial u}{\partial n}(0,y) = 0 = \frac{\partial u}{\partial n}(x,0).$$

[If we take $w = \phi_i = (1 - x^i)(1 - y^i), i = 1, \cdots, n$, then this choice satisfies the essential boundary conditions, but not the natural boundary conditions. Hence, we assume the first approximate solution as $u_1 = a\phi_1, \ \phi_1 = (1 - x^2)(1 - y^2)$. Alternatively, we can take $w = \phi_i = \cos\dfrac{(2i-1)\pi x}{2}\cos\dfrac{(2i-1)\pi y}{2}, i = 1, \cdots, n$. The exact solution is

$$u(x,y) = \frac{1}{2}\Big\{(1 - y^2) +$$

$$+ \frac{32}{\pi^3}\sum_{k=1}^{\infty}\frac{(-!)^k\cos[(2k-1)\pi y/2]\cosh[(2k-1)\pi x/2]}{(2k-1)^3\cosh[(2k-1)\pi/2]}\Big\}.]$$

References and Bibliography for Additional Reading

C. A. Brebbia, *The Boundary Element Method for Engineers*, Pentech Press, London, 1980.

———(Ed.), *Recent Advances in Boundary Elements*, Pentech Press, London, 1978.

———, *New Developments in Boundary Element Methods*, Proc. 2nd Conf. on BEM, CML Publications, Southampton, 1980.

———, *Boundary Element Methods*, Proc. 3rd Conf. on BEM, Springer–Verlag, Berlin, 1981.

———, *Boundary Element Methods in Engineering*, Proc. 4th Conf. on BEM, Springer–Verlag, Berlin, 1982.

———, *Progress in Boundary Element Methods*, Vol. 1 and 2, Pentech Press, London, 1981, 1983.

J. J. Connor and C. A. Brebbia, *Finite Element Techniques for Structural Engineers*, Butterworths, London, 1973.

———, *Finite Element Techniques for Fluid Flow*, Butterworths, London, 1976.

A. J. Davies, *The Finite Element Method*, Clarendon Press, Oxford, 1980.

B. A. Finlayson, *The Method of Weighted Residuals and Variational Principles*, Academic Press, New York, 1972.

L. V. Kantorovitch and V. L. Krylov, *Approximate Methods of Higher Analysis*, Interscience, New York, 1958.

S. G. Mikhlin S.G., *Integral Equations*, Pergamon Press, New York, 1957.

———, *Variational Methods in Mathematical Physics*, MacMillan, New York, 1964.

J. N. Reddy, *An Introduction to the Finite Element Method*, McGraw–Hill, New York, 1984.

V. J. Smirnov, *A Course in Higher Mathematics*, Vol. IV, Addison–Wesley, London, 1964.

O. D. Kellogg, *Foundations of Potential Theory*, Dover, New York, 1953.

G. T. Symm, *Integral Equation Methods in Potential Theory* II, Proc. Roy. Soc. Ser. A **275** (1963), 33–46.

K. Washizu, *Variational Methods in Elasticity and Plasticity*, 2nd ed., Pergamon Press, New York, 1975.

O. C. Zienkiewicz and G. S. Holister (Eds.), *Stress Analysis,* Wiley, London, 1966.

3

One–Dimensional Problems

For one–dimensional regions the boundary consists of the two endpoints. We
will introduce the boundary element method for one–dimensional problems
by integrating the fundamental solution of the boundary value problem at the
boundary points. Although the boundary element method is more useful in
two–dimensional and three–dimensional problems, this chapter is introduced
to clarify the concepts and the method in the simplest possible geometric setting. We will consider one–dimensional potential flow problem (2nd–order
equations) and the rectangular beam problem (4th–order equations) to illustrate the method.

3.1. Potential Flow

The potential flow in a one–dimensional region is defined by the equation

$$-\frac{d}{dx}\left(a\frac{du(x)}{dx}\right) + f(x) = 0, \quad 0 \le x \le l, \tag{3.1}$$

where u denotes the potential, and $a = a(x)$. The flux $q(x)$ is then given by

$$q(x) = \frac{du(x)}{dx}. \tag{3.2}$$

Eq (3.1) is found in problems of transverse deflection of a cable, axial deformation of a bar, heat transfer along a fin in heat exchangers, flow though

pipes, laminar incompressible flow through a channel under constant pressure gradient, linear flow through porous media, and electrostatics. We introduce a function $u^*(x, x')$, sufficiently continuous to be differentiable as often as needed. Using the 3–step procedure of §2.1 with $w = u^*$, we start with

$$0 = - \int_0^l \frac{d}{dx} \left(a\frac{du(x)}{dx} \right) u^* \, dx + \int_0^l f(x)u^* \, dx,$$

and get

$$\left[a\left(u^*\frac{du}{dx} - \frac{du^*}{dx}u(x) \right) \right]_0^l + \int_0^l au\frac{d^2u^*}{dx^2} \, dx = \int_0^l f(x)u^* \, dx. \qquad (3.3)$$

Let u^* be a solution of

$$-\frac{d}{dx} \left(a\frac{du^*(x, x')}{dx} \right) = -\delta(x, x'). \qquad (3.4)$$

Then substituting (3.4) into (3.3) and using (1.55) we get

$$\left[a\left(u^*\frac{du}{dx} - u(x)\frac{du^*}{dx} \right) \right]_0^l + u(x') = \int_0^l f(x)u^* \, dx. \qquad (3.5)$$

Now, if $a = $ const (for a homogeneous isotropic medium), it is known (see Chapter 4) that the fundamental solution of (3.4) is given by

$$u^*(x, x') = \frac{1}{2a} (l - r), \quad r = |x - x'|. \qquad (3.6)$$

This yields

$$q^*(x, x') = \frac{du^*}{dx} = \frac{1}{2a} \, \text{sgn}\, r, \qquad (3.7)$$

where

$$\text{sgn}\, r = \begin{cases} 1 & \text{for } r > 0, x > x', \\ -1 & \text{for } r < 0, x < x', \\ \text{undefined} & \text{for } r = 0, x = x', \end{cases} \qquad (3.8)$$

but $r\,\text{sgn}\, r = 0$ at $r = 0$. Note that $u^* = 0$ at $r = l$. Now, using the solution

(3.6)–(3.7), we find from (3.5) the boundary integral equation (BI Eq)

$$
\begin{aligned}
u(x') &= a\big[q^*u - u^*q\big]_0^l + \int_0^l f(x)u^*\,dx \\
&= a\big[-q^*(0,x')u(0) + q^*(l,x')u(l) + u^*(0,x')q(0) - u^*(l,x')q(l)\big] \\
&\quad + \int_0^l f(x)u^*(x,x')\,dx, \\
&= a\big[-q^*(0,x') \quad q^*(l,x')\big]\left\{\begin{array}{c} u(0) \\ u(l) \end{array}\right\} \\
&\quad + a\big[u^*(0,x') \quad -u^*(l,x')\big]\left\{\begin{array}{c} q(0) \\ q(l) \end{array}\right\} \\
&\quad + \int_0^l f(x)u^*(x,x')\,dx,
\end{aligned}
$$

(3.9)

which, in view of (3.2), gives the flux as

$$
\begin{aligned}
q(x') &= a\big[Q^*(0,x') \quad -Q^*(l,x')\big]\left\{\begin{array}{c} u(0) \\ u(l) \end{array}\right\} \\
&\quad + a\big[-U^*(0,x') \quad U^*(l,x')\big]\left\{\begin{array}{c} q(0) \\ q(l) \end{array}\right\} - \int_0^l f(x)U^*(x,x')\,dx,
\end{aligned}
$$

(3.10)

where $Q^*(x,x') = -dq^*/dx'$, $U^*(x,x') = -du^*/dx'$. Eqs (3.9) and (3.10) determine the potential u and the flux q at an interior point x' provided that any two of the boundary conditions $u(0)$, $u(l)$, $q(0)$, $q(l)$ are known and the initial source function $f(x)$ is prescribed. Note that if there is a single point–source $f(x_0)$ at some point x_0 in the interval $(0,l)$, then

$$
\int_0^l f(x_0)u^*(x_0,x')\,dx = lf(x_0)u^*(x_0,x').
$$

(3.11)

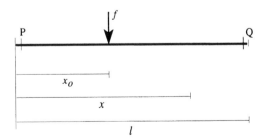

Fig. 3.1. One–Dimensional Element PQ.

Now, we take $x' = 0 + \varepsilon$ at P, and $x' = l - \varepsilon$ at Q. Then, using (3.9), we get

$$\left\{ \begin{array}{c} u(0 + \varepsilon) \\ u(l - \varepsilon) \end{array} \right\} = a \left[\begin{array}{cc} -q^*(0, 0 + \varepsilon) & q^*(l, 0 + \varepsilon) \\ -q^*(0, l - \varepsilon) & q^*(l, l - \varepsilon) \end{array} \right] \left\{ \begin{array}{c} u(0) \\ u(l) \end{array} \right\}$$

$$+ a \left[\begin{array}{cc} u^*(0, 0 + \varepsilon) & -u^*(l, 0 + \varepsilon) \\ u^*(0, l - \varepsilon) & -u^*(l, l - \varepsilon) \end{array} \right] \left\{ \begin{array}{c} q(0) \\ q(l) \end{array} \right\}$$

$$+ \left\{ \begin{array}{c} \int_0^l f(x) u^*(x, 0 + \varepsilon)\, dx \\ \int_0^l f(x) u^*(x, l - \varepsilon)\, dx \end{array} \right\}. \tag{3.12}$$

In view of (3.6) and (3.7), we find that $u^*(0, 0 + \varepsilon) = l/2a = u^*(l, l - \varepsilon)$, $u^*(l, 0 + \varepsilon) = 0 = u^*(0, l - \varepsilon)$, $q^*(0, 0 + \varepsilon) = -1/2a = q^*(0, l - \varepsilon)$, and $q^*(l, 0 + \varepsilon) = 1/2a = q^*(l, l - \varepsilon)$. Thus, as $\varepsilon \to 0$, we obtain from (3.12) the boundary element equation (BE Eq)

$$\left[\begin{array}{cc} l/2 & 0 \\ 0 & l/2 \end{array} \right] \left\{ \begin{array}{c} q(0) \\ q(l) \end{array} \right\} - \left[\begin{array}{cc} 1/2 & -1/2 \\ -1/2 & 1/2 \end{array} \right] \left\{ \begin{array}{c} u(0) \\ u(l) \end{array} \right\}$$

$$+ \left\{ \begin{array}{c} \int_0^l f(x) u^*(x, 0)\, dx \\ \int_0^l f(x) u^*(x, l)\, dx \end{array} \right\} = 0. \tag{3.13}$$

The formula (3.13) computes the unknowns as follows:

(i) If $u(0), u(l)$ and $f(x)$ are prescribed, it gives $q(0), q(l)$ (Dirichlet).

(ii) If $q(0), q(l)$ and $f(x)$ are prescribed, it gives $u(0), u(l)$ (Neumann).

(iii) If $u(0), q(l)$ or $u(l), q(0)$ and $f(x)$ are prescribed, it gives $u(l), q(0)$ or $u(0), q(l)$ (Mixed).

Example 3.1:
(Dirichlet boundary value problem) $u(0) = 0 = u(l)$, $a = 1$, and f is prescribed at $x = x_1' \in (0, l)$. Then from (3.13)

$$\left[\begin{array}{cc} l/2 & 0 \\ 0 & -l/2 \end{array} \right] \left\{ \begin{array}{c} q(0) \\ q(l) \end{array} \right\} = - \left\{ \begin{array}{c} lf(x_1') u^*(0, x_1') \\ lf(x_1') u^*(l, x_1') \end{array} \right\} = -\frac{l}{2} f(x_1') \left\{ \begin{array}{c} l - x_1' \\ x_1' \end{array} \right\},$$

which gives

$$q(0) = -(l - x_1') f(x_1'), \quad q(l) = x_1' f(x_1'). \ \blacksquare$$

Example 3.2:
(Mixed boundary value problem) $u(0) = U_0, q(l) = V$, $a = 1$, and f is prescribed at $x = x_1'$. Then, from (3.12)

$$\begin{bmatrix} l/2 & 0 \\ 0 & -l/2 \end{bmatrix} \begin{Bmatrix} q(0) \\ V \end{Bmatrix} = \begin{bmatrix} 1/2 & -1/2 \\ -1/2 & 1/2 \end{bmatrix} \begin{Bmatrix} U_0 \\ u(l) \end{Bmatrix} =$$

$$= -\frac{l}{2} f(x_1') \begin{Bmatrix} l - x_1' \\ x_1' \end{Bmatrix},$$

which gives

$$q(0) = V - f(x_1'), \quad u(l) = U_0 - lV + lx_1' f(x_1'). \ \blacksquare$$

Example 3.3:
Let the axial deformation of a bar be governed by (3.1), $0 < x < l$, with the boundary conditions $u(0) = 0$, $\left(a \dfrac{du}{dx} \right)_{x=l} = P$. The exact solution of this boundary value problem is given by

$$u = \frac{f}{2a}(2xl - x^2) + \frac{P}{a}x, \quad a\frac{du}{dx} = f(l - x) + P.$$

The BE solution from (3.13) is

$$\begin{bmatrix} l/2 & 0 \\ 0 & l/2 \end{bmatrix} \begin{Bmatrix} q(0) \\ q(l) = P/a \end{Bmatrix} - \begin{bmatrix} 1/2 & -1/2 \\ -1/2 & 1/2 \end{bmatrix} \begin{Bmatrix} u(0) = 0 \\ u(l) \end{Bmatrix}$$

$$+ \begin{Bmatrix} \int_0^l f(x)u^*(x,0)\, dx = fl^2/4a \\ \int_0^l f(x)u^*(x,l)\, dx = fl^2/4a \end{Bmatrix} = 0.$$

This yields $u(l) = fl^2/2a + Pl/a$, $q(0) = -du/dx|_{x=0} = -(fl + P)/a$, which matches with the above exact solution. $\ \blacksquare$

3.2. Bending of an Elastic Beam

This problem is defined by

$$\frac{d^2}{dx^2} \left(b \frac{d^2 u}{dx^2} \right) = f(x), \quad 0 \leq x \leq l, \tag{3.14}$$

where $b = EI$ is the flexural rigidity of the beam (E being the modulus of elasticity and I the moment of inertia). Then, using the three–step technique of §2.1, we start with

$$\int_0^l \left[\frac{d^4 u(x)}{dx^4} - f(x) \right] u^* \, dx = 0$$

where u^* contains b (see (3.20)), and by a repeated integration by parts, get

$$\left[u^* \frac{d^3 u}{dx^3} - \frac{du^*}{dx} \frac{d^2 u}{dx^2} + \frac{d^2 u^*}{dx^2} \frac{du}{dx} - \frac{d^3 u^*}{dx^3} u \right]_0^l$$

$$+ \int_0^l \frac{d^4 u^*}{dx^4} u(x) \, dx - \int_0^l f(x) u^* \, dx = 0. \tag{3.15}$$

We will use the following notation:

Displacement	$u(x)$
Rotation	$\theta(x) = \dfrac{du}{dx}$
Moment	$m(x) = -\dfrac{d^2 u}{dx^2}$
Shear	$s(x) = -\dfrac{d^3 u}{dx^3}$
Virtual displacement	$u^*(x, x')$
Virtual Rotation	$\theta^*(x, x') = \dfrac{du^*}{dx}$
Virtual Moment	$m^*(x, x') = -\dfrac{d^2 u^*}{dx^2}$
Virtual Shear	$s^*(x, x') = -\dfrac{d^3 u^*}{dx^3}$

Note that the rotation θ is also known as the angle of twist per unit length of the beam. From (3.14) we have $f^*(x, x') = b d^4 u^* / dx^4$. Using the above notation, Eq (3.15) becomes

$$\left[-u^*(x, x') s(x) + \theta^*(x, x') m(x) - m^*(x, x') \theta(x) + s^*(x, x') u(x) \right]_0^l$$

$$+ \int_0^l \frac{d^4 u^*}{dx^4} u(x) \, dx - \int_0^l f(x) u^*(x) \, dx = 0. \tag{3.16}$$

Let $u^*(x, x')$ be the fundamental solution of

$$\frac{d^4 u^*}{dx^4} = \delta(x, x'). \tag{3.17}$$

Then, since

$$\int_0^l u \frac{d^4 u^*}{dx^4} \, dx = \int_0^l u(x)\delta(x,x') \, dx = u(x'), \tag{3.18}$$

Eq (3.16) leads to the BI Eq

$$-u(x') = \left[-u^* s(x) + \theta^*(x,x')m(x) - m^*(x,x')\theta(x) + s^*(x,x')u(x) \right]_0^l$$
$$- \int_0^l f(x)u^*(x,x') \, dx. \tag{3.19}$$

This formula computes the displacement at any interior point x'.

Note that the fundamental solution of (3.17) is

$$u^*(x,x') = \lambda l^3 \left(2 + |\rho|^3 - 3|\rho|^2 \right), \tag{3.20}$$

where $\rho = r/l$, $\lambda = b/12$, and $r = x - x'$. This gives (since $\dfrac{d}{dx}|\rho| = \dfrac{|\rho| \, d\rho}{\rho \, dx}$,
$\dfrac{d\rho}{dx} = \dfrac{1}{l}$, $\dfrac{d\rho}{dx'} = -\dfrac{1}{l}$, and $\dfrac{|\rho|}{\rho} = \dfrac{|r|}{r} = \operatorname{sgn} r$)

$$u^*(x,x') = \lambda \left[|r|^3 - 3l|r|^2 + 2l^3 \right],$$
$$\theta^*(x,x') = \frac{du^*(x,x')}{dx} = 3\lambda \left[|r|^2 - 2l|r| \right] \operatorname{sgn} r,$$
$$m^*(x,x') = -\frac{d^2 u^*(x,x')}{dx^2} = 6\lambda \left[l - |r| \right], \tag{3.21}$$
$$s^*(x,x') = -\frac{d^3 u^*(x,x')}{dx^3} = -6\lambda \operatorname{sgn} r,$$

where u^* contains b (in λ), and $u^* = 0$ at $r = l$. Also, from (3.21)

$$U^*(x,x') = \frac{du^*}{dx'} = -3\lambda \left[|r|^2 - 2l|r| \right] \operatorname{sgn} r,$$
$$Q^*(x,x') = \frac{d\theta^*}{dx'} = 6\lambda \left[l - |r| \right],$$
$$P^*(x,x') = \frac{dm^*}{dx'} = 6\lambda \operatorname{sgn} r, \tag{3.22}$$
$$D^*(x,x') = \frac{ds^*}{dx'} = 0.$$

Substituting (3.21)–(3.22) into (3.19) we get the BE Eq

$$-u(x') = [\, u^*(0,x') \quad -u^*(l,x') \quad -\theta^*(0,x') \quad \theta^*(l,x')\,] \left\{ \begin{array}{c} s(0) \\ s(l) \\ m(0) \\ m(l) \end{array} \right\}$$

$$+ [\, m^*(0,x') \quad -m^*(l,x') \quad -s^*(0,x') \quad s^*(l,x')\,] \left\{ \begin{array}{c} \theta(0) \\ \theta(l) \\ u(0) \\ u(l) \end{array} \right\}$$

$$- \int_0^l f(x)u^*(x,x')\,dx.$$

$$(3.23)$$

This formula computes the transverse displacement at any interior point x' if any four of the eight values of u, θ, m, s at $x = 0, l$ are known, and the initial transverse load of intensity $f(x)$ is prescribed.

If we differentiate (3.23) with respect to x' and use (3.22), we get the formula for rotation at an interior point x':

$$-\theta(x') = [\, U^*(0,x') \quad -U^*(l,x') \quad -Q^*(0,x') \quad Q^*(l,x')\,] \left\{ \begin{array}{c} s(0) \\ s(l) \\ m(0) \\ m(l) \end{array} \right\}$$

$$+ [\, P^*(0,x') \quad -P^*(l,x') \quad -D^*(0,x') \quad D^*(l,x')\,] \left\{ \begin{array}{c} \theta(0) \\ \theta(l) \\ u(0) \\ u(l) \end{array} \right\}$$

$$- \int_0^l f(x)U^*(x,x')\,dx.$$

$$(3.24)$$

Similar expressions can be written for the evaluation of moments and shears at an interior point x'.

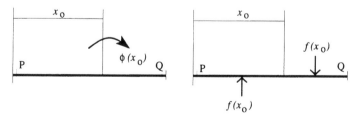

Fig. 3.2. Point–load and point–moment on a beam.

Now, consider an applied point–moment $\phi(x_0)$ at a point $x_0 \in (0, l)$. Then the problem can be solved as a limiting case of two opposite point–loads as shown in Fig. 3.2. As before, we take the point x' on the boundaries (i.e., the endpoints) such that $x' = 0 + \varepsilon$ at P and $x' = l - \varepsilon$ at Q. Then, from (3.23)

$$-\left\{ \begin{array}{c} u(0+\varepsilon) \\ u(l-\varepsilon) \\ \theta(0+\varepsilon) \\ \theta(l-\varepsilon) \end{array} \right\} =$$

$$= \begin{bmatrix} u^*(0,0+\varepsilon) & -u^*(l,0+\varepsilon) & -\theta^*(0,0+\varepsilon) & \theta^*(l,0+\varepsilon) \\ u^*(0,l-\varepsilon) & -u^*(l,l-\varepsilon) & -\theta^*(0,l-\varepsilon) & \theta^*(l,l-\varepsilon) \\ U^*(0,0+\varepsilon) & -U^*(l,0+\varepsilon) & -Q^*(0,0+\varepsilon) & Q^*(l,0+\varepsilon) \\ U^*(0,l-\varepsilon) & -U^*(l,l-\varepsilon) & -Q^*(0,l-\varepsilon) & Q^*(l,l-\varepsilon) \end{bmatrix} \times$$

$$\times \left\{ \begin{array}{c} s(0) \\ s(l) \\ m(0) \\ m(l) \end{array} \right\} +$$

$$+ \begin{bmatrix} -s^*(0,0+\varepsilon) & s^*(l,0+\varepsilon) & m^*(0,0+\varepsilon) & -m^*(l,0+\varepsilon) \\ -s^*(0,l-\varepsilon) & s^*(l,l-\varepsilon) & m^*(0,l-\varepsilon) & -m^*(l,l-\varepsilon) \\ -D^*(0,0+\varepsilon) & D^*(l,0+\varepsilon) & P^*(0,0+\varepsilon) & -P^*(l,0+\varepsilon) \\ -D^*(0,l-\varepsilon) & D^*(l,l-\varepsilon) & P^*(0,l-\varepsilon) & -P^*(l,l-\varepsilon) \end{bmatrix} \times$$

$$\times \left\{ \begin{array}{c} u(0) \\ u(l) \\ \theta(0) \\ \theta(l) \end{array} \right\} - \left\{ \begin{array}{c} \int_0^l f(x)u^*(x,0+\varepsilon)\, dx \\ \int_0^l f(x)u^*(x,l-\varepsilon)\, dx \\ \int_0^l f(x)U^*(x,0+\varepsilon)\, dx \\ \int_0^l f(x)U^*(x,l-\varepsilon)\, dx \end{array} \right\}. \tag{3.25}$$

As $\varepsilon \to 0$, Eq (3.24) determines the unknowns whenever four out of the eight boundary conditions (on u, θ, m, s at $x = 0, l$) and f are prescribed. Note that

1. Eq (3.17) is the virtual work equation for a simple loaded beam.

2. If we integrate (3.16) or (3.17) by parts twice, we find that

$$\left[-u^*(x,x')s(x) + \theta^*(x,x')m(x) \right]_0^l - \int_0^l f(x)u^*(x,x')\, dx$$

$$= \left[-u(x)s^*(x,x') + \theta(x)m^*(x,x') \right]_0^l - \int_0^l f^*(x,x')u(x)\, dx. \tag{3.26}$$

Eq (3.26) is known as the reciprocal theorem which states that the work done by the forces of a system A (the real system) on the displacements of another

system B (any admissible system) is equal to the work done by the forces of
the system B on the displacements of the system A.

Before we solve some problems, we shall evaluate various quantities in
(3.25) as $\varepsilon \to 0$. Thus, with $f =$const,

$$u^*(0, 0 + \varepsilon) \to 2\lambda l^3, \quad u^*(l, 0 + \varepsilon) \to 0,$$
$$u * (0, l - \varepsilon) \to 0, \quad u^*(l, l - \varepsilon) \to 2\lambda l^3,$$
$$\theta^*(0, 0 + \varepsilon) \to 0, \quad \theta^*(l, 0 + \varepsilon) \to -3\lambda l^2,$$
$$\theta^*(0, l - \varepsilon) \to 3\lambda l^2, \quad \theta^*(l, l - \varepsilon) \to 0,$$

$$m^*(0, 0 + \varepsilon) \to 6\lambda l, \quad m^*(l, 0 + \varepsilon) \to 0,$$
$$m^*(0, l - \varepsilon) \to 0, \quad m^*(l, l - \varepsilon) \to 6\lambda l,$$
$$s^*(0, 0 + \varepsilon) \to 6\lambda, \quad s^*(l, 0 + \varepsilon) \to -6\lambda,$$
$$s^*(0, l - \varepsilon) \to 6\lambda, \quad s^*(l, l - \varepsilon) \to -6\lambda,$$

$$U^*(0, 0 + \varepsilon) \to 0, \quad U^*(l, 0 + \varepsilon) \to 3\lambda l^2,$$
$$U^*(0, l - \varepsilon) \to -3\lambda l^2, \quad U^*(l, l - \varepsilon) \to 0,$$
$$Q^*(0, 0 + \varepsilon) \to 6\lambda l, \quad Q * (l, 0 + \varepsilon) \to 0,$$
$$Q^*(0, l - \varepsilon) \to 0, \quad Q^*(l, l - \varepsilon) \to 6\lambda l,$$
$$P^*(0, 0 + \varepsilon) \to -6\lambda, \quad P^*(l, 0 + \varepsilon) \to 6\lambda,$$
$$P^*(0, l - \varepsilon) \to -6\lambda, \quad P^*(l, l - \varepsilon) \to 6\lambda,$$
$$D^*(0, 0 + \varepsilon) \to 0, \quad D^*(l, 0 + \varepsilon) \to 0,$$
$$D^*(0, l - \varepsilon) \to 0, \quad D^*(l, l - \varepsilon) \to 0,$$

$$\int_0^l (-f)u^*(x, 0 + \varepsilon) \to \frac{5}{4}f\lambda l^4, \quad \int_0^l (-f)u^*(x, l - \varepsilon) \to \frac{5}{4}f\lambda l^4,$$
$$\int_0^l (-f)U^*(x, 0 + \varepsilon) \to 2f\lambda l^3, \quad \int_0^l (-f)U^*(x, l - \varepsilon) \to -2f\lambda l^4.$$

$$(3.27)$$

Example 3.4:
The bending of an elastic beam is defined by

$$\frac{d^4u}{dx^4} = f, \quad 0 < x < l,$$

$$u(0) = 0, \quad \frac{du(0)}{dx} = 0, \quad \left(\frac{d^2u}{dx^2}\right)_{x=l} = M, \quad \frac{d}{dx}\left(\frac{d^2u}{dx^2}\right)_{x=l} = F.$$

The exact solution of this problem is given by (with $f =$const)

$$u(x) = \frac{f}{24}(l - x)^4 - \frac{F}{6}(l - x)^3 + \frac{M}{2}x^2 + \left(\frac{fl^3}{6} - \frac{Fl^2}{2}\right)x - \frac{fl^4}{24} + \frac{Fl^3}{6},$$

thus, giving

$$\theta(x) = -\frac{f}{6}(l - x)^3 + \frac{F}{2}(l - x)^2 + Mx + \frac{fl^3}{6} - \frac{Fl^2}{2},$$

$$m(x) = -\frac{f}{2}(l - x)^2 + F(l - x) - M,$$

$$s(x) = f(l - x) - F.$$

Now, the BE Eq (3.25), with the values evaluated in (3.27) and $b = 1$, yields

$$-\left\{\begin{array}{c} u(0) = 0 \\ u(l) \\ \theta(0) = 0 \\ \theta(l) \end{array}\right\} = \begin{bmatrix} 2\lambda l^3 & 0 & 0 & -3\lambda l^2 \\ 0 & 2\lambda l^3 & -3\lambda l^2 & 0 \\ 0 & -3\lambda l^2 & -6\lambda l & 0 \\ -3\lambda l^2 & 0 & 0 & 6\lambda l \end{bmatrix} \times$$

$$\times \left\{\begin{array}{c} s(0) \\ s(l) = -F \\ m(0) \\ m(l) = -M \end{array}\right\} +$$

$$+ \begin{bmatrix} -6\lambda & -6\lambda & 6\lambda l & 0 \\ -6\lambda & -6\lambda & 0 & -6\lambda l \\ 0 & 0 & -6\lambda & -6\lambda \\ 0 & 0 & -6\lambda & -6\lambda \end{bmatrix} \left\{\begin{array}{c} u(0) = 0 \\ u(l) \\ \theta(0) = 0 \\ \theta(l) \end{array}\right\} + \left\{\begin{array}{c} 5f\lambda l^4/4 \\ -5f\lambda l^4/4 \\ 2f\lambda l^3 \\ -2f\lambda l^3 \end{array}\right\}.$$

Solving this system, we find that

$$u(l) = \frac{Ml^2}{2} + \frac{fl^4}{8} - \frac{Fl^3}{3},$$

$$\theta(l) = Ml + \frac{fl^3}{6} - \frac{Fl^2}{2},$$

$$m(0) = -\frac{fl^2}{2} + Fl - M,$$

$$m(l) = -M,$$

$$s(0) = fl - F,$$

which matches with the exact solution given above. ∎

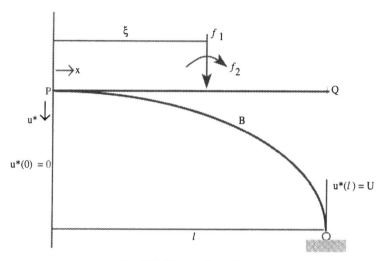

Fig. 3.3. Propped cantilever.

Example 3.5:
Problem of the propped cantilever, where

$$-EI\left(\frac{d^2u^*}{dx^2}\right)_Q = m^*(Q) = 0,$$

$$\left(\frac{du^*}{dx}\right)_P = \theta^*(P) = 0,$$

and the boundary conditions are $u(0) = \theta(0) = 0, u(l) = U, m(l) = 0$. From (3.25) with $\varepsilon \to 0$, we get after using the values (3.21) and (3.22)

$$\begin{bmatrix} 0 & -2\lambda l^3 & -3\lambda l^2 & -l/2 \\ -2\lambda l^3 & 0 & 0 & 0 \\ -3\lambda l^2 & 0 & 0 & 1/2 \\ 0 & -3\lambda l^2 & -l/2 & -1/2 \end{bmatrix} \begin{Bmatrix} s(0) \\ s(l) \\ m(0) \\ \theta(l) \end{Bmatrix} = -U \begin{Bmatrix} -1/2 \\ 1/2 \\ 0 \\ 0 \end{Bmatrix},$$

which gives

$$s(0) = \frac{U}{4\lambda l^3}, \quad s(l) = \frac{U}{4\lambda l^3}, \quad m(0) = -\frac{U}{4\lambda l^2}, \quad \theta(l) = \frac{3U}{2l}. \ \blacksquare$$

The theory developed in this chapter is the simplest in the BEM's and the results are derived under the assumption that the coefficients a, b, and f in Eq (3.1) and (3.14) are constant. Problems where these coefficients are polynomials or other types of functions in x cannot be solved by the above formulas. These cases require the derivation of related fundamental solutions before the BEM becomes applicable.

3.3. Problems

3.1. The equation for one–dimensional heat conduction in an insulated rod of cross–section area A, length l, and conductivity k, is given by

$$-\frac{d}{dx}\left(kA\frac{dT}{dx}\right) = f(x),$$

where $a = kA = \begin{cases} a_1, & 0 \le x \le l_1 \\ a_2, & l_1 \le x \le l_2 \end{cases}$ in two portions of the rod which means that either the cross–section and/or the material is different, and f is the distributed heat source. The boundary conditions are

$$T(0) = T_0, \quad \left[kA\frac{dT}{dx} + q\right]_{x=l} = 0,$$

where T_0 the prescribed temperature at $x = 0$, and q the prescribed heat flux at $x = l$. Solve for temperature at $x = 4, 10$ cm, given that $a_1 = 76\,(\text{W}\cdot\text{cm}^2)/^\circ C$, $a_2 = 96\,(\text{W}\cdot\text{cm}^2)/^\circ C$, $l_1 = 4$ cm, $l_2 = 6$ cm, and $T_0 = 100$ C.

3.2. The one–dimensional inviscid flow in a pipe is governed by

$$-\frac{d}{dx}\left(\rho A\frac{d\phi}{dx}\right) = 0, \quad 0 \le x \le l,$$

where ρ is the density, A the cross–section area of the pipe, and ϕ the velocity potential so that the velocity $u = -d\phi/dx$. Assume that $u(0) = 1, \phi(l) = 0$, and take $A = 2 - x/l$. Determine $\phi(0)$ when $l = 10$.

3.3. Solve the problem of the bending of an elastic beam with $u(0) = 0 = \theta(0)$, $s(l) = -F_l$, $m(l) = -M_l$, where $f = $ const for $x \in (0, l)$. Compute the displacements u and rotations θ at $x = 30, 60$ in, where $l = 90$ in, $EI = 60 \times 10^7$ lbin^2, $F_l = 4000$ lb, $M_l = 0$, and $f = 50$ lb/in. Find the values of $s(0)$ and $m(0)$.

3.4. Solve the problem of Example 3.3 by dividing the interval $(0, l)$ into two equal elements. Compute $u(l/2)$, $\theta(l/2)$, $m(l/2)$ and $s(l/2)$, given that $EI = 58 \times 10^8$ lbin^2, $l = 150$ in, $f = 100$ lb/in, and $s(l) = 8000$ lb, $m(l) = 0$.

3.5. Solve the beam problem of §3.2 for the Fig. 3.4 and determine the unknown quantities at the point C.

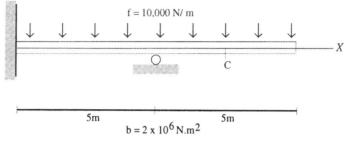

Fig. 3.4.

3.6. Solve the problem for the simple homogeneous cantilever beam ($b = 1$), shown in Fig. 3.5.

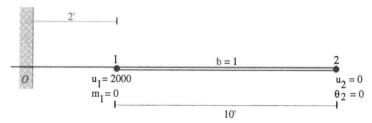

Fig. 3.5. Homogeneous Cantilever Beam.

References and Bibliography for Additional Reading

P. K. Banerjee and R. Butterfield, *Boundary Element Methods in Engineering Science*, McGraw–Hill, 1981.

J. N. Reddy, *An Introduction to the Finite Element Method*, McGraw–Hill, New York, 1984.

4

Fundamental Solutions

The weak variational formulation, studied in Chapter 2, and the fundamental solutions are the two building blocks for the direct BEM. The adage that one can formulate and develop a BEM for any boundary value problem if the fundamental solution for the governing equation is known is not without merit. We shall present some techniques to derive the fundamental solutions for the equations considered in this book. For further study the interested reader must refer to the bibliography at the end of this chapter.

4.1. Eigenpairs and Dirac Delta Function

First we shall find eigenfunctions for boundary value problems, and then represent the Dirac delta function as a sum of eigenfunctions. This will lead us to certain methods of explicitly finding fundamental solutions of some boundary value problems that are considered in the sequel.

Let a linear boundary value problem be defined by

$$L(u) = 0, \quad u = g(s) \quad \text{on } C_1; \quad \frac{\partial u}{\partial n} + k(s)u = h(s). \qquad (4.1)$$

The eigenfunctions ϕ_n corresponding to L are given by

$$L(\phi_n) = \lambda_n \phi_n, \qquad (4.2)$$

where λ_n are called the eigenvalues associated with the eigenfunctions ϕ_n. Thus, (ϕ_n, λ_n) is called the eigenpair for the boundary value problem (4.1). The form of the eigenpair depends on the coordinate system used in the boundary value problem. In the case where the coordinate system is cartesian, the eigenfunctions are of the form $\cos nx$, $\sin nx$, or e^{inx}, and the eigenvalues depend on n if the linear operator L contains constant coefficients. The eigenfunctions ϕ_n must be a complete set in the sense that for any function u there exists an integer N such that

$$\left\| u - \sum_{n=1}^{N} c_n \phi_n \right\| \leq \varepsilon \tag{4.3}$$

where $\varepsilon > 0$ is a preassigned arbitrarily small quantity and c_n are constants. The norm used in (4.3) is defined by

$$\|u\| = \sqrt{\int_V u\bar{u}\, dV}, \tag{4.4}$$

where $\bar{u}$ denotes the complex conjugate of u. In view of (4.3), a sufficiently continuous function u possesses an eigenfunction expansion to any degree of accuracy, thus

$$u = \lim_{N \to \infty} \sum_{n=0}^{N} c_n \phi_n. \tag{4.5}$$

Also, a complete set of eigenfunctions forms an orthonormal set with the property that

$$< \phi_m, \phi_n > = \delta_{mn}, \tag{4.6}$$

where δ_{mn} is the Kronecker delta (see §1.1). Now we can define the Dirac delta function on a region V in terms of the complete orthonormal set of eigenfunctions for the region V, as

$$\delta_V(\mathbf{x}, \mathbf{x}') = \sum_{n=0}^{\infty} \phi_n(\mathbf{x})\bar{\phi}_n(\mathbf{x}'), \tag{4.7}$$

where $\mathbf{x}$ and $\mathbf{x}'$ are two points in V, sometimes called the field point and the source point respectively (see Fig. B.1 in Appendix B).

Example 4.1:
Consider the following one–dimensional transient problem:

$$\frac{\partial^2 u}{\partial x^2} = \frac{1}{k}\frac{\partial u}{\partial t}, \quad -a < x < a, \tag{4.8}$$

with the initial and boundary conditions as $u(x,0) = F(x)$ for $-a < x < a$, and $u(-a,t) = 0 = u(a,t)$ for $t > 0$. We will use the method of separation of variables, and assume that $u(x,t) = X(x)T(t)$. Then we get

$$\frac{1}{X}\frac{d^2X}{dx^2} = \frac{1}{kT}\frac{dT}{dt} = -\lambda^2, \tag{4.9}$$

where λ is a real parameter. The boundary conditions and initial condition become: $X(-a) = 0 = X(a)$, and $T(0) = F(x)$. It will be found that only the choice of $\lambda > 0$ gives meaningful results, leading to the solution of the following two equations:

$$\frac{d^2X}{dx^2} + \lambda^2 X = 0,$$
$$\frac{dT}{dt} + k\lambda^2 T = 0. \tag{4.10}$$

The solution is found to be

$$u(x,t) = \sum_{-\infty}^{\infty} A_n \sin\frac{n\pi x}{a} e^{-n^2\pi^2 kt/a^2}, \tag{4.11}$$

where

$$2aA_n = \int_{-\infty}^{\infty} F(x)\sin\frac{n\pi x}{a}\,dx. \tag{4.12}$$

Thus, the orthonormal set of spatial eigenpairs for this problem is given by $\left(\frac{n\pi}{a}, \frac{1}{2a}\sin\frac{n\pi x}{a}\right)$. In complex notation the orthonormal eigenfunctions are

$$\phi_n = \frac{1}{2a}e^{in\pi x/a}.$$

Hence the Dirac delta function in the region $-a < x < a$ for the steady–state (as $t \to \infty$) one–dimensional Laplace equation over the interval $-a < x < a$ is represented by

$$\delta(x,x') = \frac{1}{4a^2}\sum_{n=-\infty}^{\infty} e^{in\pi x/a}e^{-in\pi x'/a} = \frac{1}{4a^2}\sum_{-\infty}^{\infty} e^{in\pi(x-x')/a}. \tag{4.13}$$

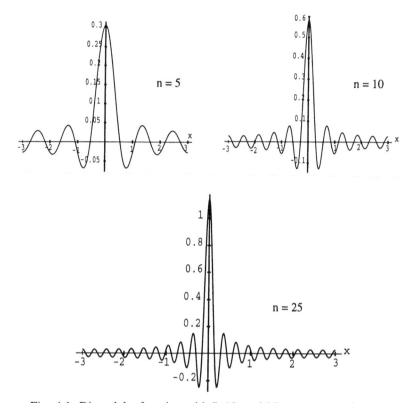

Fig. 4.1. Dirac delta function with 5, 10, and 25 terms respectively.

Note that in (4.12) we have assumed that the eigenfunction expansion is periodic with period $2a$. Hence the Dirac delta function (4.13) also has period $2a$. The solution (4.11) can also be obtained by using the Laplace transform on time. We have graphed the real part of this function for the basic interval $-a < x < a$ in Fig. 4.1, using 5, 10, and 25 terms in (4.13), with $a = 3$, and $x' = 0$. The graphs, however, repeat outside this interval with a period $2a$. From the graphs in Fig. 4.1, it is obvious that the peak becomes infinitely higher and narrower as n increases. ∎

Example 4.2:
Consider the two–dimensional Laplace equation

$$\frac{\partial^2 u}{\partial x^2} + \frac{\partial^2 u}{\partial y^2} = 0$$

over the rectangle $R = \{-a < x < a,\ -b < y < b\}$. The eigenpair for this region is given by

$$\lambda_n = -\frac{n^2\pi^2}{2\sqrt{ab}}\left(\frac{1}{a^2}+\frac{1}{b^2}\right),$$

$$\phi_n(x,y) = \frac{1}{2\sqrt{ab}}e^{in\pi x/a}e^{in\pi y/b}.$$

Note that

$$\frac{\partial^2\phi_n}{\partial x^2}+\frac{\partial^2\phi_n}{\partial y^2} = -\frac{n^2\pi^2}{2\sqrt{ab}}\left(\frac{1}{a^2}+\frac{1}{b^2}\right)\phi_n. \qquad (4.14)$$

It can be easily verified that these eigenvalues are orthonormal by showing that $\langle\phi_m,\phi_n\rangle = \int_{-a}^{a}\int_{-b}^{b}\phi_m\bar{\phi}_n\,dy\,dx = \delta_{mn}$. The solution in the complex form of a double Fourier series over the rectangle is given by

$$u(x,y) = \frac{1}{2\sqrt{ab}}\sum_{n=-\infty}^{\infty}\sum_{m=-\infty}^{\infty}c_{nm}e^{in\pi x/a}\,e^{im\pi y/b}. \qquad (4.15)$$

This representation of u is periodic both in x and y with periods $2a$ and $2b$ respectively. The Dirac delta function for this region is

$$\delta(\mathbf{x},\mathbf{x}') = \delta(x,x')\delta(y,y')$$

$$= \frac{1}{4ab}\sum_{n=-\infty}^{\infty}e^{in\pi(x-x')/a}e^{in\pi(y-y')/b}, \qquad (4.16)$$

where $\mathbf{x} = (x,y)$ is the field point and $\mathbf{x}' = (x',y')$ the source point. The solution of the problem is given by

$$u(\mathbf{x}) = u(x,y) = \sum_{n_1=-\infty}^{\infty}\sum_{n_2=-\infty}^{\infty}c_{n_1 n_2}\phi_{n_1 n_2}. \qquad (4.17)$$

Again,

$$\left\langle u(\mathbf{x}),\delta_R(\mathbf{x},\mathbf{x}')\right\rangle_R = \int_{-a}^{a}\int_{-b}^{b}u(x,y)\bar{\delta}_R(\mathbf{x},\mathbf{x}')\,dy\,dx$$

$$= \frac{1}{2\sqrt{ab}}\sum_{n=0}^{\infty}e^{in\pi(x'/a+y'/b)}\times$$

$$\times\frac{1}{2\sqrt{ab}}\int_{-a}^{a}\int_{-b}^{b}u(x,y)\,e^{-in\pi(x/a+y/b)}\,dy\,dx$$

$$= \frac{1}{2\sqrt{ab}}\sum_{n=0}^{\infty}c_n e^{in\pi(x'/a+y'/b)},$$

$$(4.18)$$

where

$$c_n = \frac{1}{2\sqrt{ab}} \int_{-a}^{a} \int_{-b}^{b} u(x,y) \, e^{-in\pi(x/a+y/b)} \, dy \, dx. \tag{4.19}$$

Note that (4.18) is the Fourier series for $u(x,y)$. Hence

$$\left\langle u(x,y), \delta_R(\mathbf{x},\mathbf{x}') \right\rangle_R = u(x',y'). \tag{4.20}$$

Also

$$\begin{aligned}
\langle u, \phi_m \rangle &= \left\langle \sum_{n=-\infty}^{\infty} c_n \phi_n, \phi_m \right\rangle_R \\
&= \sum_{n=-\infty}^{\infty} c_n \left\langle \phi_n, \phi_m \right\rangle_R \\
&= \sum_{n=-\infty}^{\infty} c_n \delta_{nm} = c_n.
\end{aligned} \tag{4.21}$$

Thus, finally

$$\begin{aligned}
\left\langle u(\mathbf{x}), \delta_R(\mathbf{x},\mathbf{x}') \right\rangle_R &= \left\langle \sum_{n=0}^{\infty} c_n \phi_n(\mathbf{x}), \sum_{m=0}^{\infty} \phi_m(\mathbf{x}) \bar\phi_m(\mathbf{x}') \right\rangle_R \\
&= \sum_{n=0}^{\infty} c_n \sum_{m=0}^{\infty} \phi_m(\mathbf{x}') \left\langle \phi_n(\mathbf{x}), \phi_m(\mathbf{x}) \right\rangle_R \\
&= \sum_{n=0}^{\infty} c_n \sum_{m=0}^{\infty} \phi_m(\mathbf{x}') \delta_{nm} \\
&= \sum_{n=0}^{\infty} c_n \phi_n(\mathbf{x}') = u(\mathbf{x}'). \ \blacksquare
\end{aligned} \tag{4.22}$$

The following two examples use a direct method of finding fundamental solutions in a three–dimensional free–space:

Example 4.3:
The fundamental solution for the free-space three–dimensional Laplace equation

$$-\nabla^2 u^* = \delta(x) \tag{4.23}$$

can be obtained directly as follows: Since the operator $-\nabla^2$ is invariant under a rotation of coordinate axes, we shall seek a solution that depends only on

$r = |x|$. For $r > 0$, $u^*(r)$ will satisfy the homogeneous equation $\nabla^2 u^* = 0$, i.e., in spherical coordinates

$$\frac{1}{r}\frac{\partial}{\partial r}\left(r^2\frac{\partial u^*}{\partial r}\right) = 0,$$

which has a solution $u^* = \dfrac{A}{r} + B$. If we require the potential to vanish at infinity, then $B = 0$, and $u^* = A/r$. In order to determine A, we take into account the magnitude of the source at $x = 0$. Integrating (4.23) over a small sphere S_ε of radius ε and center at $x = 0$, we obtain

$$-\int_{S_\varepsilon} \nabla^2 u^* \, dx = 1,$$

which, by using (1.5) (divergence theorem), gives

$$-\int_{\partial S_\varepsilon} \frac{\partial u^*}{\partial r}\bigg|_{r=\varepsilon} dS = 1, \tag{4.24}$$

where ∂S_ε is the surface of the sphere S_ε. Physically, Eq (4.24) expresses the conservation of charge, i.e., the flux of the electric field through the closed surface ∂S_ε (of area $4\pi\varepsilon^2$) is equal to the charge in the interior of S_ε. Now, substituting $u^* = A/r$ in (4.24), we find that $A = 1/(4\pi)$, and hence the fundamental solution for the three–dimensional Laplace equation is

$$u^* = \frac{1}{4\pi r} = \frac{1}{4\pi|\mathbf{x} - \mathbf{x'}|}. \tag{4.25}$$

For the fundamental solution of the two–dimensional Laplace equation (4.23), we have

$$\frac{1}{r}\frac{\partial}{\partial r}\left(r\frac{\partial u^*}{\partial r}\right) = 0, \quad r = |x| = \sqrt{x^2 + y^2},$$

which has a solution $u^* = C\ln r + D$. We arbitrarily set $D = 0$ and use a result similar to (4.24) for the flux of an electric charge through the boundary ∂C_ε (of length $2\pi\varepsilon$) of a circle of radius ε. Then $C = -1/2\pi$, and

$$u^* = \frac{1}{2\pi}\ln\frac{1}{r} = \frac{1}{2\pi}\ln\frac{1}{|x - x'|}. \tag{4.26}$$

For the fundamental solution of the one–dimensional Laplace equation, we note that (4.23) becomes $-\dfrac{d^2 u^*}{dx^2} = \delta(x)$, whose general solution is

$$u^*(x) = -\frac{1}{2}|x| + Ax + B.$$

If we require spherical symmetry, i.e., $u^* = u^*(|x|)$, then $A = 0$, and we set $B = 0$. Then the fundamental solution is given by

$$u^*(x) = -\frac{1}{2}|x|. \tag{4.27}$$

Note that the fundamental solution for the free–space Laplace equation $-\nabla^2 u^* = \delta(x)$ in R^n is given by

$$u^*(r) = \frac{1}{(n-2)S_n(1)r^{n-2}}, \quad n > 2, \tag{4.28}$$

where $S_n(1)$ is the surface area of a sphere of radius unity, and $r = |\mathbf{x} - \mathbf{x}'|$.

Let R_+ denote the upper half of the x, y–plane and $\mathbf{x}' = (x', y')$ be a fixed point in R_+. Then both

$$u_1^*(x, y, x', y') = u^*(x, y, x', y') + x^2 - y^2, \tag{4.29}$$
$$u_2^*(x, y, x', y') = u^*(x, y, x', y') + u^*(x, y, x', -y'), \tag{4.30}$$

where u^* is defined by (4.26), are the fundamental solutions of the Laplace equation in R_+. ∎

Example 4.4:
The fundamental solution for the Helmholtz equation in R^n

$$-(\nabla^2 + \mu)u^* = \delta(x), \quad \text{or} \quad -(\nabla^2 - h^2)u^* = \delta(x), \tag{4.31}$$

where $h^2 = -\mu$, and $\sqrt{\mu}$ is defined such that it has a nonnegative imaginary part, i.e., $\sqrt{\mu} = \alpha + i\beta$, with $\beta \geq 0$, and $\beta = 0$ iff $\mu \in [0, \infty)$. We will therefore take $h = -i\sqrt{\mu}$, so that h is real positive when μ is real negative. We will assume that u^* is spherically symmetric. Then for $x \neq 0$, the function u^* must satisfy in spherical coordinates

$$\frac{d}{dr}\left(r^{n-1}\frac{du^*}{dr}\right) + \mu r^{n-1}u^* = 0. \tag{4.32}$$

If we substitute $u^* = wr^{1-(n/2)}$, then Eq (4.32) can be reduced to an equation of the Bessel type of order $(n/2) - 1$ with parameter μ, i.e.,

$$\frac{d}{dr}\left(r\frac{dw}{dr}\right) - \frac{w}{r}\left(1 - \frac{n}{2}\right)^2 + \mu rw = 0, \tag{4.33}$$

whose general solution can be written in terms of the Hankel functions as

$$w(r) = C_1 H^{(1)}_{(n/2)-1}(\sqrt{\mu}\,r) + C_2 H^{(2)}_{(n/2)-1}(\sqrt{\mu}\,r), \quad n \geq 2. \qquad (4.34)$$

If $\mu \notin [0, \infty)$, then $\sqrt{\mu}$ has positive imaginary part and the Hankel function $H^{(2)}_{(n/2)-1}(\sqrt{\mu}\,r)$ becomes exponentially large as $r \to \infty$, but $H^{(1)}_{(n/2)-1}(\sqrt{\mu}, r)$ is exponentially small. Since u^* vanishes at $r = \infty$, we must have $C_2 = 0$, and then from (4.34) we get

$$u^* = C_1 H^{(1)}_{(n/2)-1}(\sqrt{\mu}\,r). \qquad (4.35)$$

As in the previous example, we apply (1.5) (divergence theorem) to (4.31) and obtain

$$-\int_{S_\epsilon} \frac{\partial u^*}{\partial r}\, dS = 1,$$

or

$$\lim_{r \to \infty} r^{n-1} S_n(1) \frac{\partial u^*}{\partial r} = -1, \qquad (4.36)$$

where $S_n(1)$ is the surface area of a sphere of unit radius. For small r, we have the asymptotic expansion

$$H_n^{(1)}(r) \sim -\frac{i 2^n (n-1)!}{\pi} r^{-n}. \qquad (4.37)$$

Thus, substituting (4.35) into (4.36) and using (4.37) we find that

$$C_1 = \frac{i\pi 2^{-n/2}(\sqrt{\mu})^{(n/2)-1}}{[(n/2)-1]! S_n(1)} = \frac{i}{4}\left(\frac{\sqrt{\mu}}{2\pi}\right)^{(n/2)-1}. \qquad (4.38)$$

Hence, for $n \geq 2$ and $\mu \notin [0, \infty)$, the required fundamental solution for (4.31) is given by

$$u^*(r, \mu) = \frac{i}{4}\left(\frac{\sqrt{\mu}}{2\pi}\right)^{(n/2)-1} H^{(1)}_{(n/2)-1}(\sqrt{\mu}\,r), \quad n \geq 2, \qquad (4.39)$$

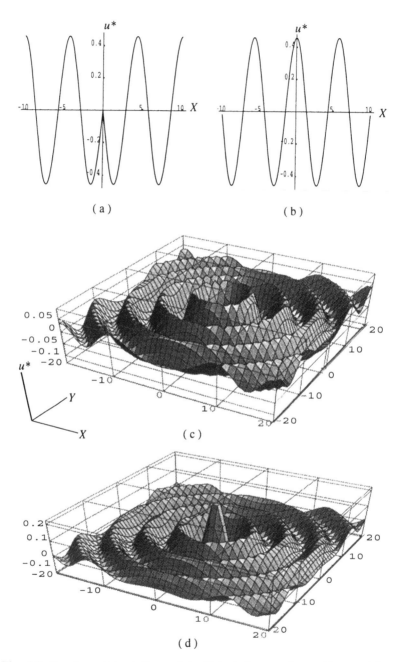

Fig. 4.2. Fundamental solutions of the Helmholtz equation, with $\mu = 1.2$:
(a), (b): Real and imaginary parts of one–dimensional solution;
(c), (d): Real and imaginary parts of two–dimensional solution.

or, writing $h^2 = -\mu$, i.e., $h = -i\sqrt{\mu}$, or $\sqrt{\mu} = ih$, we have

$$H^{(1)}_{(n/2)-1}(\sqrt{\mu}\,r) = H^{(1)}_{(n/2)-1}(ihr) = \frac{2}{i\pi}K_{(n/2)-1}(hr), \qquad (4.40)$$

where $K_{(n/2)-1}$ are the modified Bessel functions, thereby yielding

$$u^*(r, -h^2) = \frac{1}{2\pi}\left(\frac{h}{2\pi r}\right)^{(n/2)-1}K_{(n/2)-1}(hr), \quad n \geq 2, \qquad (4.41)$$

which holds whenever $-h^2 \notin [0, \infty)$, i.e., for all h with $\Re h > 0$. For $n = 2$, the fundamental solutions (4.39) and (4.41) become

$$u^*(r, \mu) = \frac{i}{4}H^{(1)}_0(\sqrt{\mu}\,r) = \frac{1}{2\pi}K_0(hr). \qquad (4.42)$$

For $n = 3$, by using $H^{(1)}_{1/2}(z) = \frac{1}{i}\left(\frac{2}{\pi}\right)^{1/2}\frac{e^{iz}}{z^{1/2}}$, we get

$$u^*(r, \mu) = \frac{e^{i\sqrt{\mu}\,r}}{4\pi r} = \frac{e^{-hr}}{4\pi r}. \qquad (4.43)$$

The one–dimensional fundamental solution is found directly, and it is

$$u^*(x, \mu) = \frac{ie^{i\sqrt{\mu}\,|x|}}{2\sqrt{\mu}} = \frac{e^{-h|x|}}{2h}. \qquad (4.44)$$

Note that if the Helmholtz equation is taken as $\left(\nabla^2 + k^2\right)u = 0$, where $k \neq 0$ is the wave number, then the fundamental solution in a two–dimensional region is given by

$$u^*(r, k) = -\frac{i}{4}H^{(2)}_0(\sqrt{\mu}\,r), \qquad (4.42*)$$

and in a three–dimensional region by

$$u^*(r, k) = \frac{e^{-ikr}}{4\pi r}. \qquad (4.43*)$$

It is obvious from Fig. 4.2 that the real and the imaginary parts of the fundamental solutions of the Helmholtz equation in one– and two–dimensional cases exhibit wave structure. ∎

Example 4.5:
Consider the three–dimensional diffusion equation

$$\nabla^2 u \equiv \frac{\partial^2 u}{\partial x^2} + \frac{\partial^2 u}{\partial y^2} + \frac{\partial^2 u}{\partial z^2} = \frac{1}{k}\frac{\partial u}{\partial t}, \qquad (4.45)$$

where k is the thermal conductivity. This equation is satisfied by

$$u = \frac{1}{8(\pi kt)^{3/2}} e^{-[(x-x')^2+(y-y')^2+(z-z')^2]/4kt}. \qquad (4.46)$$

As $t \to t'$, this solution tends to zero at all points except the source point $\mathbf{x}' = (x', y', z')$ where it becomes infinite. If there is an instantaneous line source of unit strength at $t = t'$, parallel to z–axis and passing through the point (x', y'), then we consider a distribution of point sources of strength dz' at z' along the line. This gives the fundamental solution of Eq (4.45) as

$$u^* = \frac{1}{8[\pi k(t - t')]^{3/2}} \int_{-\infty}^{\infty} dz' \, e^{-[(x-x')^2+(y-y')^2+(z-z')^2]/4k(t-t')}$$

$$= \frac{1}{4\pi k(t - t')} e^{-[(x-x')^2+(y-y')^2]/4k(t-t')} = \frac{1}{4\pi k(t - t')} e^{-r^2/4k(t-t')},$$

$$(4.47)$$

where $r^2 = (x - x')^2 + (y - y')^2 + (z - z')^2.$ ∎

4.2. Green's Functions

In a region V with boundary S, let

$$\begin{aligned} L(u) &= f(\mathbf{x}) \quad \text{in } V, \\ B(u) &= 0 \quad \text{on } S, \end{aligned} \qquad (4.48)$$

represent a linear second–order partial differential equation and a linear boundary condition, respectively, such that the problem (4.48) has a unique solution for each continuous f. Then $G(\mathbf{x}, \mathbf{x}')$ is the Green's function for this problem if the unique solution of (4.48) is given by

$$u(\mathbf{x}) = \int_V G(\mathbf{x}, \mathbf{x}') f(\mathbf{x}') \, d_{x'} S, \qquad (4.49)$$

where $d_{x'} S$ denotes integration with respect to $\mathbf{x}'$. If $f(\mathbf{x})$ is replaced by the Dirac delta function $\delta(\mathbf{x}, \mathbf{x}')$, then the Green's function $G(\mathbf{x}, \mathbf{x}')$ represents the fundamental solution $u^*(\mathbf{x}, \mathbf{x}')$ for the problem (4.48).

A useful results in deriving the fundamental solutions is the **Duhamel's Theorem.** Let $F(\mathbf{x}, t) \in C(R^n)$ for $t > 0$. For a fixed positive parameter t', let $v_F(\mathbf{x}, t, t')$ denote the solution of the boundary value problem

$$\frac{\partial^2 v}{\partial t^2} = \nabla^2 v(\mathbf{x}, t, t'), \quad \mathbf{x} \in R^n, t > t',$$

$$v(\mathbf{x}, t') = 0, \quad \mathbf{x} \in R^n, \tag{4.50}$$

$$\frac{\partial v}{\partial t} = F(\mathbf{x}, t'), \quad \mathbf{x} \in R^n.$$

Then the function

$$u(\mathbf{x}, t) \equiv \int_0^t v_F(\mathbf{x}, t, t') \, dt' \tag{4.51}$$

is the solution of

$$\frac{\partial^2 v}{\partial t^2} = \nabla^2 u(\mathbf{x}, t) + F(\mathbf{x}, t), \quad \mathbf{x} \in R^n, t > 0,$$

$$u(\mathbf{x}, 0) = 0 = \frac{\partial u}{\partial t}(\mathbf{x}, 0), \quad \mathbf{x} \in R^n. \tag{4.52}$$

Example 4.6:
We shall consider the following one–dimensional initial value problem:

$$\frac{\partial v}{\partial t} = \kappa \frac{\partial^2 v}{\partial t^2}, \quad -\infty < x < \infty, \ t > 0,$$

$$v(x, 0) = f(x), \quad -\infty < x < \infty, \tag{4.53}$$

where $v(x, t)$ is of exponential growth in x. If we take the Fourier transform with respect to x, the problem (4.53) in the transform domain becomes

$$\frac{d}{dt}\tilde{v}(\alpha, t) = -\kappa \alpha^2 \tilde{v}(\alpha, t), \quad t > 0,$$

$$\tilde{v}(\alpha, 0) = \tilde{F}(\alpha), \tag{4.54}$$

where $\tilde{v}$ and $\tilde{F}$ denote the Fourier transforms of v and F, respectively, defined by

$$\tilde{v}(\alpha, t) = \frac{1}{\sqrt{2\pi}} \int_{-\infty}^{\infty} v(x, t) e^{-i\alpha x} \, dx, \quad \tilde{F}(\alpha) = \frac{1}{\sqrt{2\pi}} \int_{-\infty}^{\infty} f(x) e^{-i\alpha x} \, dx, \tag{4.55}$$

α being the variable of the transform. The solution of (4.54) is

$$\tilde{v}(\alpha t) = F(\alpha) e^{-\kappa \alpha^2 t}, \tag{4.56}$$

which on inversion gives the unique solution

$$u(x,t) = \frac{1}{2\sqrt{\pi \kappa t}} \int_{-\infty}^{\infty} e^{-(x+y)^2/4\kappa t} f(y)\, dy. \tag{4.57}$$

Alternatively, by differentiating (4.57) under the integral sign it can be easily verified that (4.56) is the solution of (4.53). Now, by applying the Duhamel's theorem it follows that the problem

$$\frac{\partial u}{\partial t} - \kappa \frac{\partial^2 v}{\partial x^2} = f(x), \quad -\infty < x < \infty,\ t > 0,$$
$$u(x,0) = 0, \quad -\infty < x < \infty, \tag{4.58}$$

has the unique solution

$$u(x,t) = \int_0^t v(x, t - t')\, dt'$$
$$= \int_0^t \int_{-\infty}^{\infty} \frac{1}{2\sqrt{\pi \kappa(t-t')}} e^{-(x-x')^2/4\kappa(t-t')} f(x')\, dx'\, dt'. \tag{4.59}$$

From this solution we obtain the Green's function for the problem (4.58) as

$$G(x,t,x',t') = \frac{1}{2\sqrt{\pi \kappa(t-t')}} e^{-(x-x')^2/4\kappa(t-t')}, \quad t > t'. \tag{4.60}$$

This Green's function becomes singular as (x,t) approaches (x',t'). Now, if we replace the function $f(x)$ in (4.58) by $\delta(x,x')$, then the fundamental solution for the problem (4.58) is given by

$$u^* = \frac{1}{2\sqrt{\pi \kappa(t-t')}} e^{-(x-x')^2/4\kappa(t-t')}, \quad t > t'. \tag{4.61}$$

The function u^* also exhibits a similar singular behavior as (x,t) approaches (x',t'). ∎

The fundamental solution for the diffusion equation

$$\frac{\partial u^*}{\partial t} - \kappa \nabla^2 u^* = \delta(x)\delta(t), \tag{4.62}$$

where $\delta(x) = \delta(x,0)$, $\delta(t) = \delta(t,0)$, in R^n is given by

$$u^*(r,t) = H(t - t')(4\pi\kappa t)^{-n/2} e^{-r^2/4at}, \quad t > 0, \tag{4.63}$$

where $H(t - t')$ is the Heaviside function, and $r = |\mathbf{x} - \mathbf{x}'|$.

If V^* is a bounded region to which the divergence theorem (see §1.1) applies, then the fundamental solution of the diffusion equation (4.62) in V^* is given by

$$u_1^*(\mathbf{x}, t, \mathbf{x}', t') = u^*(\mathbf{x} - \mathbf{x}', t - t') + \phi(\mathbf{x}, t, \mathbf{x}', t'), \qquad (4.64)$$

where u^* is given by (4.63) and ϕ is the solution of the dual problem

$$\frac{\partial \phi}{\partial t} - \kappa \nabla_x^2 \phi = 0, \quad t > t' \qquad -\frac{\partial \phi}{\partial t} - \kappa \nabla_{x'}^2 \phi = 0, \quad t > t' \quad \atop 2 \quad (4.65)$$
$$\phi \equiv 0, \quad t < t' \qquad\qquad\qquad \phi \equiv 0, \quad t < t'.$$

Thus, e.g., in one–dimensional case, let V^* be the interval $(0, 1)$. Then for $0 < x, x' < 1$ the fundamental solution for the one–dimensional diffusion equation is

$$u_2^*(x, t, x', t') = u^*(x - x', t - t') + H(t - t')e^{-\kappa(t-t')}\cos(x - x'), \quad (4.66)$$

where u^* is given by (4.63) with $n = 1$.

4.3. Fundamental Solutions on a Finite Domain

The fundamental solution $u^*(\mathbf{x}, \mathbf{x}')$ for a linear operator L in a region V is defined by

$$L\left(u_V^*(\mathbf{x}, \mathbf{x}')\right) \overset{\text{def}}{=} \delta_V(\mathbf{x}, \mathbf{x}'), \qquad (4.67)$$

where the operator L contains differentials in the variable $\mathbf{x}$. If we expand u_V^* and δ_V in terms of the eigenfunctions, i.e.,

$$u_V^*(\mathbf{x}, \mathbf{x}') = \sum_{n=0}^{\infty} c_n(\mathbf{x}')\phi_n(\mathbf{x}'), \qquad (4.68)$$

then we find that

$$L\left(\sum_{n=0}^{\infty} c_n(\mathbf{x}')\phi_n(\mathbf{x})\right) = \sum_{n=0}^{\infty} \phi_n(\mathbf{x})\bar{\phi}_n(\mathbf{x}'), \qquad (4.69)$$

where

$$\sum_{n=0}^{\infty} c_n(\mathbf{x'}) \, L(\phi_n) = \sum_{n=0}^{\infty} c_n(\mathbf{x'}) \lambda_n \phi_n(\mathbf{x}). \qquad (4.70)$$

Thus,

$$\sum_{n=0}^{\infty} c_n(\mathbf{x'}) \lambda_n \phi_n(\mathbf{x}) = \sum_{n=0}^{\infty} \phi_n(\mathbf{x}) \bar{\phi}_n(\mathbf{x'}). \qquad (4.71)$$

This is an identity which holds for all $\mathbf{x}$ and $\mathbf{x'}$. It gives

$$c_n(\mathbf{x'}) = \frac{\bar{\phi}_n(\mathbf{x'})}{\lambda_n}. \qquad (4.72)$$

Since L has constant coefficients, the eigenvalues λ_n in the above formula will not be a function of $\mathbf{x}$. Hence from (4.68) we find that the fundamental solution $u_V^*(\mathbf{x}, \mathbf{x'})$ for a finite domain V is given by

$$u_V^*(\mathbf{x}, \mathbf{x'}) = \sum_{n=0}^{\infty} \frac{\bar{\phi}_n(\mathbf{x'})}{\lambda_n} \phi_n(\mathbf{x}). \qquad (4.73)$$

The fundamental solution represents the field due to a point source $\mathbf{x'}$ together with the 'image' sources generated by the boundary of the region V. In a finite region this solution corresponds to finitely many sources reflected according to the geometry of the region.

Example 4.7:
For the steady–state Laplace equation on the rectangle R in Example 4.2, we have

$$\lambda_n = -\frac{n^2 \pi^2}{2\sqrt{ab}} \left(\frac{1}{a^2} + \frac{1}{b^2} \right), \quad \phi_n(\mathbf{x}) = \frac{1}{2\sqrt{ab}} e^{in\pi(x/a + y/b)}. \qquad (4.74)$$

Hence the fundamental solution for this problem is

$$u_R^*(\mathbf{x}, \mathbf{x'}) = -\frac{(ab)^{3/2}}{2(a^2 + b^2)\pi^2} \sum_{\substack{n=-\infty \\ n \neq 0}}^{\infty} \frac{1}{n^2} e^{in\pi\{(x-x')/a + (y-y')/b\}} + c_0.$$

$$(4.75)$$

Note that the term corresponding to $n = 0$ is excluded in the above sum, since it makes the constant term infinite. In potential problems, however, the term $\partial u/\partial \mathbf{x}$ is physically more significant, and therefore such an infinite term becomes less important and may be ignored.

Example 4.8:

Consider the two–dimensional Helmholtz equation

$$\frac{\partial^2 u}{\partial x^2} + \frac{\partial^2 u}{\partial y^2} + h^2 u = 0, \tag{4.76}$$

on the square region $R = \{|x| \leq a, |y| \leq a\}$. The eigenvectors and the orthonormal eigenfunctions are given by

$$\lambda_n = \frac{1}{2a}\left(h^2 - \frac{2n^2\pi^2}{a^2}\right), \quad \phi_n = \frac{1}{2a}e^{in\pi(x+y)/a}, \quad n = 0, \pm1, \pm2, \cdots .$$

$$\tag{4.77}$$

Hence the fundamental solution for this region is

$$u_R^*(\mathbf{x} - \mathbf{x}') = \frac{1}{2a} \sum_{n=0}^{\infty} \frac{e^{in\pi(x'+y')/a}}{h^2 - 2n^2\pi^2/a^2} . \quad \blacksquare \tag{4.78}$$

4.4. Fundamental Solutions on an Infinite Domain

In order to study the fundamental solution in an infinite space, we shall consider the Cauchy–Navier equation of elastostatics in the absence of body forces. This equation is given by (see §6.1 for notation)

$$\mu u_{j,kk} + \frac{\mu}{1-2\nu}u_{k,kj} = 0, \tag{4.79a}$$

or

$$\mu\nabla^2\mathbf{u} + \frac{\mu}{1-2\nu}\nabla(\nabla \cdot \mathbf{u}) = 0, \tag{4.79b}$$

where ν $(0 < \nu < 0.5)$ is the Poisson ratio. The fundamental solution $u^*(\mathbf{x}, \mathbf{x}')$ satisfies the equation

$$\mu u_{j,kk}^* + \frac{\mu}{1-2\nu}u_{k,kj}^* = -\delta(\mathbf{x}, \mathbf{x}')e_j, \tag{4.80a}$$

or

$$\mu\nabla_{x'}^2\mathbf{u}^*(\mathbf{x}, \mathbf{x}') + \frac{\mu}{1-2\nu}\nabla_{x'}(\nabla_{x'} \cdot \mathbf{u}^*(\mathbf{x}, \mathbf{x}')) = -\delta(\mathbf{x}, \mathbf{x}')I_n, \tag{4.80b}$$

where e_j or I_n is the $n \times n$ identity matrix. The following results hold: For $n = 2$ (two–dimensional case):

$$u_{kl}^*(\mathbf{x}, \mathbf{x}') = \frac{1}{8\pi\mu(1-\nu)r} \left[(3-4\nu)\ln\left(\frac{1}{r}\right) \delta_{kl} + r_{,k}r_{,l} \right], \qquad (4.81)$$

and for $n = 3$ (three–dimensional case):

$$u_{kl}^*(\mathbf{x}, \mathbf{x}') = \frac{1}{16\pi\mu(1-\nu)r} \left[(3-4\nu)\delta_{kl} + r_{,k}r_{,l} \right], \qquad (4.82)$$

Example 4.9:
We shall use Lord Kelvin's method to derive the fundamental solution (4.82). Let

$$\mathbf{u}(\mathbf{x}) = \mathbf{h}(\mathbf{x}) - c\nabla\left[\mathbf{x}\cdot\mathbf{h}(\mathbf{x}) + f(\mathbf{x})\right], \qquad (4.83)$$

where $\mathbf{h}(\mathbf{x})$ is a harmonic vector function, $f(\mathbf{x})$ a harmonic scalar function, and $c = 1/4(1-\nu)$. Then it can be easily verified that this function $\mathbf{u}$ is the solution of Eq (4.80b). If $\mathbf{u}$ is sufficiently smooth, we can write

$$\mathbf{u} = \nabla g + \nabla \times \mathbf{F}, \qquad \nabla\cdot\mathbf{F} = 0, \qquad (4.84)$$

where g is a scalar and $\mathbf{F}$ a vector potential. We will now show that g is a harmonic function: Substituting (4.84) into (4.79b) we get

$$\nabla^2\left[\nabla g + \nabla\times\mathbf{F} + \frac{1}{1-2\nu}\nabla g\right] = 0, \qquad (4.85)$$

since $\nabla\cdot\nabla\times\mathbf{F} = 0$. Thus, since $\mathbf{h}$ is harmonic (i.e., $\nabla^2\mathbf{h} = 0$),

$$\mathbf{h} \equiv \nabla g + \nabla\times\mathbf{F} + \frac{1}{1-2\nu}\nabla g = \frac{2(1-\nu)}{1-2\nu}\nabla g + \nabla\times\mathbf{F}. \qquad (4.86)$$

Taking divergence of (4.86) we find, in view of (4.85), that

$$\nabla^2 g = 2c(1-2\nu)\nabla\cdot\mathbf{h} = c(1-2\nu)\nabla^2\left(\mathbf{x}\cdot\mathbf{h}\right),$$

or

$$g - c(1-2\nu)\left(\mathbf{x}\cdot\mathbf{h}\right) \equiv c(1-2\nu)f(\mathbf{x}), \qquad (4.87)$$

i.e., g is a harmonic scalar function (since f is). Next, we substitute into (4.84) the expression for $\nabla\times\mathbf{F}$ from (4.86). Then, using (4.87), we get

$$\begin{aligned}
\mathbf{u} &= \nabla g + \mathbf{h} - \frac{2(1-\nu)}{1-2\nu}\nabla g = \mathbf{h} - \frac{1}{1-2\nu}\nabla g \\
&= \mathbf{h} - \frac{1}{1-2\nu}\frac{2(1-\nu)}{1-2\nu}\nabla\left[f + (\mathbf{x}\cdot h)\right] \qquad (4.88) \\
&= \mathbf{h} - c\nabla\left[f + (\mathbf{x}\cdot h)\right].
\end{aligned}$$

Let us assume that the singularity of $u^*(\mathbf{x}, \mathbf{x}')$ is of the form $\dfrac{1}{r} = \dfrac{1}{|\mathbf{x} - \mathbf{x}'|}$

(as in the case of the Laplace equation), and write $\mathbf{h}$ as the vector

$$\mathbf{h}(\mathbf{x}, \mathbf{x}') = \begin{bmatrix} \frac{1}{r} & 0 & 0 \end{bmatrix}^T = \begin{bmatrix} \frac{1}{|\mathbf{x} - \mathbf{x}'|} & 0 & 0 \end{bmatrix}^T. \qquad (4.89)$$

Let us assume, for the sake of simplicity, that the singularity is at $\mathbf{x}' = 0$. Then

$$\mathbf{h}(\mathbf{x}) = \begin{bmatrix} \frac{1}{r} & 0 & 0 \end{bmatrix}^T = \begin{bmatrix} \frac{1}{|\mathbf{x}|} & 0 & 0 \end{bmatrix}^T. \qquad (4.90)$$

We shall further take $f = 0$. Then from (4.88) and (4.90) we obtain

$$\mathbf{u}^*(\mathbf{x}) = \begin{bmatrix} \frac{1}{r} & 0 & 0 \end{bmatrix}^T - c\nabla \left(\frac{\mathbf{x}}{r} \right)$$

$$= \begin{bmatrix} \dfrac{1-c}{r} + \dfrac{cx_1^2}{r^3} & \dfrac{cx_1 x_2}{r^3} & \dfrac{cx_3 x_1}{r^3} \end{bmatrix}^T, \qquad (4.91)$$

where, as before, $c = 1/4(1 - 2\nu)$ and $\mathbf{x} = (x_1, x_2, x_3)$. Since $\mathbf{u}^*$ satisfies Eq (4.80b) everywhere except at the origin where it is singular because of a point force in the x_1 direction, we find that the magnitude of this force is given by

$$\int_{|\mathbf{x}|<\varepsilon} \left[\mu\nabla^2\mathbf{u}^*(\mathbf{x}) + \frac{\mu}{1-2\nu}\nabla(\nabla \cdot \mathbf{u}^*(\mathbf{x})) \right] d\mathbf{x} \equiv I_x + I_y + I_z, \qquad (4.92)$$

where $|\mathbf{x}| < \varepsilon$ is a ball of radius ε about the origin and I_x, I_y, I_z denote the three components of the above integral. Note that from (4.91)

$$\nabla \cdot \mathbf{u}^* = \frac{\partial}{\partial x_1}\left(\frac{1-c}{r} + \frac{cx_1^2}{r^3} \right) + \frac{\partial}{\partial x_2}\left(\frac{cx_1 x_2}{r^3} \right) + \frac{\partial}{\partial x_3}\left(\frac{cx_3 x_1}{r^3} \right),$$

whence

$$\nabla(\nabla \cdot \mathbf{u}^*) =$$

$$\begin{bmatrix} \dfrac{\partial}{\partial x_1}\left(\dfrac{(2c-1)x_1}{r^3} \right) & \dfrac{\partial}{\partial x_2}\left(\dfrac{(2c-1)x_1}{r^3} \right) & \dfrac{\partial}{\partial x_3}\left(\dfrac{(2c-1)x_1}{r^3} \right) \end{bmatrix}^T.$$

Then the first component

$$I_{x_1} = \mu\int_{|\mathbf{x}|<\varepsilon} \left\{ \nabla^2\left[\frac{1-c}{r} + \frac{cx_1^2}{r^3} \right] + \frac{1}{1-2\nu}\frac{\partial}{\partial x_1}\left(\frac{(2c-1)x_1}{r^3} \right) \right\} d\mathbf{x}$$

$$= \mu\int_{S_\varepsilon} \left\{ \frac{\partial}{\partial r}\left(\frac{1-c}{r} + \frac{cx_1^2}{r^3} \right) + \frac{2c-1}{1-2\nu}\frac{n_1 x_1}{r^3} \right\} dS,$$

$$(4.93)$$

where S_ε is the boundary of the ball $|\mathbf{x}| < \varepsilon$, $\mathbf{n} = (n_1, n_2, n_3)$, and we have applied the Green's formula (1.6) to the second term in the integrand. Using the spherical coordinates $x_1 = \rho \sin \theta \cos \phi$, $x_2 = \rho \sin \theta \sin \phi$, $x_3 = \rho \cos \theta$, we get

$$
\begin{aligned}
I_{x_1} &= \mu \int_0^{2\pi} \int_0^{\pi} \left[(c - 1) - c \sin^2 \theta \cos^2 \phi \right] \sin \theta \, d\theta \, d\phi + \\
&\quad + \mu \frac{2c - 1}{1 - 2\nu} \int_0^{2\pi} \int_0^{\pi} \sin^3 \theta \cos^2 \phi \, d\theta \, d\phi \\
&= \mu \left[4(c - 1)\pi - \frac{4c\pi}{3} + \frac{2c - 1}{1 - 2\nu} \frac{4\pi}{3} \right] \\
&= -4\pi\mu,
\end{aligned}
\tag{4.94}
$$

which is independent of ε. Similarly, it can be found that $I_{x_2} = 0 = I_{x_3}$. Since the magnitude of the concentrated force defined by (4.92) must be -1, we should normalize $\mathbf{u}^*$ in (4.90) by a factor of $1/4\pi\mu$. Hence (4.91) becomes

$$
\begin{aligned}
\mathbf{u}^*(\mathbf{x}) &= \frac{1}{4\pi\mu} \left[\frac{1 - c}{r} \delta_{ij} + \frac{c x_i x_j}{r^3} \right] \\
&= \frac{1}{16\pi(1 - \nu)r} \left[(3 - 4\nu)\delta_{ij} + \frac{\partial r}{\partial x_i} \frac{\partial r}{\partial x_j} \right], \quad i, j = 1, 2, 3,
\end{aligned}
$$

which becomes (4.82) if we carry out the translation from $\mathbf{x}$ to $\mathbf{x} - \mathbf{x}'$. ∎

In the next example we shall derive the result (4.82) by using the Fourier transform method.

Example 4.10:
As in the above example, we shall assume that $\mathbf{x} = 0$. Then the k–th row of Eq (4.80b) becomes

$$
\mu \nabla^2 u_k^* + \frac{\mu}{1 - 2\nu} \nabla \left[\nabla \cdot \mathbf{u}^*(\mathbf{x}) \right] = -\delta(\mathbf{x}) \mathbf{e}_k,
\tag{4.95}
$$

where $\mathbf{e}_k$ is the unit vector in the direction of x_k–axis. The Fourier transform of Eq (4.95) is

$$
-\mu 4\pi^2 |\alpha|^2 \tilde{u}_{kj}^*(\alpha) + \frac{\mu}{1 - 2\nu} (2\pi i \alpha_j)(2\pi i) \sum_{m=1}^{3} \alpha_m \tilde{u}_{km}^* = -\delta_{kj}, \quad j = 1, 2, 3,
\tag{4.96}
$$

where $\alpha = (\alpha_1, \alpha_2, \alpha_3)$ is the variable of the transform (see Example 4.6 for notation and definition). If we multiply (4.96) by α_j and carry out summation over j in (4.96), we get

$$4\pi^2 \left[-\mu|\alpha|^2 \sum_{j=1}^{3} \alpha_j \tilde{\mathbf{u}}^*_{kj} - \frac{\mu}{1-2\nu}|\alpha|^2 \sum_{m=1}^{3} \alpha_m \tilde{\mathbf{u}}^*_{km} \right] = -\delta_k,$$

which gives

$$\sum_{j=1}^{3} \alpha_j \tilde{\mathbf{u}}^*_{kj} = \frac{\alpha_k(1-2\nu)}{8\mu(1-\nu)\pi^2|\alpha|^2}, \quad k = 1, 2, 3. \tag{4.97}$$

If we substitute (4.97) into (4.96) we find that

$$\tilde{\mathbf{u}}^*_{kj}(\alpha) = \frac{1}{4\pi^2\mu|\alpha|^2}\delta_{kj} - \frac{\alpha_j\alpha_k}{8\mu(1-\nu)\pi^2|\alpha|^2}, \tag{4.98}$$

which on inversion yields (4.82). ∎

We shall not derive Mindlin's fundamental solutions for the half–space here. Interested readers can find literature on this topic in the bibliography given below.

References and Bibliography for Additional Reading

C. A. Brebbia, *The Boundary Element Method for Engineers*, Pentech Press, London, 1980.

———, J. Telles and L. Wrobel, *Boundary Element Techniques – Theory and Applications in Engineering*, Springer–Verlag, Berlin, 1984.

H. S. Carslaw and J. C. Jaeger, *Operational Methods in Applied Mathematics*, Dover, New York, 1963.

———, *Conduction of Heat in Solids*, Clarendon Press, Oxford, 1959.

Y. P. Chang, C. S. Kang and D. J. Chen, *The use of fundamental Green's functions for the solution of heat conduction in anisotropic media*, Int. J. Heat and Mass Transfer **16** (1973), 1905–1918.

G. Chen and J. Zhou, *Boundary Element Methods*, Academic Press, New York, 1992.

W. F. Donoghue, *Distributions and Fourier Transforms*, Academic Press, New York, 1969.

A. Erdélyi et al, Tables of Integral Transforms (1954), McGraw–Hill, New York.

P. R. Garabedian, *Partial Differential Equations*, Wiley, New York, 1964.

I. M. Gel'fand, G. E. Shilov, M. I. Graev and N. Ya. Vilenkin, *Generalized Functions*, Vol. I, Academic Press, New York, 1964.

A. E. H. Love, *A Treatise on the Mathematical Theory of Elasticity*, Dover, New York, 1944.

R. D. Mindlin, *Force at a point in the interior of a semi–infinite solid*, Physics **7** (1936), 195–202.

H. Neuber, *Eine neuere Ansatz zur Lösung räumlicher Probleme der Elastizitätstheorie*, ZAMM **14** (1934), 203–212.

F. J. Rizzo and D. J. Shippy, *A method of solution of certain problems of transient heat conduction*, AIAA J. **8** (1970), 2004–2009.

G. E. Shilov, *Generalized Functions and Partial Differential Equations*, Gordon and Breach, New York, 1968.

I. Stakgold, *Green's Functions and Boundary Value Problems*, Wiley, New York, 1979.

⎯⎯⎯, *Boundary Value Problems of Mathematical Physics*, Vol. 2, Macmillan, London, 1968.

A. N. Tychonov and A. A. Samarski, *Partial Differential Equations of Mathematical Physics*, Vol. I and II, Holden–Day, San Francisco, 1964, 1967.

V. S. Vladimirov, *Equations of Mathematical Physics*, Dekker, New York, 1971.

R. E. Williamson, R. H. Crowell and H. F. Trotter, *Calculus of Vector Functions*, Prentice–Hall, Englewood Cliffs, NJ, 1962.

5

Potential Problems

We shall study Laplace and Poisson equations which represent a wide range of problems in applied mathematics, physics and engineering. Some of the physical situations which have models involving these equations are: (i) steady–state heat conduction problems, (ii) torsion problems in solid mechanics, (iii) diffusion flow in porous media, (iv) incompressible inviscid fluid flow, (v) electrostatic potential problems, (vi) Newtonian potentials, and (vii) magnetostatics. We will introduce the boundary element method for two– and three–dimensional steady–state potential problems. In a very general form, this method starts by subdividing the boundary of the region into finitely many elements (hence the name boundary element method, or BEM). This method then becomes a particular case of the weighted residual technique, except that it usually produces a singularity when the equation using the fundamental solution is discretized on the boundary. A simple way to introduce this method is to integrate the fundamental solution of boundary value problems on the boundary.

5.1. Laplace Equation

We will solve the mixed Laplace boundary value problem

$$\nabla^2 u = 0, \quad u = u_0 \text{ on } C_1, \quad \frac{\partial u}{\partial n} \equiv q = q_0 \text{ on } C_2, \qquad (5.1)$$

where $C = C_1 \cup C_2$ is the boundary of a region R. In view of (B.1), the potential boundary value problem with a concentrated charge acting at a point i can be written as

$$\nabla^2 u^* = -\delta(i); \quad u^* = u_0^* \text{ on } C_1, \quad \frac{\partial u^*}{\partial n} \equiv q^* = q_0^* \text{ on } C_2. \qquad (5.2)$$

The solution of this problem is called the fundamental solution for the potential problem. In view of the translation property (1.55) of the delta function,

$$0 = \iint_R u \big[\nabla^2 u^* + \delta(i) \big] \, dx \, dy = \iint_R u \nabla^2 u^* \, dx \, dy + u(i), \qquad (5.3)$$

where $u(i)$ denotes the value of the unknown potential u at the point i where the charge is applied. Note that we are writing $u(i)$ for $u(x_i)$, where x_i is the source point, thus $r = |x_j - x_i|$ (see Fig. B.1). Now, the weak variational form (2.8) for the boundary value problem (5.1) becomes

$$0 = \iint_R \big(\nabla^2 u \big) w \, dx \, dy$$
$$= - \iint_R \Big(\frac{\partial u}{\partial x} \frac{\partial w}{\partial x} + \frac{\partial u}{\partial y} \frac{\partial w}{\partial y} \Big) \, dx \, dy + \int_C w \frac{\partial u}{\partial n} \, ds,$$

which by applying (1.5c) again gives

$$0 = \iint_R u \nabla^2 w \, dx \, dy - \int_C u \frac{\partial w}{\partial n} \, ds + \int_C w \frac{\partial u}{\partial n} \, ds.$$

Thus,

$$- \iint_R u \nabla^2 w \, dx \, dy = \int_C w \frac{\partial u}{\partial n} \, ds - \int_C u \frac{\partial w}{\partial n} \, ds. \qquad (5.4)$$

Since $u = u_0$ on C_1 and $\partial u / \partial n \equiv q = q_0$ on C_2, where $C = C_1 \cup C_2$, we replace w by u^* (and hence q by q^*) in (5.4), and from (5.3) and (5.4) we obtain

$$u(i) = \int_{C_1} u^* q \, ds + \int_{C_2} u^* q_0 \, ds - \int_{C_1} u_0 q^* \, ds - \int_{C_2} u q^* \, ds, \qquad (5.5)$$

where $q^* = \partial u^* / \partial n$. Recall that for an isotropic two–dimensional region, the fundamental solution is given by (B.3) or (4.26), where r is the distance from the point of application of the delta function to the point under consideration. The symmetric form of the two–dimensional Laplace equation in polar cylindrical coordinates is

$$\frac{\partial^2 u^*}{\partial r^2} + \frac{1}{r} \frac{\partial u^*}{\partial r} = -\delta(i). \qquad (5.6)$$

Substituting (4.26) into (5.6) yields $\delta(i) = 0$ for $r \neq 0$. Thus, this equation is satisfied for any $r \neq 0$. Since at $r = 0$ the fundamental solution u^* has a logarithmic singularity, we proceed as follows: Integrate on a circle K surrounding the boundary point i where the charge is applied. This gives

$$\iint_R \nabla^2 u^* \, dx \, dy = -\iint_R \delta(i) \, dx \, dy = -1. \qquad (5.7)$$

To show that the first integral in (5.7) is also equal to -1, we find from (1.6b) that

$$\iint_R \nabla^2 u^* \, dx \, dy = \int_K q^* \, ds = \int_K \frac{\partial u^*}{\partial r} \, ds. \qquad (5.8)$$

Substituting (4.26) into (5.8) we get

$$\int_K \frac{\partial u^*}{\partial r} \, ds = \frac{1}{2\pi} \int_0^{2\pi} \left(-\frac{1}{r}\right) r \, d\theta = -1. \qquad (5.9)$$

Note that this result (-1) is independent of r. Thus, the left side goes to -1 as $r \to 0$.

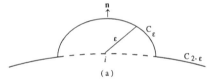

(a)

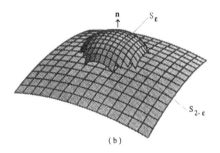

(b)

Fig. 5.1.

Eq (5.5), where $u(i)$ is the value of the unknown potential u at the point i of the application of the charge, is valid at any point of the region. However, in order to solve it by the boundary element method, we will formulate it on the boundary. One of the simplest ways to do so, as explained in Example 4.3, is as follows: Consider a semicircle C_ε of radius ε on the boundary of a two–dimensional region R, as in Fig. 5.1(a). Assume that the boundary point

i is at the center of this semicircle. As $\varepsilon \to 0$, the semicircle will reduce to the boundary point i. Further, assume that the boundary C of the region R is smooth, and that $C = C_1 \cup C_2$. Let the boundary point i be on the C_2 portion of C (similar considerations apply if it is on C_1). Divide the boundary C_2 into two parts: C_ε and $C_{2-\varepsilon}$. Then

$$\int_{C_2} uq^* \, ds = \int_{C_{2-\varepsilon}} uq^* \, ds + \int_{C_\varepsilon} uq^* \, ds. \qquad (5.10)$$

Substitute the fundamental solution (4.26) into the second integral on the right side and take the limit. This integral becomes

$$\lim_{\varepsilon \to 0} \int_{C_\varepsilon} uq^* \, ds = \lim_{\varepsilon \to 0} \int_{C_\varepsilon} u \left(-\frac{1}{2\pi\varepsilon} \right) ds = -\lim_{\varepsilon \to 0} \frac{u}{2\pi\varepsilon} \int_{C_\varepsilon} ds = -\frac{u}{2}, \qquad (5.11)$$

where $\int_{C_\varepsilon} ds = \pi\varepsilon$ (circumference of the semicircle). Now since ε is zero, the boundary $C_{2-\varepsilon}$ again becomes C_2. Also, note that the right side of (5.5) gives

$$\lim_{\varepsilon \to 0} \int_{C_\varepsilon} qu^* \, ds = \lim_{\varepsilon \to 0} \frac{q}{2\pi} \ln \frac{1}{\varepsilon} \int_{C_\varepsilon} ds = -\frac{q}{2} \lim_{\varepsilon \to 0} \varepsilon \ln \varepsilon = 0. \qquad (5.12)$$

Thus, this limiting process does not introduce any new terms in (5.5).

Now, substituting (5.11) into (5.5), we obtain the following two–dimensional BI Eq for a node i on the boundary C_2:

$$\frac{u(i)}{2} + \int_{C_1} u_0 q^* \, ds + \int_{C_2} uq^* \, ds = \int_{C_1} u^* q \, ds + \int_{C_2} u^* q_0 \, ds. \qquad (5.13)$$

We will obtain the same result if we consider the point i on the C_1 portion of the boundary instead of C_2.

In the three–dimensional case, consider the Laplace equation $\nabla^2 u = 0$ with the boundary conditions $u = u_0$ on S_1 and $q = q_0$ on S_2, where $S = S_1 \cup S_2$ is the boundary (surface) of an isotropic region. We shall take a hemisphere S_ε (of radius ε and center at i, see Fig. 5.1(b)) such that $S_2 = S_{2-\varepsilon} + S + \varepsilon$. Then Eq (5.10) becomes

$$\iint_{S_2} uq^* \, dS = \iint_{S_{2-\varepsilon}} uq^* \, dS + \iint_{S_\varepsilon} uq^* \, dS. \qquad (5.10')$$

We shall substitute the fundamental solution (B.2) or(4.25) into the second integral in (5.10'). Then

$$\lim_{\varepsilon \to 0} \iint_{S_\varepsilon} uq^* \, dS = \lim_{\varepsilon \to 0} \iint_{S_\varepsilon} u \left(-\frac{1}{4\pi\varepsilon^2} \right) dS$$

$$= -\lim_{\varepsilon \to 0} \frac{u}{4\pi\varepsilon^2} \iint_{S_\varepsilon} dS = -\frac{u}{2}, \qquad (5.11')$$

where $\iint_{S_\epsilon} dS = 2\pi\epsilon^2$ (surface area of the hemisphere). In this limit process the boundary $S_{2-\epsilon}$ becomes S_2, and

$$\lim_{\epsilon \to 0} \iint_{S_\epsilon} qu^* \, dS = \lim_{\epsilon \to 0} \frac{q}{2\pi\epsilon} \iint_{C_\epsilon} dS = \lim_{\epsilon \to 0} \frac{q}{2\pi\epsilon} \left(2\pi\epsilon^2\right) = 0. \quad (5.12')$$

Thus, the three-dimensional BI Eq for a node i on the boundary S_2 is given by

$$\frac{u(i)}{2} + \iint_{S_1} u_0 q^* \, dS + \iint_{S_2} uq^* \, dS = \iint_{S_1} u^* q \, dS + \iint_{S_2} u^* q_0 \, dS. \quad (5.13')$$

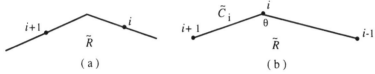

Fig. 5.2.

In general, BI Eq (5.13) or (5.13') can be written as

$$c(i)u(i) + \int_C uq^* \, ds = \int_C u^* q \, ds, \quad C = C_1 \cup C_2, \quad (5.14)$$

or

$$c(i)u(i) + \iint_S uq^* \, dS = \iint_S u^* q \, dS, \quad S = S_1 \cup S_2, \quad (5.14')$$

respectively, under the essential boundary conditions $u = u_0$ on C_1 (or on S_1) and the natural boundary condition $\partial u/\partial n \equiv q = q_0$ on C_2 (or on S_2), where in two–dimensional case

$$c(i) = \begin{cases} 0 & \text{if } i \text{ is outside } R \cup C \\ 1 & \text{if } i \text{ is inside } R \\ 1/2 & \text{if } i \text{ is on a smooth portion of } C \\ \theta/2\pi & \text{if } i \text{ is at a corner node,} \end{cases} \quad (5.15)$$

θ being the internal angle (in radian) at the corner at node i (see Fig. 5.2(b)). In Fig. 5.2(a), the value of $c(i) = 1/2$. The coefficient $c(i)$ can be evaluated analytically, or by considering different cases of the values of potential and flux 'before' and 'after' a corner node (see §5.5). In the case of a three–dimensional region, $c(i)$ has the same values relative to the boundary surface S except at a corner node on S where it has the value $\theta/4\pi$, θ being the solid (internal) angle at that node.

5.2. Boundary Elements

We shall discretize (partition) the smooth boundary C of a two–dimen sional region R into N segments C_j, $j = 1, \ldots, N$ (Fig. 5.3). The chords joining the partition points are called the *boundary elements* and will be denoted by $\tilde{C}_j, j = 1, \ldots, N$; the partition points are called the *extreme points* of the boundary elements. The discretization of the boundary produces, in general, an approximate region $\tilde{R}$ and an approximate (polygonal) boundary $\tilde{C} = \cup_{j=1}^{N}\tilde{C}_j$. The portion between the boundary C and the approximate boundary $\tilde{C}$ will produce a discretization error. The choice of boundary elements should always minimize the discretization error. If the boundary conditions are mixed, i.e., if the essential and natural boundary conditions are applied on two portions C_1 and C_2 ($C = C_1 \cup C_2$), the two points common to these portions are taken as extreme points. In the case of zero discretization error, we shall have $\tilde{R} = R$ and $\tilde{C} = C$.

The points where both known and unknown values of u and q are considered according to the prescribed boundary conditions are called *nodes*. Three types of nodes are explained in Fig. 5.4:

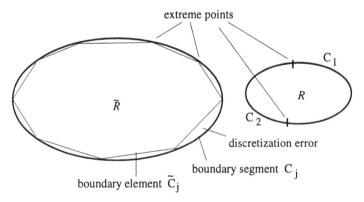

Fig. 5.3.

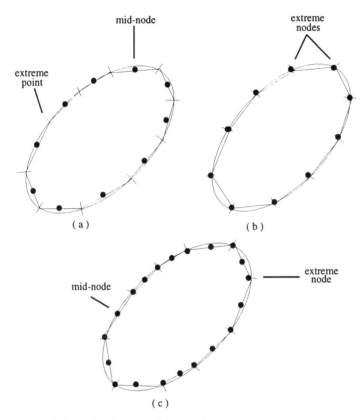

Fig. 5.4. Elements, Mid–nodes, and extreme nodes.

1. Constant elements have mid–nodes, which are taken at the mid–point of each element, as in Fig. 5.4.(a);

2. Linear elements have extreme nodes, which are at the intersection between two elements, as in Fig. 5.4.(b); and

3. Quadratic elements have both mid– and extreme nodes, as in Fig. 5.4(c).

The nomenclature of linear, quadratic, or cubic elements resembles the one used in finite element analysis (see §1.2 on line elements). We shall now discuss these types separately.

5.3. Constant Elements

Let the boundary C of the region R be smooth, and let it be discretized into N elements, of which N_1 elements belong to C_1 and N_2 to C_2. This discretization produces an approximate region $\tilde{R}$ and an approximate boundary $\tilde{C}$. Assume that the values of u and $q \equiv \partial u / \partial n$ are constant on each element and equal to the value at the mid–node of the respective element. Eq (5.14) for a given node i becomes in the discretized form

$$\frac{u(i)}{2} + \sum_{j=1}^{N} u_j \int_{\tilde{C}_j} q^* \, ds = \sum_{j=1}^{N} q_j \int_{\tilde{C}_j} u^* \, ds, \qquad (5.16)$$

where $q^* \equiv \partial u^* / \partial n$. This BE Eq applies to a particular node i. Note that the terms with $\int_{\tilde{C}_j}$ relate to the node i with segment j over which the integral is evaluated. Denote the integrals $\int_{\tilde{C}_j} q^* \, ds$ on the left side of (5.16) by $\hat{H}_{ij}$, and the integrals $\int_{\tilde{C}_j} u^* \, ds$ on the right side by G_{ij}. Then Eq (5.16) becomes

$$\frac{u(i)}{2} + \sum_{j=1}^{N} u_j \hat{H}_{ij} = \sum_{j=1}^{N} q_j G_{ij}. \qquad (5.17)$$

The integrals $\hat{H}_{ij}$ and G_{ij} are easy to evaluate for the constant element case. However, for higher order elements they are more difficult to evaluate analytically and will be computed by using the Gauss quadrature formulas (see Appendix C).

Eq (5.17) relates the value of u at the mid–node i with the value of u and q at all the nodes on the boundary including i. If we write Eq (5.17) for each mid–node i, we get a system of N equations:

$$\sum_{j=1}^{N} H_{ii} u_j = \sum_{j=1}^{N} G_{ij} q_j, \qquad (5.18)$$

where

$$H_{ij} = \begin{cases} \hat{H}_{ij} & \text{for } i \neq j \\ \hat{H}_{ij} + \frac{1}{2} & \text{for } i = j. \end{cases} \qquad (5.19)$$

Eq (5.17) can be written as in matrix form as

$$HU = GQ. \tag{5.20}$$

Note that $N = N_1 + N_2$, and that the N_1 values of u and N_2 values of q are known (prescribed). So we have a set of N unknowns in (5.20).

The terms H_{ii} include the coefficients $c(i)$ ($= 1/2$ for smooth boundary, see Fig. 5.2(a)). These terms are evaluated under the condition that when a uniform potential u is applied over the entire finite region the value of $q = \partial u / \partial n$ must be zero. Then (5.20) implies that

$$HI = 0, \tag{5.21}$$

where I is the unit column vector. Eq (5.21) means that the sum of all elements of H in a row should be zero. Thus, the values of the diagonal coefficients can be easily computed once all the off-diagonal coefficients are known, i.e.,

$$H_{ii} = -\sum_{\substack{j=1 \\ i \neq j}}^{N} H_{ij}, \quad i = 1, \cdots, N. \tag{5.22}$$

Let us denote the N_2 unknown values of u by $\hat{u}$ and the N_1 values of q by $\hat{q}$. We can reorder Eq (5.20) such that all the unknowns (N_2 of $\hat{u}$ and N_1 of $\hat{q}$) are on the right side. Then Eq (5.20) can be written as

$$AX = F, \tag{5.23}$$

where X is the vector of unknowns u and q. Hence, we can determine all the values of u and q on the entire boundary $\tilde{C}$ from (5.23). Once this is done, we can compute the value of u at any interior point by using (5.5) which in the discretized form is

$$u(i) = \sum_{j=1}^{N} q_j G_{ij} - \sum_{j=1}^{N} u_j \hat{H}_{ij}. \tag{5.24}$$

The internal fluxes $q_x = \partial u / \partial x$ and $q_y = \partial u / \partial y$ can be computed by differentiating (5.5); thus, at the node i

$$q_x(i) = \int_{\tilde{C}} q \frac{\partial u^*}{\partial x} ds - \int_{\tilde{C}} u \frac{\partial q^*}{\partial x} ds,$$
$$= \sum_{j=1}^{N} q_j \left(\int_{\tilde{C}_j} \frac{\partial u^*}{\partial x} ds \right) - \sum_{j=1}^{N} u_j \left(\int_{\tilde{C}_j} \frac{\partial q^*}{\partial x} ds \right), \tag{5.25a}$$
$$q_y(i) = \int_{\tilde{C}} q \frac{\partial u^*}{\partial y} ds - \int_{\tilde{C}} u \frac{\partial q^*}{\partial y} ds$$
$$= \sum_{j=1}^{N} q_j \left(\int_{\tilde{C}_j} \frac{\partial u^*}{\partial y} ds \right) - \sum_{j=1}^{N} u_j \left(\int_{\tilde{C}_j} \frac{\partial q^*}{\partial y} ds \right), \tag{5.25b}$$

where

$$\frac{\partial u^*}{\partial x} = \frac{1}{2\pi}\frac{\partial}{\partial x}(-\ln r) = -\frac{1}{2\pi r}\frac{\partial r}{\partial x},$$

$$\frac{\partial u^*}{\partial y} = \frac{1}{2\pi}\frac{\partial}{\partial y}(-\ln r) = -\frac{1}{2\pi r}\frac{\partial r}{\partial y},$$

$$\frac{\partial q^*}{\partial x} = \frac{1}{2\pi}\left[\frac{1}{r}\left(\frac{\partial r}{\partial x}n_1 + \frac{\partial r}{\partial y}n_2\right)\right],$$

$$\frac{\partial q^*}{\partial y} = \frac{1}{2\pi}\left[\frac{1}{r}\left(\frac{\partial r}{\partial x}n_1 + \frac{\partial r}{\partial y}n_2\right)\right],$$

(5.26)

and n_1, n_2 are the components of the unit normal $\hat{n}$. The integrals in (5.25) are evaluated numerically by the Gauss quadrature. So also are the integrations for $\hat{H}_{ij}$ and G_{ij} done numerically by Gauss quadrature for all elements $j \neq i$. For the node i, note that $\hat{H}_{ii} = 0$ (due to the orthogonality of r and $\hat{n}$), and

$$G_{ii} = \int_{C_i} u^* \, ds = \frac{1}{2\pi}\int_{C_i}\ln\left(\frac{1}{r}\right)ds = \frac{L_i}{2\pi}\left[1 - \ln\left(\frac{L_i}{2}\right)\right], \qquad (5.27)$$

where $L_i = \sqrt{(x_{i+1} - x_i)^2 + (y_{i+1} - y_i)^2}$ is the length of the element i (see Fig. 5.5.).

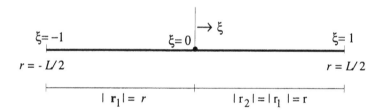

Fig. 5.5. A constant element.

For the derivation of (5.27), we take $r = \xi L_i/2$, so $r = 0$ at $\xi = 0$ and $r = \pm L_i/2$ at $\xi = \pm 1$. Then, since $\ln(1/r)$ has a logarithmic singularity at $r = \xi = 0$,

$$\begin{aligned}
G_{ii} &= \frac{1}{2\pi}\int_{\tilde{C}_i}\ln\left(\frac{1}{r}\right)ds = \frac{1}{2\pi}2\int_0^1\ln\left(\frac{2}{\xi L_i}\right)\frac{L_i}{2}\,d\xi \\
&= \frac{L_i}{2\pi}\left[\xi\ln\frac{2}{L_i} - \xi\ln\xi + \xi\right]_0^1 \\
&= \frac{L_i}{2\pi}\left[1 + \ln\left(\frac{2}{L_i}\right)\right],
\end{aligned}$$

which gives (5.27).

Program Be1

For computer implementation in this case, the program Be1.c (see Appendix D) solves isotropic potential problems with constant u and q at the mid–nodes and computes them at the required interior points of the region. The input file is created in the format explained below. It can be named Be1.in, or any other name, not to exceed 10 characters including the extension which may be .in or .dat. The output file, named Be1.out, or any other name not to exceed 10 characters, can be typed (on screen), printed (hard copy), or used as input for graphics.

Dictionary of Variables:

N	Number of boundary elements (same as the number of mid–nodes in this case)
L	Number of interior points where the results are to be computed
Code	Indicator for the type of boundary conditions at the nodes.
	Code$= 0$: Only the value of u is known at the node.
	Code$= 1$: Only the value of q is known at the node.
X	x–coordinate of extreme–points.
Y	y–coordinate of extreme–points.
Xm	x–coordinate of mid–nodes.
Xm(j)	x–coordinate of mid–node j.
Ym	y–coordinate of mid–nodes.
Ym(j)	y–coordinate of mid–node j.
G	Matrix defined in (5.20).
	After the boundary conditions are imposed, the matrix A of (5.23) is stored in this location.
H	Matrix defined in (5.20).
Bc(j)	Prescribed values of boundary conditions for node j.
	If Code$= 0$, then Bc(j) contains prescribed values of u.
	If Code$= 1$, then Bc(j) contains prescribed values of q.
F	Right side vector in (5.23).
	After solution, the values of unknown u and q are returned in this location.
Xi	x–coordinate of the interior point where the value of u is required.
Yi	y–coordinate of the interior point where the

	value of u is required.
u	Vector of potential values at interior points.
Dim	Maximum dimension of the system of Eqs (5.20).
Perp	Perpendicular distance from the point (x_p, y_p) to the element j (Fig. 5.8a).
Xg, Yg	(x, y)–coordinates of Gauss points ζ_i, $i = 1, 2, 3, 4$
HL	Half–length of the element $\tilde{C}_i$ $(= L_i/2)$.
nx, ny	n_x, n_y (components of the unit normal vector $\hat{n}$).
rx, ry, rn	$r_{,x}, r_{,y}, r_{,n}$ (note that $r_{,n} = r_{,x}n_x + r_{,y}n_y$).

Input Format: The input file contains the data in the following order:

Entry #	Variable	Explanation
1	Title	Must enter problem title (max 80 chars).
2	N	Number of boundary elements.
3	L	Number of interior points where solution is to be computed.
next N–pairs	X, Y	x, y coordinates of extreme points (see Note below).
next N–pairs	Code, Bc	N–pairs of Code (0 or 1) and Bc at each mid–node, starting with node 1 and ending with node N.
next L–pairs	Xi, Yi	x and y coordinates of interior points where the solution is to be computed.

Note: The N–pairs of entries for the coordinates of the extreme points of boundary elements are read in counterclockwise order if the region R is interior to the boundary C, and in clockwise order if R is exterior to the boundary C (see Fig. 5.6). The convention is to traverse the boundary in such a manner that the region R remains to the left.

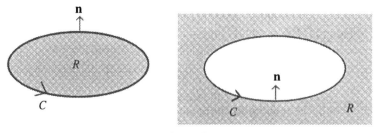

Fig. 5.6.

The program Bel calls the following functions: Sys1, Quad1, Diag1, In-
ter1, and Solve.

Sys1 can be summarized by the following algorithm:

1. Compute the coordinates $(X_m, Y_m), m = 1, \cdots, N$, of the mid–nodes
   from the extreme nodes (x_j, y_j):

$$x_{n+1} = x_1, \; y_{n+1} = y_1 \quad \text{(see Fig. 5.7.)};$$

$$X_m = \frac{x_j + x_{j+1}}{2}, \quad Y_m = \frac{y_j + y_{j+1}}{2} \quad \text{for } j = 1 \text{ to } N.$$

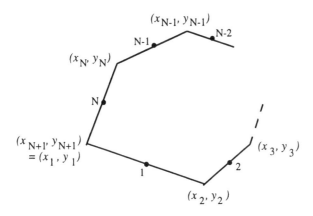

Fig. 5.7.

2. Compute the matrices H and G:

$$\text{for } j = 1 \text{ to } N$$
$$\text{for } k = 1 \text{ to } N$$
$$j \neq k: \quad \text{call Quad1}$$
$$j = k: \quad \text{call DIAG1}$$

3. Produce the matrix Eq (5.23) $F = AX$:

   for $k = 1$ to N

   if $\text{Code}(k) = 1$, then for $j = 1$ to N

   $$\text{temp} = G_{jk}; \quad G_{jk} = -H_{jk}; H_{jk} = -\text{temp} = -G_{jk},$$

   where temp is a temporary memory location. As a result, the matrix A
   is in location G; F is not yet evaluated, but all known terms are in the

location of H and U. The location H contains both known G and H, and the location U contains both known U and Q.

4. Finally,

$$F_j = 0.0 \quad \text{for } j = 1 \text{ to } N$$
$$F_j = F_j + H_{jk} \cdot \text{Bc}_k \quad \text{for } k = 1 \text{ to } N.$$

The left side now stores F (all known values). The rearrangement of the system of equations to form the matrix A (now stored in G) and the left side vector F (stored as F) of Eq (5.23) is explained as follows: Suppose, Eq (5.19) (or in its matrix form (5.20)) is written, in the expanded form, as

$$\text{Code 0} \qquad \text{Code 1}$$

$$H_{11}\hat{u}_1 + H_{12}\hat{u}_2 + H_{13}u_3 + \cdots + H_{1n}u_n = G_{11}q_1 + G_{12}q_2 + G_{13}\hat{q}_3 +$$
$$+ \cdots + G_{1n}\hat{q}_n$$
$$H_{21}\hat{u}_1 + H_{22}\hat{u}_2 + H_{23}u_3 + \cdots + H_{2n}u_n = G_{21}q_1 + G_{22}q_2 + G_{23}\hat{q}_3 +$$
$$+ \cdots + G_{2n}\hat{q}_n$$

$$\vdots$$

$$H_{n1}\hat{u}_1 + H_{n2}\hat{u}_2 + H_{n3}u_3 + \cdots + H_{nn}u_n = G_{n1}q_1 + G_{n2}q_2 + G_{n3}\hat{q}_3 +$$
$$+ \cdots + G_{nn}\hat{q}_n,$$
$$(5.28)$$

where $\hat{u}$, $\hat{q}$ denote the unknown values. In view of part 3 of the above algorithm, we note that G_{jk} is first stored in the location temp, then $-H_{jk}$ is moved to G_{jk}, and finally replace the values G_{jk}, stored at temp, to the location $-H_{jk}$. The result is the matrix equation (5.23), thus rendering the system (5.28) into the form

$$-G_{11}q_1 - G_{12}q_2 + H_{13}u_3 + \cdots + H_{1n}u_n = -H_{11}\hat{u}_1 - H_{12}\hat{u}_2 + G_{13}\hat{q}_3 +$$
$$+ \cdots + G_{1n}\hat{q}_n$$
$$-G_{21}q_1 - G_{22}q_2 + H_{23}u_3 + \cdots + H_{2n}u_n = -H_{21}\hat{u}_1 - H_{22}\hat{u}_2 + G_{23}\hat{q}_3 +$$
$$+ \cdots + G_{2n}\hat{q}_n$$

$$\vdots$$

$$-G_{n1}q_1 - G_{n2}q_2 + H_{n3}u_3 + \cdots + H_{nn}u_n = -H_{n1}\hat{u}_1 - H_{n2}\hat{u}_2 + G_{n3}\hat{q}_3 +$$
$$+ \cdots + G_{nn}\hat{q}_n,$$

which is the expanded form of Eq (5.23).

As mentioned above, Sys1 uses the two functions: Quad1 and Diag1. The function Quad1 computes the off–diagonal elements of H and G by using the Gauss quadrature formula (Appendix C). The variable Ra is explained in Fig. 5.8(a). In order to compute H_{ij} and G_{ij}, $i \neq j$, we consider two different nodes i and j in Fig. 5.8(a), where (x_p, y_p) are the coordinates of the node i under consideration; ζ_k, $k = 1, 2, 3, 4$ are the Gauss points marked on the element with mid–node j; $\zeta = 1$ and $\zeta = -1$ are the extreme points with coordinates (x_j, y_j) and (x_{j+1}, y_{j+1}) respectively of the node j; m denotes the slope of the boundary element with node j; and Ra is the distance from node i to a Gauss point ζ_k, $k = 1, 2, 3, 4$ (Fig. 5.8(a)). The coordinates of the Gauss points are denoted by (Xg, Yg). If we denote $\mathtt{Ax} = (x_{j+1} - x_j)/2$, $\mathtt{Ay} = (y_{j+1} - y_j)/2$, $\mathtt{Bx} = (x_{j+1} + x_j)/2$, and $\mathtt{By} = (y_{j+1} + y_j)/2$, then $m = (y_{j+1} - y_j)/(x_{j+1} - x_j) = \mathtt{Ay/Ax} = \mathtt{slope}$ of the element $\tilde{C}_j$. Also, the equation of the element with node j is

$$m(x_j - x) - (y_j - y) = 0,$$

the distance

$$\mathtt{Ra} = \sqrt{(x_p - Xg)^2 + (y_p - Yg)^2},$$

and the half–length of the element $\tilde{C}_i$

$$\frac{L_i}{2} = \sqrt{\mathtt{Ax}^2 + \mathtt{Ay}^2} = \mathtt{HL}.$$

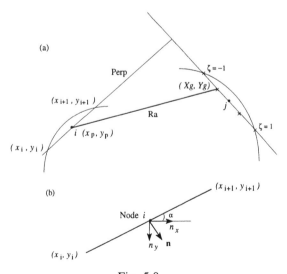

Fig. 5.8.

Also, the directional derivative of $\ln(1/r) = -\frac{1}{2}\ln(x^2 + y^2)$ in the direction of $\hat{n}$ is given by

$$D_{\hat{n}} \ln\left(\frac{1}{r}\right) = \nabla f \cdot \hat{n} = -\frac{y}{x^2 + y^2}$$

$$= -\frac{r_{,x} n_x + r_{,y} n_y}{r^2} = -\frac{\text{rx} \cdot \text{nx} + \text{ry} \cdot \text{ny}}{(\text{Ra})^2},$$

where $\text{rx} = (\text{Xg} - x_p)/\text{Ra} = \cos\alpha$, $\text{ry} = (\text{Yg} - x_p)/\text{Ra} = \sin\alpha$ (see Fig. 5.8(a),(b)). Then, by an 4–point Gauss quadrature, we have

$$G_{ij} = \int_{\tilde{C}_j} u^* \, ds$$

$$= \sum_{i=1}^{4} \ln\left(\frac{1}{\text{Ra}}\right) W_i \sqrt{\text{Ax}^2 + \text{Ay}^2}, \tag{5.29}$$

and

$$H_{ij} = \int_{\tilde{C}_j} q^* \, ds$$

$$= \sum_{i=1}^{4} D_{\hat{n}} \ln\left(\frac{1}{r}\right)_i W_i \frac{\sqrt{(x_j - x_{j+1})^2 + (y_j - y_{j+1})^2}}{2} \tag{5.30}$$

$$= -\sum_{i=1}^{4} \frac{1}{(\text{Ra})_i^2} (\text{rx} \cdot \text{nx} + \text{ry} \cdot \text{ny}) W_i \sqrt{\text{Ax}^2 + \text{Ay}^2}.$$

We can use a Gauss quadrature formula with different Gauss points. An empirical rule to decide which one of the Gauss quadrature formulas can be used is as follows: Let

$$s = \frac{1}{2L_j} \sqrt{\left(2x_p - X(1) - X(2)\right)^2 + \left(2y_p - Y(1) - Y(2)\right)^2}, \tag{5.31}$$

where (x_p, y_p) are the coordinates of the node i, and $X(j), Y(j), j = 1, 2$, are the coordinates of the extreme points of the node j of the element of length L_j. Then, use 6–point Gauss quadrature formula if $s \leq 1.5$, 4–point formula if $1.5 < s < 5.5$, and 2–point formula if $s \geq 5.5$. However, the 4–point formula gives good results in the constant and linear cases, whereas a 10–point formula is used with advantage in the quadratic and higher order boundary elements.

Diag1 computes the diagonal elements G_{ii} of the matrix G given by (5.27).

Solve uses the Gauss elimination method to solve the system of equations $AX = F$ by providing interchange of rows when a zero diagonal element is present.

Note that since the fundamental solution in the program is taken as $\ln(1/r)$, and not $\dfrac{1}{2\pi}\ln(1/r)$, all elements of H and G are divided by 2π in the end before the output is produced.

Inter1 computes the values of u at interior points by using (5.24). It reorders Bc (boundary condition vector) and F (unknown vector) such that all values of u are stored in Bc and all values of q in F. Note that since all H and G terms appear multiplied by 2π, the solution for interior points is also finally divided by 2π.

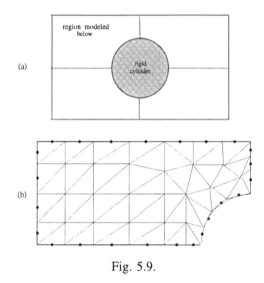

Fig. 5.9.

Example 5.1:

This example deals with a region shown in Fig. 5.9. Although this region is used in various heat transfer, fluid flow and elasticity problem, we have given it here for the purpose of showing how the boundary element method compares with the finite element method in terms of the number of elements and nodes. Imagine a confined potential flow about a circular cylinder with its axis perpendicular to the plane of the flow between two long horizontal walls. We consider only the quarter region because of symmetry, and find that for the mesh in Fig. 5.9(b) the finite element method requires 58 triangular elements and 42 nodes, whereas the boundary element method needs only 24 constant elements (or mid–nodes); see §8.2 and Problem 8.17.∎

Example 5.2:
Heat flow problem in the region shown in Fig. 5.10 is the simplest case of a linear one–directional flow.

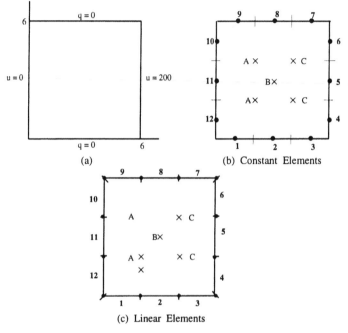

Fig. 5.10.

The input file is created in this order:

```
Example 5.2 (Constant elements, Fig. 5.10(b) )
12 5
0. 0. 2. 0. 4. 0. 6. 0. 6. 2. 6. 4.
6. 6. 4. 6. 2. 6. 0. 6. 0. 4. 0. 2.
1  0.  1 0.  1 0.  0 200.  0 200.  0 200.
1  0.  1 0.  1 0.  0 0.  0 0.  0 0.
2. 2. 2. 4. 3. 3. 4. 2. 4. 4.
```

The output is:

Mid-nodes (X, Y)	u	q
(1.0000, 0.0000)	31.8337	0.0
(3.0000, 0.0000)	100.0125	0.0

(5.0000, 0.0000)	168.1661	0.0
(6.0000, 1.0000)	200.0	35.3129
(6.0000, 3.0000)	200.0	32.4913
(6.0000, 5.0000)	200.0	35.3129
(5.0000, 6.0000)	168.1660	0.0
(3.0000, 6.0000)	100.0123	0.0
(1.0000, 6.0000)	31.8335	0.0
(0.0000, 5.0000)	0.0	-35.3076
(0.0000, 3.0000)	0.0	-32.5141
(0.0000, 1.0000)	0.0	-35.3076

Interior Points	u	q_x	q_y
(2.0000, 2.0000)	66.4933	33.5370	0.0971
(2.0000, 4.0000)	66.4933	33.5370	-0.0971
(3.0000, 3.0000)	100.0063	33.4765	0.0
(4.0000, 2.0000)	133.5199	33.5356	-0.0998
(4.0000, 4.0000)	133.5199	33.5356	0.0998

The exact solution for the problem $\nabla^2 u = 0$ with $u(0, y) = 0, u(a, y) = T_0$ is given by $u(x, y) = \dfrac{T_0}{a}x$. In this example, $a = 6, T_0 = 200$ gives the exact $u = 100x/3$, and $q_x = 100/3, q_y = 0$ ∎.

Example 5.3:
Consider the potential/heat transfer problem represented in Fig. 5.11. The input data are as follows:

```
Example 5.3 (Constant Elements)
12  5
0. 0. 2. 0. 4. 0. 6. 0. 6. 2. 6. 4.
6. 6. 4. 6. 2. 6. 0. 6. 0. 4. 0. 2.
0  0.  0  0.  0  0.  1  48.9  1  43.7  1  48.9
0  45.3  0  148.9  0  231.6  1  -48.9  1  -43.7  1  -48.9
2. 2. 2. 4. 3. 3. 4. 2. 4. 4.
```

The output is:

Interior Points	u	q_x	q_y
(2.0000, 2.0000)	31.7247	18.0850	22.1574
(2.0000, 4.0000)	92.1874	6.8650	39.4603
(3.0000, 3.0000)	71.1492	11.9264	24.0222

(4.0000, 2.0000)   62.6037   17.7637   25.4824
(4.0000, 4.0000)   97.8296   5.5206   8.8485

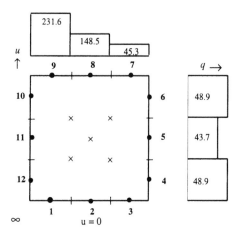

Fig. 5.11.

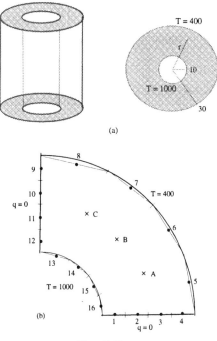

Fig. 5.12.

Example 5.4:

Consider the heat transfer in a hollow circular cylinder of radii 10 and 30 (Fig. 5.12(a)). Because of the axial symmetry we shall discretize the quarter region into 16 constant elements (Fig. 5.12(b)). The input file is:

```
Example 5.4 (Constant Elements)
16  3
10. 0. 15. 0. 20. 0. 25. 0. 30. 0.
27.7164 11.4805 21.2132 21.2132 11.4805 27.7164
0. 30. 0. 25. 0. 20. 0. 15. 0. 10.
3.82683 9.2388 7.07107 7.07107 9.2388 3.82683
1  0.  1  0.  1  0.  1  0.
0  400.  0  400.  0  400.  0  400.
1  0.  1  0.  1  0.  1  0.
0  1000.  0  1000.  0  1000.  0  1000.
18.4776 7.65367 14.1421 14.1421 7.65367 18.4776
```

The output is:

```
Ex 5.4:
  Mid-nodes (X, Y)       u            q
(12.5000, 0.0000)    876.1199      0.0000
(17.5000, 0.0000)    688.5445      0.0000
(22.5000, 0.0000)    548.9433      0.0000
(27.5000, 0.0000)    438.1204      0.0000
(28.8582, 5.7403)    400.0000    -18.8493
(24.4648, 16.3468)   400.0000    -18.4573
(16.3468, 24.4648)   400.0000    -18.4573
(5.7403, 28.8582)    400.0000    -18.8494
(0.0000, 27.5000)    438.1204      0.0000
(0.0000, 22.5000)    548.9433      0.0000
(0.0000, 17.5000)    688.5446      0.0000
(0.0000, 12.5000)    876.1201      0.0000
(1.9134, 9.6194)    1000.0000     58.3131
(5.4489, 8.1549)    1000.0000     53.7123
(8.1549, 5.4489)    1000.0000     53.7123
(9.6194, 1.9134)    1000.0000     58.3130


Interior Points                    u          q_x q_y
(Xi,Yi)
(   18.4776,      7.6537)    613.8828   -25.3482   -10.6362
(   14.1421,     14.1421)    613.4034   -19.3797   -19.3797
(    7.6537,     18.4776)    613.8827   -10.6362   -25.3482
```

Exact solution for a hollow cylinder of inner and outer radius a and b respectively is given by

$$T = T_a + \frac{T_a - T_b}{\ln(a/b)} \ln \frac{r}{a}, \quad 10 \le r \le 30,$$

where T_a and T_b are the prescribed temperatures on the circular boundaries $r = a$ and $r = b$ respectively. Exact $T(20) = 621.1442$. A better accuracy is achieved by taking more elements on the circular boundaries. See also Example 5.9. ∎

5.4. Linear Elements

The variation of u and q is assumed to be linear within each element. The nodes are at the intersection of straight elements, and hence called extreme nodes (Fig. 5.4(b)). In this case the BI Eq (5.14) leads to the BE Eq

$$c(i)u(i) + \sum_{j=1}^{N} \int_{\tilde{C}_j} uq^* \, ds = \sum_{j=1}^{N} \int_{\tilde{C}_j} qu^* \, ds \qquad (5.32)$$

for N elements. Unlike Eq (5.14), we cannot take u_j and q_j out of the integral sign since they vary linearly within each element. Also, $c(i) = 1/2$ only for smooth boundaries. For nonsmooth boundaries, we will discuss a method to determine $c(i)$ later (see Discontinuous Elements at a Corner Node in the sequel).

Consider an arbitrary segment as shown in Fig. 5.13(a). The values of u and q at any point of this segment can be determined in terms of their nodal values and two linear interpolation functions ϕ_1 and ϕ_2 such that, for the case of both u and q varying linearly, we have

$$
\begin{aligned}
u(\xi) &= \phi_1 u_1 + \phi_2 u_2 = [\phi_1\ \phi_2]\,[u_1\ u_2]^T, \\
q(\xi) &= \phi_1 q_1 + \phi_2 q_2 = [\phi_1\ \phi_2]\,[q_1\ q_2]^T,
\end{aligned}
\qquad (5.33)
$$

(a) Linear

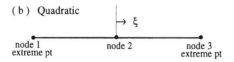

(b) Quadratic

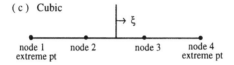

(c) Cubic

Fig. 5.13.

where ξ is the dimensionless coordinate $\xi = x/(l/2) = 2x/l$, and ϕ_1, ϕ_2 are defined in (1.27). The integrals along the element j on the left side of Eq (5.32) is

$$\int_{\tilde{C}_j} u q^* \, ds = \int_{\tilde{C}_j} [\phi_1 \ \phi_2] \, [u_1 \ u_2]^T q^* \, ds$$

$$= \int_{\tilde{C}_j} [\phi_1 \ \phi_2] q^* \, ds \, [u_1 \ u_2]^T \qquad (5.34)$$

$$= [h_{i1} \ h_{i2}] \, [u_1 \ u_2]^T,$$

where

$$h_{i1} = \int_{\tilde{C}_j} \phi_1 q^* \, ds \equiv a_1, \quad h_{i2} = \int_{\tilde{C}_j} \phi_2 q^* \, ds \equiv a_2. \qquad (5.35)$$

For the right side of Eq (5.32)

$$\int_{\tilde{C}_j} q u^* \, ds = \int_{\tilde{C}_j} [\phi_1 \ \phi_2] u^* \, ds \, [q_1 \ q_2]^T$$

$$= \int_{\tilde{C}_j} [g_{i1} \ g_{i2}] \, [q_1 \ q_2]^T, \qquad (5.36)$$

where

$$g_{i1} = \int_{\tilde{C}_j} \phi_1 u^* \, ds \equiv b_1, \quad g_{i2} = \int_{\tilde{C}_j} \phi_2 u^* \, ds \equiv b_2. \qquad (5.37)$$

Substituting (5.34) and (5.36) for all j elements into (5.32), we get for node i

$$c(i)u(i) + \begin{bmatrix} \hat{H}_{i1} & \hat{H}_{i2} & \cdots & \hat{H}_{in} \end{bmatrix} \begin{bmatrix} u_1 & u_2 & \cdots & u_n \end{bmatrix}^T =$$
$$= \begin{bmatrix} G_{i1} & G_{i2} & \cdots & G_{in} \end{bmatrix} \begin{bmatrix} q_1 & q_2 & \cdots & q_n \end{bmatrix}^T, \quad (5.38)$$

where

$$\hat{H}_{ij} = h_{i1} \text{ term of element } j + h_{i2} \text{ term of element } (j-1) = a_1 + a_2.$$
$$(5.39)$$

Similarly

$$G_{ij} = g_{i1} \text{ term of element } j + g_{i2} \text{ term of element } (j-1) = b_1 + b_2. \quad (5.40)$$

Formulas (5.38) and (5.40) represent the assembled equation for the collocation point i. For $i = j$,

$$G_{ii} = \frac{1}{2\pi} \int_{\tilde{C}_i} u^* \, ds = \frac{1}{2\pi} \int_{\tilde{C}_i} \left[u_1 \phi_1 + u_2 \phi_2 \right] ds$$
$$= \frac{1}{2\pi} \left(u_1 G_{ii}^1 + u_2 G_{ii}^2 \right). \quad (5.41)$$

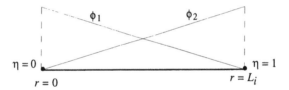

Fig. 5.14.

But in this case we find that although the interpolation functions $\phi_{1,2}$, defined by (1.27), are used in the evaluation of H_{ii}, they fail to remove the singularity at $\xi = 0$ when used to compute G_{ii}. We shall, therefore, modify the function $\phi_{1,2}$ by using the translation $\xi = 1 - 2(1 - \eta)$, which yields $\phi_1 = 1 - \eta$ and $\phi_2 = \eta$, $0 \le \eta \le 1$. Then, using the substitution $r = L_i\eta$, so that $ds = dr = L_i d\eta$ (see Fig. 5.14), we get

$$G_{ii}^1 = \int_{\tilde{C}_i} \phi_1 \ln \left(\frac{1}{r} \right) ds = \int_0^1 (1 - \eta) \ln \left(\frac{1}{L_i\eta} \right) L_i \, d\eta$$
$$= L_i \int_0^1 (1 - \eta) \left(-\ln L_i - \ln \eta \right) d\eta \quad (5.42)$$
$$= \frac{L_i}{2} \left[\frac{3}{2} - \ln L_i \right],$$

$$G_{ii}^2 = \int_{\tilde{C}_i} \phi_2 \ln\left(\frac{1}{r}\right) ds = \int_0^1 \eta \ln\left(\frac{1}{L_i\eta}\right) L_i \, d\eta$$

$$= L_i \int_0^1 \eta \left(-\ln L_i - \ln \eta\right) d\eta \tag{5.43}$$

$$= \frac{L_i}{2} \left[\frac{1}{2} - \ln L_i\right].$$

We can now write Eq (5.38) as

$$c(i)u(i) + \sum_{j=1}^n \hat{H}_{ij}u_j = \sum_{j=1}^n G_{ij}q_j, \tag{5.44a}$$

or

$$\sum_{j=1}^n H_{ij}u_j = \sum_{j=1}^n G_{ij}q_j, \tag{5.44b}$$

or, in matrix form, as

$$HU = GQ. \tag{5.44c}$$

Note that the value $-u/2$ obtained in (5.11) is now not valid unless the curve/surface is smooth. However, we can always compute the diagonal terms of H by using the fact that when a uniform potential is applied over the entire boundary, the normal derivative (i.e., q values) must be zero. Hence, Eq (5.44c) becomes $HU = 0$. It means that the sum of all elements of H in any row must be zero, and the value of the diagonal elements can be easily evaluated once all off–diagonal elements are known, i.e.,

$$H_{ii} = 1 - \sum_{j=1, j\neq i}^N H_{ij}. \tag{5.45}$$

Therefore, we need not compute the value of $c(i)$ explicitly. Also, since the fundamental solution in the program is taken as $-\ln r$, and not as $-\ln r/(2\pi)$, Eq (5.45) is written in the program as

$$H_{ii} = 2\pi - \sum_{j=1, j\neq i}^N H_{ij}, \quad i = 1, \ldots, N. \tag{5.46}$$

The results are then finally divided by 2π before the output is produced. The solution at the interior points is also finally divided by 2π in Inter2 (see program Be2 in the next section). This technique, simple as it is, is maintained throughout all computer programs to ensure uniformity.

5.5. Discontinuous Elements

A corner node becomes significant in linear and higher elements where the second node (marked i) of the $(i-1)$–st element is the same as the first node of the i-th element $\tilde{C}_i$ (Fig. 5.15(a)). If the potential is the same throughout the boundary, then the value of u on the $\tilde{C}_{i-1}$ is the same as on $\tilde{C}_i$. But this is not, in general, true for the flux q at a corner since the normals to the adjacent elements may be different, and hence not unique, at a corner; or the prescribed value of q along an element may possess discontinuities at some points. The former situation which occurs in most physical problems is solved by rearranging the terms in (5.38)–(5.40), which leads to the BE Eq

$$c(i)u(i) + \sum_{j=1}^{N} \hat{H}_{ij}u_j = \sum_{j=1}^{2N} G_{ij}q_j, \tag{5.47}$$

where the upper limit $2N$ of right sum corresponds to the case when the value of q at node i for the element $\tilde{C}_i$ is different from that for the element $\tilde{C}_{i-1}$ (see Fig. 5.15(b)). Defining, as in (5.19),

$$H_{ij} = \begin{cases} \hat{H}_{ij} & \text{for } i \neq j \\ \hat{H}_{ij} + c(i) & \text{for } i = j, \end{cases} \tag{5.48}$$

we get the same matrix form of the BE Eq as in (5.20), where G is now a $N \times 2N$ matrix.

The following observations are worth noting: At a corner node the flux 'before' and 'after' the node (i.e., the prenodal and postnodal fluxes) may be the same, or they may be prescribed differently. A similar situation may occur for the prescribed potential. However, only one of these variables will be unknown at a node. The following four cases arise at a corner node depending on adjacent boundary conditions:

(i) The prenodal and postnodal values of q are known (unknown: values of u);

(ii) The value of u and the prenodal value of q are known (unknown: postnodal value of q);

(iii) The value of u and the postnodal value of q are known (unknown: prenodal value of q); and

(iv) The values of u are known (unknown: both pre– and postnodal values of q).

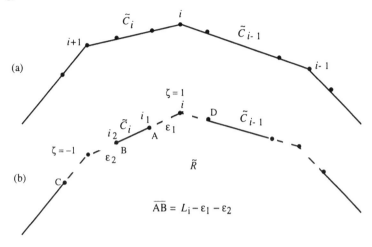

Fig. 5.15.

Except for the case (iv), there is only one unknown in the three cases which can be computed by using (5.47) and solving the subsequent matrix equation $HU = GQ$. The case (iv) can be computed by making the two adjacent elements at a corner into two discontinuous elements, by replacing the corner node i into two nodes i_1 and i_2 arbitrarily close to i such that $|i - i_1| = \varepsilon_1$ and $|i - i_2| = \varepsilon_2$, and $\xi_{i_1} = 2\varepsilon_1/L_i - 1$, and $\xi_{i_2} = 1 - 2\varepsilon_2/L_i$ are the local coordinates of the nodes i_1 and i_2, and the values of u and q prescribed at node i are now assigned to the nodes i_1 and i_2 (Fig. 5.15(b)). Since Eqs (5.33) hold, we have

$$\left\{ \begin{matrix} u_{i_1} \\ u_{i_2} \end{matrix} \right\} = \left[\begin{matrix} \phi_1(\xi_{i_1}) & \phi_2(\xi_{i_1}) \\ \phi_1(\xi_{i_2}) & \phi_2(\xi_{i_2}) \end{matrix} \right] \left\{ \begin{matrix} u_1 \\ u_2 \end{matrix} \right\}. \tag{5.49}$$

Substituting (5.49) into (5.33) we find that

$$u(\xi) = [\phi_1 \quad \phi_2] \, \mathbf{Q} \, [u_{i_1} \quad u_{i_2}]^T, \tag{5.50}$$

$$q(\xi) = [\phi_1 \quad \phi_2] \, \mathbf{Q} \, [q_{i_1} \quad q_{i_2}]^T, \tag{5.51}$$

where

$$\mathbf{Q} = \frac{1}{L_i - \varepsilon_1 - \varepsilon_2} \left[\begin{matrix} L_i - \varepsilon_2 & -\varepsilon_1 \\ -\varepsilon_2 & L_i - \varepsilon_1 \end{matrix} \right], \tag{5.52}$$

where $L_i - \varepsilon_1 - \varepsilon_2$, denoted by $\overline{AB}$ in Fig. 5.15(b), is the length of the discontinuous element $\tilde{C}_i$. It should be noted that the coefficient $c(i) = 1/2$

for discontinuous elements. The integrals $h_{i_1 1}$, $h_{i_2 1}$, $g_{i_1 1}$, $g_{i_2 1}$ along the discontinuous elements are then given by

$$h_{i_k 1} = \int_{\tilde{C}_j} \phi_k \mathbf{Q} q^* \, ds, \quad g_{i_k 1} = \int_{\tilde{C}_j} \phi_k \mathbf{Q} u^* \, ds, \ k = 1, 2, \qquad (5.53)$$

which are evaluated by Gauss quadrature for the case when the node i does not belong to the element. In the case when the node i belongs to the element, $h_{i_1 j} = 0 = h_{i_2 j}$, and $g_{i_1 j}$, $g_{i_2 j}$ are obtained by integration as in (5.41). For the corner nodes, as shown in Fig. 5.15(b), the segments AD, BC must be treated like additional linear elements. The linear elements of the type AD, BC may be further discretized to smoothen the corners out, which are thus replaced by polygonal curves.

Besides corner nodes, discontinuous elements also arise in certain cases where the potential u is undefined (and hence discontinuous) at a point on the boundary C. Example 5.8 discusses this type of discontinuous elements.

For linear elements we shall use the program Be2 (see Appendix D) which is described below.

Program Be2

This program carries out the computer implementation of the case of linear and discontinuous elements. It deals with the case of Figs. 5.4.(b) and 5.15, and solves the orthotropic potential boundary value problems where u and q vary linearly along the boundary elements, with the nodes same as the extreme points. The Input file is created as, e.g., Be2.in, and it follows the same format as in Be1.in. However, it should be noted that in this case the dimension of the array Bc becomes $2N$, since there are two boundary conditions for each element (one at each node). The dimension of the array F remains N. The Output file is created as Be2.out which can be typed (on the monitor screen), printed (as hard copy), or used as input for a graphics subroutine. The program calls the following functions: Sys2, Quad2, Diag2, Inter2, and Solve.

Quad2 differs from Quad1 in that instead of computing only one value for the boundary element $\tilde{C}_i$ along which the integration is carried out, it computes the elements of the matrices H and G corresponding to the adjacent nodes i and $(i \pm 1)$.

Diag2 is a little different from Diag1; it computes the elements of the matrix G along the boundary elements which include the node under consideration.

Inter2 computes the potential at interior points as given by (5.24), but the values of G_{ij} and $\hat{H}_{ij}$ to be used in (5.24) are defined by (5.38)–(5.43).

Dictionary of Variables: In addition to the variable defined in the program Be1, the following variables are also used:

ux, uy $u_{,x}$, $u_{,y}$ defined analogous to (5.26).
qx, qy $q_{,x}$, $q_{,y}$ defined analogous to (5.26).

Example 5.5:

We shall solve Example 5.2 with linear elements. The input file, from Fig. 5.10(c), is:

```
Example 5.5 (Linear Elements)
12 5
0. 0. 2. 0. 4. 0. 6. 0. 6. 2. 6. 4.
6. 6. 4. 6. 2. 6. 0. 6. 0. 4. 0. 2.
1  0.  1 0.  1 0.  1 0.  1 0.  1 0.
0 200.  0 200.  0 200.  0 200.  0 200.  0 200.
1  0.  1 0.  1 0.  1 0.  1 0. 1 0.
0 0.  0 0.  0 0.  0 0.  0 0.  0 0.
2. 2. 2. 4. 3. 3. 4. 2. 4. 4.
```

The output is:

Boundary Nodes

	(X,Y)		u	Prenodal q	Postnodal q
(	0.0000,	0.0000)	0.0000	-33.3332	0.0000
(	2.0000,	0.0000)	66.6665	0.0000	0.0000
(	4.0000,	0.0000)	133.3334	0.0000	0.0000
(	6.0000,	0.0000)	200.0000	0.0000	33.3333
(	6.0000,	2.0000)	200.0000	33.3334	33.3334
(	6.0000,	4.0000)	200.0000	33.3334	33.3334
(	6.0000,	6.0000)	200.0000	33.3333	0.0000
(	4.0000,	6.0000)	133.3335	0.0000	0.0000
(	2.0000,	6.0000)	66.6665	0.0000	0.0000
(	0.0000,	6.0000)	0.0000	0.0000	-33.3333
(	0.0000,	4.0000)	0.0000	-33.3334	-33.3334
(	0.0000,	2.0000)	0.0000	-33.3334	-33.3334

Interior Points

	(Xi,Yi)		u	q_x	q_y
(	2.0000,	2.0000)	66.6667	33.3332	-0.0002

(	2.0000,	4.0000)	66.6667	33.3332	0.0002
(	3.0000,	3.0000)	100.0000	33.3333	0.0000
(	4.0000,	2.0000)	133.3339	33.3339	-0.0005
(	4.0000,	4.0000)	133.3339	33.3339	0.0005

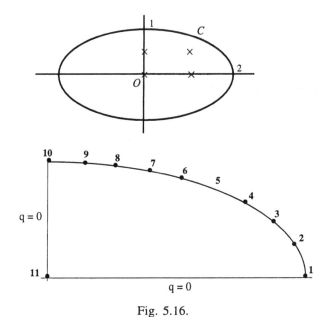

Fig. 5.16.

Example 5.6:

Solve the Laplace equation $\nabla^2 u = 0$ for an elliptic region R, shown in Fig. 5.16, with the semi–axes 2 and 1 respectively, and the boundary condition $u = (x^2 + y^2)/2$ on C. Because of the symmetry about both x and y axes, we shall consider the quarter region with 11 linear elements (11 nodes). The input file is as follows:

```
Example 5.6 (Linear Elements)
11  1
2. 0. 1.9696 0.1736 1.8794 0.342 1.732 0.5
1.532 0.6428 1.2856 0.766 1. 0.866 0.684 0.9397
0.3473 0.9848 0. 1. 0. 0.
0 2.   0 1.9547  0 1.9547  0 1.8245  0 1.8245  0 1.6249
0 1.6249  0 1.38  0 1.38  0 1.1197  0 1.1197  0 0.8749
0 0.8749  0 0.6754  0 0.6754  0 0.5422  0 0.5422  0 0.5
1 0.  1 0.  1 0.  1 0.
1. 0.5
```

Note that the first 10 extreme points, marked 1 through 10 in Fig. 5.16, are given by $x_j = 2\cos j\pi/18$, $y_j = \sin j\pi/18$, $j = 0, 1, \cdots, 9$. The results are:

```
Boundary Nodes
          (X,Y)                 u       Prenodal q  Postnodal q
(    2.0000,    0.0000)     2.0000        0.0000       1.2138
(    1.9696,    0.1736)     1.9547        0.7349       0.7349
(    1.8794,    0.3420)     1.8245        0.4469       0.4469
(    1.7320,    0.5000)     1.6249        0.1941       0.1941
(    1.5320,    0.6428)     1.3800       -0.1510      -0.1510
(    1.2856,    0.7660)     1.1197       -0.4201      -0.4201
(    1.0000,    0.8660)     0.8749       -0.5929      -0.5929
(    0.6840,    0.9397)     0.6754       -0.7533      -0.7533
(    0.3473,    0.9848)     0.5422       -0.7199      -0.7199
(    0.0000,    1.0000)     0.5000       -1.5266       0.0000
(    0.0000,    0.0000)     0.9506        0.0000       0.0000


Interior Points
          (Xi,Yi)                u         q_x          q_y
(    1.0000,    0.5000)     1.0755        0.5645      -0.3547
```

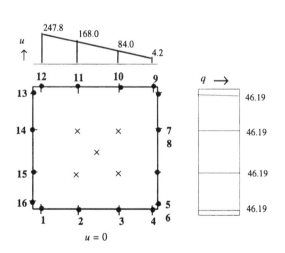

Fig. 5.17.

Example 5.7:

This is an example of discontinuous elements at corner nodes. The input file from Fig. 5.17 is:

Example 5.7 (Linear Discontinuous Elements)
16 5
0.001 0.   2. 0.   4. 0.   5.999 0.
6. 0.001 6. 2.   6. 4.   6. 5.999
5.999 6.   4. 6.   2. 6.   0.001 6.
0. 0.599 0. 4.   0. 2.   0. 0.001
0 0. 0 0. 0 0. 0 0. 0 0. 0 0. 0 0. 1 46.19
1 46.19 1 46.19 1 46.19 1 46.19 1 46.19 1 46.19
1 46.19 0 4.2   0 4.2 0 84.   0 84.   0 168.   0 168.
0 247.8 0 247.8 1 -46.19
1 -46.19 1 -46.19 1 -46.19 1 -46.19 1 -46.19
1 -46.19 1 -46.19 0 0.
2. 2. 2. 4. 3. 3. 4. 2. 4. 4.

The results are:

Boundary Nodes

(X,Y)		u	Prenodal q	Postnodal q
(0.0010,	0.0000)	0.0000	156.5302	156.5302
(2.0000,	0.0000)	0.0000	-16.8707	-16.8707
(4.0000,	0.0000)	0.0000	-17.7104	-17.7104
(5.9990,	0.0000)	0.0000	-103.2299	-103.2299
(6.0000,	0.0010)	6.1784	46.1900	46.1900
(6.0000,	2.0000)	114.0960	46.1900	46.1900
(6.0000,	4.0000)	131.4527	46.1900	46.1900
(6.0000,	5.9990)	12.4991	46.1900	46.1900
(5.9990,	6.0000)	4.2000	-151.6348	-151.6348
(4.0000,	6.0000)	84.0000	11.2953	11.2953
(2.0000,	6.0000)	168.0000	44.0098	44.0098
(0.0010,	6.0000)	247.8000	71.9776	71.9776
(0.0000,	0.5990)	-89.3071	-46.1900	-46.1900
(0.0000,	4.0000)	52.9450	-46.1900	-46.1900
(0.0000,	2.0000)	-116.3532	-46.1900	-46.1900
(0.0000,	0.0010)	-8.5014	-46.1900	-46.1900

Interior Points

(Xi,Yi)		u	q_x	q_y
(2.0000,	2.0000)	16.6571	25.3384	20.5395
(2.0000,	4.0000)	84.4682	-1.0487	41.9869
(3.0000,	3.0000)	59.1773	12.0620	23.8646
(4.0000,	2.0000)	52.5295	17.4905	22.7641

(4.0000,   4.0000)   83.9397   4.0897   7.2784

Example 5.8:

Solve the Laplace equation $\nabla^2 u = 0$ in the upper–half circular region of radius 10 (Fig. 5.18(a)). The problem has a discontinuous potential at $x = 0$ where the flux has a singularity. We shall treat the first and the last element as discontinuous elements, although the point $x = 0$ does not qualify as a corner (Fig. 5.18(b)).

Input:

```
Example 5.8 (Discontinuous linear elements)
26  6
0.02 0. 0.5 0. 1. 0. 2.5 0. 4. 0. 5. 0. 7.5 0. 9. 0 10. 0.
9.2388 3.8268   7.07106 7.07106  3.8268 9.2388   0. 10.
-3.8268 9.2388 -7.07106 7.07106  -9.2388 3.8268  -10. 0.
-9. 0. -7.5 0. -5. 0. -4. 0. -2.5 0. -1. 0. -0.5 0. -0.02 0.
0. 0.02
0 100. 0 100.   0 100. 0 100.   0 100. 0 100.   0 100. 0 100.
0 100. 0 100.   0 100. 0 100.   0 100. 0 100.   0 100. 0 100.
0 100.   1 0.   1 0.   1 0.   1 0.   1 0.   1 0.   1 0.   1 0.
1 0.   1 0.   1 0.   1 0.   1 0.   1 0.   1 0.   0 0.   0 0.   0 0.
0 0.   0 0.   0 0.   0 0.   0 0.   0 0.   0 0.   0 0.
0 0.   0 0.   0 0.   0 0.   0 0.   0 0.   0 0.   0 0.
-4.9497 4.9497  -2.2121 2.2121   0. 3.  0. 7.
4.9497 4.9497   2.2121 2.2121
```

Output:

Boundary Nodes

	(X,Y)		u	Prenodal q	Postnodal q
(	0.0200,	0.0000)	100.0000	1317.7314	1317.7314
(	0.5000,	0.0000)	100.0000	-181.4680	-181.4680
(	1.0000,	0.0000)	100.0000	41.6827	41.6827
(	2.5000,	0.0000)	100.0000	6.1767	6.1767
(	4.0000,	0.0000)	100.0000	7.6256	7.6256
(	5.0000,	0.0000)	100.0000	5.6411	5.6411
(	7.5000,	0.0000)	100.0000	3.7710	3.7710
(	9.0000,	0.0000)	100.0000	3.3227	3.3227
(	10.0000,	0.0000)	100.0000	0.4944	0.4944
(	9.2388,	3.8268)	88.3650	0.0000	0.0000

(7.0711,	7.0711)	76.0512	0.0000	0.0000
(3.8268,	9.2388)	63.7051	0.0000	0.0000
(0.0000,	10.0000)	51.2231	0.0000	0.0000
(-3.8268,	9.2388)	38.6032	0.0000	0.0000
(-7.0711,	7.0711)	25.8645	0.0000	0.0000
(-9.2388,	3.8268)	13.0275	0.0000	0.0000
(-10.0000,	0.0000)	1.3482	0.0000	0.0000
(-9.0000,	0.0000)	0.0000	-5.0253	-5.0253
(-7.5000,	0.0000)	0.0000	-4.3527	-4.3527
(-5.0000,	0.0000)	0.0000	-6.7003	-6.7003
(-4.0000,	0.0000)	0.0000	-8.4800	-8.4800
(-2.5000,	0.0000)	0.0000	-12.6731	-12.6731
(-1.0000,	0.0000)	0.0000	-33.2685	-33.2685
(-0.5000,	0.0000)	0.0000	-87.2868	-87.2868
(-0.0200,	0.0000)	0.0000	-226.3264	-226.3264
(0.0000,	0.0200)	0.0000	-4947.9248	-4947.9248

Interior Points

(Xi,Yi)		u	q_x	q_y
(-4.9497,	4.9497)	25.8223	3.2881	3.2935
(-2.2121,	2.2121)	26.0080	7.5758	7.3495
(0.0000,	3.0000)	51.6792	10.7238	-0.3675
(0.0000,	7.0000)	51.1955	4.5655	-0.0131
(4.9497,	4.9497)	75.9859	3.1497	-3.1411
(2.2121,	2.2121)	76.4182	6.7149	-7.2596

The exact solution is

$$u(x, y) = \frac{100}{\pi} \left(\pi - \arctan \frac{y}{x} \right), \quad 0 < |x| \le 10, 0 \le y \le 10.$$

The flux along the x–axis is given by $q = -100/\pi x$, $0 < |x| \le 10$, whereas on the circumference by

$$q = \frac{100}{(x^2 + y^2)^2} (n_1 y - n_2 x) = 0.$$

The graphs for u, and q along the x–axis are shown in Fig. 5.18(c) and (d) respectively (see Example 8.10 also). ■

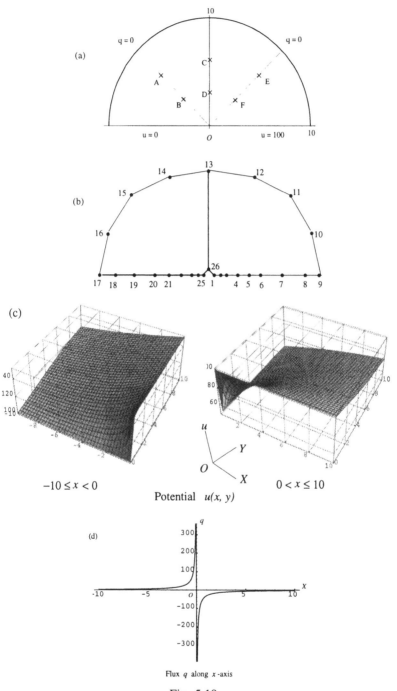

Fig. 5.18.

Example 5.9:

We shall solve the problem of Example 5.4 by treating the four corner nodes
as in Fig. 5.19.

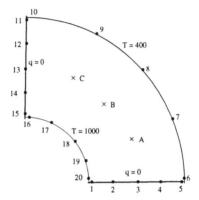

Fig. 5.19. Corner nodes.

Input:

```
Example 5.9 (Corner nodes: linear elements)
20  3
10.02 0.   15. 0.   20. 0.   25. 0.   29.98 0.   29.993 0.628
27.716 11.48   21.213 21.213   11.48 27.716   0.628 29.993
0. 29.98   0. 25.   0. 20.   0. 15.   0. 10.02   0.209 9.997
3.826 9.238   7.071 7.071   9.238 3.826   9.997 0.209
1 0.   1 0.   1 0.   1 0.   1 0.   1 0.   1 0.   1 0.   1 0.   0 400.
0 400.   0 400.   0 400.   0 400.   0 400.   0 400.   0 400.   0 400.
0 400.   1 0.   1 0.   1 0.   1 0.   1 0.   1 0.   1 0.   1 0.   1 0.
1 0.   0 1000.   0 1000.   0 1000.   0 1000.   0 1000.
0 1000.   0 1000.   0 1000.   0 1000.   0 1000.   1 0.
18.477 7.653   14.142 14.142   7.653 18.477
```

Output

```
Boundary Nodes
          (X,Y)                 u      Prenodal q  Postnodal q
(   10.0200,     0.0000)   994.2380     0.0000      0.0000
(   15.0000,     0.0000)   775.7899     0.0000      0.0000
(   20.0000,     0.0000)   617.7275     0.0000      0.0000
(   25.0000,     0.0000)   495.1680     0.0000      0.0000
(   29.9800,     0.0000)   404.1531     0.0000      0.0000
```

(29.9930,	0.6280)	400.0000	-20.6881	-20.6881
(27.7160,	11.4800)	400.0000	-17.7511	-17.7511
(21.2130,	21.2130)	400.0000	-18.4223	-18.4223
(11.4800,	27.7160)	400.0000	-17.7511	-17.7511
(0.6280,	29.9930)	400.0000	-20.6881	-20.6881
(0.0000,	29.9800)	404.1530	0.0000	0.0000
(0.0000,	25.0000)	495.1682	0.0000	0.0000
(0.0000,	20.0000)	617.7275	0.0000	0.0000
(0.0000,	15.0000)	775.7903	0.0000	0.0000
(0.0000,	10.0200)	994.2382	0.0000	0.0000
(0.2090,	9.9970)	1000.0000	61.6049	61.6049
(3.8260,	9.2380)	1000.0000	53.1925	53.1925
(7.0710,	7.0710)	1000.0000	55.4601	55.4601
(9.2380,	3.8260)	1000.0000	53.1924	53.1924
(9.9970,	0.2090)	1000.0000	61.6049	61.6049

Interior Points

(Xi,Yi)		u	q_x	q_y
(18.4770,	7.6530)	618.6964	-25.3328	-10.5420
(14.1420,	14.1420)	618.4969	-19.3454	-19.3454
(7.6530,	18.4770)	618.6965	-10.5420	-25.3328

These results can be compared with Example 5.4.

5.6. Quadratic and Higher Elements

For quadratic elements (Fig. 5.4(c)) we use the interpolation functions ϕ_1, ϕ_2, ϕ_3 defined by (1.28). Then the functions u and q are written as

$$u(\xi) = \phi_1 u_1 + \phi_2 u_2 + \phi_3 u_3 = [\phi_1 \ \phi_2 \ \phi_3][u_1 \ u_2 \ u_3]^T,$$
$$q(\xi) = \phi_1 q_1 + \phi_2 u_2 + \phi_3 u_3 = [\phi_1 \ \phi_2 \ \phi_3][q_1 \ q_2 \ q_3]^T. \tag{5.54}$$

The functions ϕ_1, ϕ_2, ϕ_3 which vary quadratically give the nodal values of the functions when specified for the nodes (see Fig. 5.13(b)). Using these interpolation functions, the evaluation of the integrals along the boundary element j gives (compare with (5.34) and (5.36))

$$\int_{\tilde{C}_j} u q^* \, ds = [h_{j1} \ h_{j2} \ h_{j3}][u_1 \ u_2 \ u_3]^T, \tag{5.55}$$

where

$$h_{jk} = \int_{\tilde{C}_j} \phi_k q^* \, ds, \quad k = 1, 2, 3. \tag{5.56}$$

which, using the Jacobian

$$|J| = \frac{ds}{d\xi} = \sqrt{\left(\frac{dx}{d\xi}\right)^2 + \left(\frac{dy}{d\xi}\right)^2} \tag{5.57}$$

yields

$$\int_{\tilde{C}_j} u q^* \, ds = \int_{(1)}^{(2)} u(\xi) |J| q^* \, ds. \tag{5.58}$$

Similarly,

$$\int_{\tilde{C}_j} q u^* \, ds = \int_{\tilde{C}_j} [g_{j1} \ g_{j2} \ g_{j3}][q_1 \ q_2 \ q_3]^T \, ds, \tag{5.59}$$

where

$$g_{jk} = \int_{\tilde{C}_j} \phi_k u^* \, ds. \tag{5.60}$$

Hence, for node i

$$
\begin{aligned}
c(i)u(i) &= [\hat{H}_{i1} \ \hat{H}_{i2} \ \cdots \ \hat{H}_{in}][u_1 \ u_2 \ \cdots \ u_n]^T \\
&= [G_{i1} \ G_{i2} \ \cdots \ G_{in}][q_1 \ q_2 \ \cdots \ q_n]^T,
\end{aligned} \tag{5.61}
$$

where

$$\hat{H}_{ij} = h_{i1} \text{ term of element } (j-1) + h_{i2} \text{ term of element } (j+1) +$$
$$+ \, h_{i3} \text{ terms of element } j, \quad (5.62)$$

and

$$G_{ij} = g_{i1} \text{ term of element } (j-1) + g_{i2} \text{ term of element } (j+1) +$$
$$+ \, g_{i3} \text{ terms of element } j, \quad (5.63)$$

(compare with (5.38)).

It is not easy to evaluate G_{ii} for quadratic elements, as we have done in the cases of constant and linear elements. For better accuracy in numerically computing both H_{ij} and G_{ij}, a 10–point Gauss quadrature is recommended. The value of the Jacobian $|J|$, defined by (5.57), is needed for evaluating (5.58) and (5.59) numerically. It can be computed by defining x and y in terms of the quadratic interpolation functions ϕ_1, ϕ_2, ϕ_3, defined by (1.28), as

$$x = \phi_1 x^{\text{node 1}} + \phi_2 x^{\text{node 2}} + \phi_3 x^{\text{node 3}},$$
$$y = \phi_1 y^{\text{node 1}} + \phi_1 y^{\text{node 2}} + \phi_1 y^{\text{node 3}}, \tag{5.64}$$

where the superscript refers to the local node number of a quadratic element. Then the derivatives $\partial x/\partial \xi$ and $\partial y/\partial \xi$ can be easily computed from (5.64).

For the cubic elements (see Fig. 5.13(c)) the interpolation functions are defined by (1.29). The integrals for H and G can be developed as above (see Prob. 5.13).

5.7. Poisson Equation

We shall consider the Poisson boundary value problem

$$\nabla^2 u = b, \quad \text{in } R \tag{5.65}$$

$$u = u_0 \quad \text{on } C_1; \quad \frac{\partial u}{\partial n} \equiv q = q_0 \quad \text{on } C_2, \tag{5.66}$$

where $b = b(x, y)$ and, as before, $C = C_1 \cup C_2$ is the boundary of a two-dimensional region R. As in §5.1, we will start by taking the test function as u^*, which, defined by (B.3) or (4.26), is the fundamental solution of (5.2). This, in view of (1.5c), leads to

$$0 = \iint_R (\nabla^2 u - b) u^* \, dx \, dy = \iint_R (u\nabla^2 u^* - bu^*) \, dx \, dy + \int_{C_1} u^* q \, ds +$$
$$+ \int_{C_2} u^* q_0 \, ds - \int_{C_1} u_0 q^* \, ds - \int_{C_2} u q^* \, ds. \tag{5.67}$$

Using $\nabla^2 u^* = \delta(i)$, so that $\iint_R u\nabla^2 u^* \, dx \, dy = -u(i)$, as in (5.3), we obtain

$$-\iint_R bu^* \, dx \, dy - u(i) + \int_{C_1} u^* q \, ds + \int_{C_2} u^* q \, ds - \int_{C_1} u_0 q^* \, ds -$$
$$- \int_{C_2} u q^* \, ds = 0,$$

or

$$\iint_R bu^* \, dx \, dy + u(i) + \int_{C_1} u_0 q^* \, ds + \int_{C_2} u q^* \, ds = \int_{C_1} u^* q \, ds + \int_{C_2} u^* q_0 \, ds,$$
(5.68)

or, as in the derivation of (5.14), we obtain the two–dimensional BI Eq as

$$c(i)u(i) + \iint_R bu^* \, dx \, dy + \int_C u q^* \, ds = \int_C u^* q \, ds, \qquad (5.69)$$

where $c(i)$ is defined by (5.15). Discretization of (5.69) using constant elements, as in the case of Fig. 5.4(a), leads to the BE Eq

$$c(i)u(i) + \iint_{\tilde{R}} bu^* \, dx \, dy + \sum_{j=1}^{N} u_j \int_{\tilde{C}_j} q^* \, ds = \sum_{j=1}^{N} q_j \int_{\tilde{C}_j} u^* \, ds, \quad (5.70\text{a})$$

or

$$c(i)u(i) + B_i + \sum_{j=1}^{N} \hat{H}_{ij} u_j = \sum_{j=1}^{N} G_{ij} q_j, \qquad (5.70\text{b})$$

or

$$B_i + \sum_{j=1}^{N} H_{ij} u_j = \sum_{j=1}^{N} G_{ij} q_j, \qquad (5.70\text{c})$$

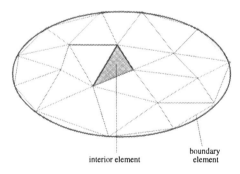

interior element　　　boundary element

Fig. 5.20. Boundary elements and interior cells.

where H_{ij} is defined by (5.19), and B_i is obtained by integrating the domain integral

$$B_i = \iint_{\tilde{R}} bu^* \, dx \, dy \qquad (5.71)$$

numerically as follows: Partition the region $\tilde{R}$ into a mesh of finite elements which are called interior cells (see Fig. 5.20). Note that these interior cells are conceptually different from the usual finite elements although they appear similar. Then use the formula

$$B_i = \sum_{m=1}^{M} \left(\sum_{j=1}^{k} W_j b_j u_j^* \right) A_M, \tag{5.72}$$

where M is the number of interior cells, k the number of integration points on each cell, W_j the weight function, b_j the value of b at integration point j, u_j^* the value of u^* at integration point j, and A_M the area of the cell. Thus, in matrix form, the BE Eq for the N nodes of the type as in Fig. 5.4(a) finally becomes

$$B + HU = GQ. \tag{5.73}$$

Note that N_1 values of u and N_2 values of q are known since they are prescribed on the boundary. We can then reorder (5.73) such that all unknown quantities (called vector X) are on the left side. Then Eq (5.73) becomes

$$AX = F. \tag{5.74}$$

After solving (5.74), we obtain all values of u and q on the boundary nodes. Then we compute the value of u and q at an interior point i by using the formulas (5.24) and (5.25).

Example 5.10:
Solve the Poisson equation $-\nabla^2 u = 2$ on the elliptic region R shown in Fig. 5.16 with the semi–axes 2 and 1 respectively, and the boundary condition $u = 0$ on C. The solution of this problem can be divided into two parts: $u = u_1 + u_2$, where $u_1 = -(x^2 + y^2)/2$ is a particular solution and u_2 is the complementary function. Since $-\nabla^2 u_1 = 2$, the problem reduces to the Laplace equation $\nabla^2 u_2 = 0$ with the boundary condition $u_2 = -u_1$ on C, which can be solved as in Example 5.6. A similar problem is discussed in the next example. ∎

Example 5.11:
The steady Poiseuille flow in a pipe of circular cross–section in the direction of z–axis is defined by

$$\mu \left(\frac{\partial u}{\partial x^2} + \frac{\partial u}{\partial y^2} \right) = \frac{\partial p}{\partial z},$$

where u is the fluid velocity in the z–direction, μ its viscosity, and $\partial p/\partial z = -G$ a constant pressure gradient. The flow is then governed by the Poisson equation $-\nabla^2 u = G/\mu$. The exact solution is

$$u(x, y) = \frac{G a^2 b^2}{\mu(a^2 + b^2)} \left(1 - \frac{x^2}{a^2} - \frac{y^2}{b^2} \right).$$

Because of the axial symmetry we will use the quarter region shown in Fig.
5.16. If we take $G/\mu = 2$, we solve $\nabla^2 u_2 = 0$ with the same input file as in
Example 5.6. ∎

5.8. Non–convex Surfaces

The above analysis can be extended to problems on non–convex surfaces (i.e.,
a region with more than one boundary). An example of such a region is given
in Fig. 5.21.

Fig. 5.21. A non–convex surface with one hole

On non–convex surfaces (with holes), the direction of the normal derivatives on the exterior and interior boundaries is determined by the following
rule: For exterior boundaries, the numbering scheme for boundary elements is
carried out in the counterclockwise direction, whereas for interior boundaries
it is defined in clockwise direction. This rule ia analogous to the convention
used in contour integration in complex analysis, maintaining the direction on
the boundary so that the region always remains to the left. This rule will enable
us define the normal derivatives in computer programs.

Program Be5

This program (see Appendix D) is used for the computer implementation
of potential problems on nonconvex surfaces with constant elements (Fig.
5.4(a)). It starts in the same manner as Be1 and adds two new variables:
M (≥ 2): Number of different boundaries; if M $= 1$, use Be1 or Be2.
Last: Number of last node on each different boundary.

Be5 calls the functions Sys5, Quad1, Diag1, Inter5 and Solve.

Example 5.12:

Solve the Poisson equation on an annular region with boundaries as two concentric circles of radii 1 and 2. We take 8 boundary elements on each circle (see Fig. 5.22).

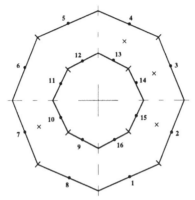

Fig. 5.22. Concentric 8–gon annular region

Thus, there are 16 (constant) boundary elements and 2 different boundaries. The last numbers of the nodes on two surfaces are 8 and 16. As boundary conditions, we assume that $u = 0$ on the exterior boundary and $u = 100$ on the interior boundary. The extreme points are numbered in Fig. 5.22 according to the convention mentioned above. This boundary value problem can be viewed as that of heat transfer between the two concentric circles. Note that

N   Number of boundary elements   16
L   Number of interior points   4
M   Number of different surfaces   2
Last   Last node numbers on these surfaces 8, 16

Input:

```
Example 5.12 (multiply-connected region, constant elements)
16  4  2  8  16
0.0  -2.0  1.4142  -1.4142  2.0  0.0  1.4142  1.4142
0.0  2.0  -1.4142  1.4142  -2.0  0.0  -1.4142  -1.4142
-0.7071  -0.7071  -1.0  0.0  -0.7071  0.7071  0.0  1.0
0.7071  0.7071  1.0  0.0  0.7071  -0.7071  0.0  -1.0
0 0.  0 0.  0 0.  0 0.  0 0.  0 0.  0 0.  0 0.
0 100.  0 100.  0 100.  0 100.
0 100.  0 100.  0 100.  0 100.
1.38582 -0.574025  1.38582  0.574025
```

```
    0.574025 1.38582   -1.38582 -0.574025
```

Output:

```
Boundary Nodes


                    (X,Y)              u          q
    (    0.7071,    -1.7071)      0.0000   -72.9385
    (    1.7071,    -0.7071)      0.0000   -72.9385
    (    1.7071,     0.7071)      0.0000   -72.9384
    (    0.7071,     1.7071)      0.0000   -72.9386
    (   -0.7071,     1.7071)      0.0000   -72.9386
    (   -1.7071,     0.7071)      0.0000   -72.9385
    (   -1.7071,    -0.7071)      0.0000   -72.9386
    (   -0.7071,    -1.7071)      0.0000   -72.9386
    (   -0.8535,    -0.3535)    100.0000   150.2988
    (   -0.8535,     0.3535)    100.0000   150.2992
    (   -0.3535,     0.8535)    100.0000   150.2989
    (    0.3535,     0.8535)    100.0000   150.2989
    (    0.8535,     0.3535)    100.0000   150.2989
    (    0.8535,    -0.3535)    100.0000   150.2991
    (    0.3535,    -0.8535)    100.0000   150.2989
    (   -0.3535,    -0.8535)    100.0000   150.2990


Interior Points
                   (Xi,Yi)             u         q_x        q_y
    (    1.3858,    -0.5740)     31.5684   -89.7677    37.1834
    (    1.3858,     0.5740)     31.5684   -89.7676   -37.1834
    (    0.5740,     1.3858)     31.5683   -37.1834   -89.7677
    (   -1.3858,    -0.5740)     31.5683    89.7677    37.1834
```

For the same problem with 32 boundary elements and Last = 16, 32 (Fig. 5.23), the input file becomes

```
Example 5.12 (multiply-connected region, constant elements)
32  4  2  16  32
0. -2.  0.7653 -1.8478  1.4142 -1.4142  1.8478 -0.7653
2. 0.  1.8478 0.7653  1.4142 1.4142  0.7653 1.8478
0. 2.  -0.7653 1.8478  -1.4142 1.4142  -1.8478 0.7653
-2. 0.  -1.8478 -0.7653  -1.4142 -1.4142  -0.7653 -1.8478
```

```
-0.3826 -0.9238   -0.7971 -0.7971   -0.9238 -0.3826   -1. 0.
-0.9238 0.3826   -0.7971 0.7971   -0.3826 0.9238   0. 1.
0.3826 0.9238   0.7971 0.7971   0.9238 0.3826   1. 0.
0.9238 -0.3826   0.7971 -0.7971   0.3826 -0.9238   0. -1.
0 0. 0 0. 0 0. 0 0. 0 0. 0 0. 0 0. 0 0.
0 0. 0 0. 0 0. 0 0. 0 0. 0 0. 0 0. 0 0.
0 100. 0 100. 0 100. 0 100. 0 100. 0 100. 0 100. 0 100.
0 100. 0 100. 0 100. 0 100. 0 100. 0 100. 0 100. 0 100.
1.3858 -0.574 1.3858 0.574 0.574 1.3858 -1.3858 -0.574
```

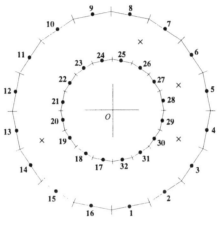

Fig. 5.23.

Note that the coordinates of the outer extreme points are given by

$$x_i = 2\cos\left(\frac{i\pi}{8} - \frac{\pi}{2}\right), \quad y_i = 2\sin\left(\frac{i\pi}{8} - \frac{\pi}{2}\right),$$

and of the inner extreme points by

$$x_i = \cos\left(\frac{i\pi}{8} + \frac{\pi}{2}\right), \quad y_i = -\sin\left(\frac{i\pi}{8} + \frac{\pi}{2}\right), \quad i = 1, \ldots, 16.$$

5.9. Domain Integral

The presence of the domain integral term B_i, defined by (5.71), in the BE Eq (5.70) introduces numerical integration on interior cells. This means more lengthy computations as compared to those performed for integration on the boundary elements. A method that transform the domain integral into a boundary integral will be helpful in maintaining the computational simplicity of the BEM.

A very simple case is when the function b is constant, linear or harmonic. Then $\nabla^2 b = 0$. Let $v^*(x,y)$ denote a Galerkin–type function such that $\nabla^2 v^* = u^*$. Then, using Green's second identity (1.8), we get

$$\iint_R \left(b\nabla^2 v^* - v^*\nabla^2 b \right) \, dx \, dy = \int_C \left(b\frac{\partial v^*}{\partial n} - v^*\frac{\partial b}{\partial n} \right) ds, \qquad (5.75)$$

which gives

$$B_i = \int_{\tilde{C}} \left(b\frac{\partial v^*}{\partial n} - v^*\frac{\partial b}{\partial n} \right) ds, \qquad (5.76)$$

Since $u^* = \dfrac{1}{2\pi} \ln \left(\dfrac{1}{r} \right)$, a choice of v^* is obtained by solving $\nabla^2 v^* = u^*$, i.e.,

$$\nabla^2 v^* = \frac{1}{r}\frac{\partial}{\partial r}\left(r\frac{\partial v^*}{\partial r} \right) = \frac{1}{2\pi}\ln\frac{1}{r}, \qquad (5.77)$$

which yields

$$v^* = \frac{r^2}{8\pi}\left[1 + \ln\frac{1}{r} \right]. \qquad (5.78)$$

If we assume that a source of strength Q_i is concentrated at an interior point i, then

$$b = Q_i\delta(i). \qquad (5.79)$$

If finitely many sources are situated at interior points, then the BI Eq (5.69) becomes

$$c(i)u(i) + \int_C uq^* \, ds + B_i + \sum_i Q_i u_i^* = \int_C u^*q \, ds, \qquad (5.80)$$

where B_i is now defined by (5.76) as a line integral, and the concentrated sources are easy to compute.

Other methods to transform the domain integral B_i into boundary integrals for different types of the function b will be discussed in detain in Chapter 9.

Example 5.13:
Consider the Poisson equation $-\nabla^2 u = 2$ in R with the Dirichlet boundary

condition $u = 0$ on C, where (a) R is the elliptic region of semi–major axis a_1 and semi–minor axis a_2. In this case the exact solution is given by

$$u = \frac{1}{a_1^{-2} + a_2^{-2}} \left(1 - \frac{x^2}{a_1^2} - \frac{y^2}{a_2^2} \right)$$

(see Example 5.10). Taking $a_1 = 2$ and $a_2 = 1$, we get

Interior point	u (Be2)	u (exact)
$(1.5, 0.0)$	0.345	0.35
$(0.6, 0.45)$	0.563	0.566
$(0.0, 0.45)$	0.634	0.638 ∎

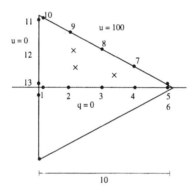

Fig. 5.24.

(b) R is the equilateral triangle of altitude a, (see Fig. 5.24), where, because of the axial symmetry, we have taken 11 linear elements (three discontinuous, one at each corner) in the upper half triangle. The input data is as follows:

```
Example 5.13(b) (Discontinuous linear elements)
13   3
0.02 0.   2.5 0.   5. 0.   7.5 0.   9.98 0.   0.98 0.01   7.5 1.25
5.  2.5   2.5 3.75   0.02 4.99   0. 4.98   0. 2.5   0. 0.02
1 0. 1 0.   1 0. 1 0.   1 0. 1.0   1 0. 1.0   1 0.   0 100.
0 100.   0 100.   0 100.   0 100.   0 100.   0 100.   0 100.   0 100.
0 100.   0 0.   0 0.   0 0.   0 0.   0 0.   0 0.   1 0.
3. 2.   3. 3.   6. 1.5
```

The exact solution is given by

$$u = \frac{2}{27} a^2 - \frac{1}{2} \left(x^2 + y^2 \right) + \frac{1}{2} a \left(x^3 - 3xy^2 \right) . \blacksquare$$

5.10. Unbounded Regions

The BI Eq (5.14) is also valid for unbounded regular domains in the following sense: Let R^* be the region exterior to a finite domain D (obstacle) with boundary C, and let K be a circle in R^* of radius r and center at a point i on C (see Fig. 5.25)). Eq (5.14) for the region outside the boundary C and inside the circle K becomes

$$c(i)u(i) + \int_C u(x)q^*(\xi, x)\, ds(x) + \int_K u(x)q^*(\xi, x)\, ds(x) =$$

$$= \int_C q(x)u^*(\xi, x)\, ds(x) + \int_K q(x)u^*(\xi, x)\, ds(x). \qquad (5.81)$$

Now let $r \to \infty$ in (5.81). Then the BI Eq for an unbounded region with a cavity should satisfy the condition

$$\lim_{r \to \infty} \int_C [q(x)u^*(\xi, x)\, ds(x) - u(x)q^*(\xi, x)]\, ds(x) = 0. \qquad (5.82)$$

The regularity conditions for Eq (5.82) at infinity are as follows: The function u^* behaves like $\ln r$, its derivative q^* is of order $O(1/r)$ as $r \to \infty$, and $ds(x) = |J|\, d\theta = O(r)$, where J is the Jacobian of the transformation from the cartesian to polar cylindrical coordinates. Hence, the two terms in the integral (5.82) do not approach zero separately as $r \to \infty$, but they must cancel each other. Thus, if we apply the condition (5.82) to (5.81) as $r \to \infty$, we obtain the same BE Eq as (5.14), i.e., the BI Eq for an unbounded domain with a cavity is the same as that for finite domains. The same is also true for points inside an unbounded domain.

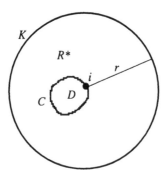

Fig. 5.25. Unbounded region with a finite obstacle

However, in the case of an unbounded domain, the matrix H is not defined like (5.17). Instead, the regularity conditions at infinity will not hold if u is assumed to be constant everywhere in the unbounded domain Ω, where

$$\lim_{r \to \infty} \int_K q^* \, ds = \lim_{r \to \infty} \int_0^{2\pi} \left(-\frac{1}{2\pi r}\right) r \, d\theta = -1.$$

Hence, the coefficients H_{ii} in the case of an unbounded domain are again given by (5.46).

5.11. Mixed Boundary Conditions

The mixed boundary conditions, mentioned in §1.4, play an important role in some boundary value problems. In heat transfer problems, for example, this type of condition has the form $q + \beta(T - T_a) = q_0$, where T is the temperature field, T_a the ambient temperature, β the heat transfer coefficient (also called the film coefficient), and q_0 a prescribed heat flux.

Example 5.14:
Consider a column of rectangular cross–section; a portion of the boundary is subjected to an internal condition, another portion is subjected the exterior weather conditions, and the remaining portion is in contact with the abutting wall which separates both portions (see Fig. 5.26(a)). The prescribed boundary conditions on the internal face ($x = 0$) are: $T_a = 100°F$ and $\beta = 0.5$ Btu/(hft^2 $°F$), and on the exterior face ($x = l$): $T_a = 0°F$ and $\beta = 6.0$ Btu/(hft^2 $°F$). The temperature and the heat transfer coefficient on the faces $y = \pm 6$ are noted in Fig. 5.26(a). We shall assume that $q_0 = 0$, and the thermal conductivity is 1.0 Btu/(hft $°F$). Because of the symmetry about x–axis, only the half of the structure is discretized into 21 linear elements, and 21 boundary nodes are marked (Fig. 5.26(b)). We shall assign Code= 2 for mixed boundary conditions. The input file is created in the following order:

Entry #	Variable	Explanation
1	Title	Must enter problem title.
2	N	Number of boundary elements.
3	L	Number of interior points where solution is to be computed.
1	M	Number of different boundaries

4	Last	Number of the last node on each boundary.
next N–pairs	X, Y	x, y coordinates of extreme points
next N–triplets	Code,β,T_a	N–triplets of Code (2), β and T_a each node, starting with node 1 and ending with node N.
next L–pairs	Xi, Yi	coordinates of interior points where the solution is to be computed.

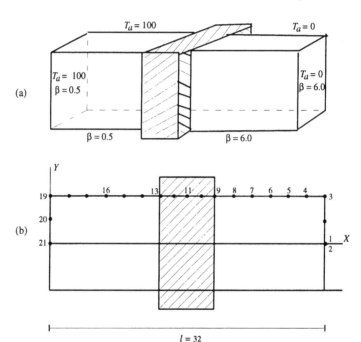

Fig. 5.26. A column with an abutting wall.

Input:

```
21  15  1  21
32. 0. 32. 3. 32. 6. 30. 6. 28. 6. 26. 6. 24. 6. 22. 6.
20. 6. 18. 6. 16. 6. 14. 6. 12. 6. 10. 6. 8. 6. 6. 6.
4. 6. 2. 6. 0. 6. 0. 3. 0. 0.
2  6. 0.  2  6. 0.  2  6. 0.  2  6. 0.  2  6. 0.  2  6. 0.
2  6. 0.  2  6. 0.  2  6. 0.  2  3. 0.  2  0. 0.
2  0.25 25.  2  0.5 50.  2  0.5 50.
2  0.5 50.  2  0.5 50.  2  0.5 50.  2  0.5 50.
2  0.5 50.  2  0.5 50.  2  0.5 50.
```

2. 0. 4. 0. 6. 0. 8. 0. 10. 0. 12. 0. 14. 0. 16. 0.
18. 0. 20. 0. 22. 0. 24. 0. 26. 0. 28. 0. 30. 0.

Note that β is assumed to be zero at node 13 as this node is approached from both sides. The value of β is also halved at nodes 12 and 14 respectively from each side. In the above input file, Code$= 2$ defines a linear relation between the potential and the flux. An algorithm for generating the boundary conditions for Code $= 0, 1$, or 2 can be written as follows: Let F_j denote the value of the potential u if Code$= 0$, the value of the flux q if Code$= 1$, and the value of the coefficient β if Code$= 2$; also, let S_j denote the value of q if Code$= 2$; here $j = 1, \ldots, N$, where N is the total number of boundary elements (hence the total number of boundary conditions). Then

$$j = 1$$
$$\text{For } i = 1 \text{ to } N \text{ Do}$$
$$i_1 := j$$
$$\text{Read } j, Code_j, F_j, S_j$$
$$j_1 := j - i_1$$
$$j_2 := j_1 - 1$$
$$a_1 := \text{Float}(j_1)$$
$$a_x := (F_j - F_{i_1})/a_1; \; a_y := (S_j - S_{i_1})/a_1$$
$$\text{For } n = 1 \text{ to } j_2 \text{ Do}$$
$$i_2 := i_1 + n; i_3 := i_2 - 1$$
$$Code_{i_2} := Code_{i_3}$$
$$F_{i_2} := F_{i_3} + a_x; \; S_{i_2} := S_{i_3} + a_y.$$

Program Be5 can then be modified accordingly (see Problem 5.22).

5.12. Indirect Method

Although we have studied the direct BEM, it will be instructive to see how the indirect method deals with the boundary value problems. One important feature of the indirect method is that the physical variables of the problem do not remain the unknown quantities. Moreover, this method leads to integral equations of the Fredholm type which must be solved, usually by numerical methods. For example, if we consider the Laplace equation $\nabla^2 u = 0$ in R with Dirichlet or Neumann boundary conditions (i.e., the Dirichlet or the neumann problem), then u is a harmonic function in R. From complex analysis we know that every harmonic function can be represented by a potential, and conversely; and that the real and imaginary parts of an analytic function are harmonic and, conversely, given any harmonic function, we can find an analytic function which has that function as its real (or imaginary) part, i.e., if $u(x, y)$ is harmonic, then a conjugate harmonic function $v(x, y)$ exists such that $u(x, y) + iv(x, y)$ is an analytic function of $z = x + iy$. This conjugate function is always unique with an arbitrary constant. This result also holds even if the region R is multiply–connected, except that the conjugate function v can be multiple–valued, since the integration paths taken on opposite side of a 'hole' may give different results. For example, $u = \ln r$, where $r = \sqrt{x^2 + y^2}$, defined on the region R which surrounds the origin but excludes it by a circle centered at the origin, has the multiple–valued conjugate $v(x, y) = \arctan(y/x) \pm 2n\pi + \text{const}, \ n = 1, 2, \ldots$.

The solution of the Dirichlet or Neumann boundary value problem is called a real potential and its level curves $u(x, y) = \text{const}$ are called equipotential lines. In hydrodynamic interpretation, the function u is known as the velocity potential and its conjugate v the stream function. The corresponding analytic function $f(z) = u + iv$ is called the complex potential which (or its real part u) describes a field in the region R; the derivative $f'(z) = u_x - iu_y$ is termed the complex intensity. Thus the results for a single–layer or double–layer potential, given by (B.7)–(B.12), hold and they are Fredholm integral equations of the first or second kind.

Example 5.15:
To obtain a solution for the Dirichlet problem, we represent u as a double–layer potential defined by (B.10)–(B.11) with an unknown source density τ given by (B.12); or represent u as a single–layer potential with an unknown source density σ, i.e.,

$$u(x') = \int_C \sigma(x)u^*(x, x')\, ds(x), \quad x \in R, \tag{5.83}$$

which is a Fredholm integral equation of the first kind. Taking the limit in (5.83) as x' approaches the boundary C, we get

$$u(x') = \int_C \sigma(x)u^*(x, x')\, ds(x), \quad x \in C, \tag{5.84}$$

and thus u becomes known, although the density σ is still unknown. ∎

Example 5.16:
In order to obtain a solution for the Neumann problem, we shall assume that the unknown harmonic function u is expressed as a single–layer potential with an unknown source density σ, i.e.,

$$u(x') = \int_C \sigma(x')u^*(x, x')\, ds(x), \qquad (5.85)$$

where u^* is the logarithmic potential given by (B.3) for two–dimensional problems; thus, Eq (5.85) yields (B.7)–(B.8) which have a solution if the Gaussian condition $\int_C q(x)\, ds(x) = 0$ is satisfied ($q = \partial u/\partial n$). This solution is unique up to an arbitrary additive constant.

From these examples it is obvious that the indirect method results in integral equations which are, in general, difficult to solve even by numerical techniques.

Alternatively, if we represent u as a single–layer potential with an unknown density σ, i.e., if

$$u(x) = \int_C \sigma(x)u^*(x, x')\, ds(x'), \qquad x \in R, \qquad (5.86)$$

then by taking the limit in (5.100) as $x \to C$, we get

$$u(x) = \int_C \sigma(x)u^*(x, x')\, ds(x'), \qquad x \in C; \qquad (5.87)$$

and thus u becomes a known quantity although the density σ is still unknown. Note that Eq (5.87) is a Fredholm integral equation of first kind which must be solved numerically. ∎

Example 5.17:
In order to obtain an integral equation solution of the Neumann boundary value problem, we shall assume that the unknown harmonic function u is expressed as a single–layer potential with an unknown density σ, i.e.,

$$u(P) = \int_C \sigma(P)u^*(P, Q)\, ds(Q), \qquad (5.88)$$

where $u^* = \dfrac{1}{2\pi} \ln \dfrac{1}{r}$ (logarithmic potential) for 2–D problems or $u^* = \dfrac{1}{4\pi r}$ (Newtonian potential) for 3–D problems. From (5.88) we get (B.7)–(B.8) which have a solution if the Gauss condition $\int_C q(x)\,ds(x) = 0$ is satisfied, and the solution is then unique up to an arbitrary additive constant. ∎

5.13. Problems

5.1. Use Bel to solve the heat conduction problem in the square plate as shown in Fig. 5.27, where (a) $u_0 = 1$, and (b) $u_0 = \sin \pi x$.

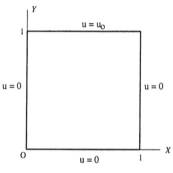

Fig. 5.27.

Exact solution:

$$u(x,y) = \frac{4}{\pi} \sum_{n=0}^{\infty} \frac{\sin(2n+1)\pi x \sinh(2n+1)\pi y}{(2n+1)\sinh[(2n+1)\pi]}.$$

5.2. Determine the temperature distribution $T(x,y)$ over the cross–section of a long rectangular bar of uniform thermal conductivity as shown in Fig. 5.28, where $\nabla^2 T = 0$, $T(0,y) = 0 = T(a,y)$ for $0 < y < b$, and $T(x,0) = f(x)$, $0 < x < a$.
[Exact solution:

$$T(x,y) = \frac{2}{a} \sum_{n=1}^{\infty} \frac{\sin(n\pi x/a)\sinh(n\pi y/b)}{\sin(n\pi/a)b} \int_0^a f(x)\sin\frac{n\pi t}{a}\,dt.]$$

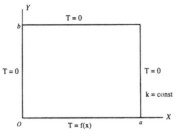

Fig. 5.28.

5.3. Three surfaces of a long bar of square cross–section are maintained at $0°C$, while the fourth surface is at $100°C$. Use Be1 to compute the centerline temperature distribution under steady–state conditions.

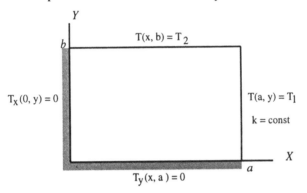

Fig. 5.29.

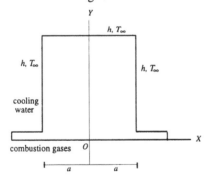

Fig. 5.30.

5.4. Develop the BE Eq and modify the program Be1 for the steady–state temperature distribution $T(x, y)$ in a long bar of rectangular cross–section s for the following boundary conditions: $T(x, b) = T_1$, $T(x, 0) = T_2$, $T(a, y) = T_3$, $T(0, y) = T_4$, where T_1, T_2, T_3 and T_4 are all prescribed constant and uniform temperatures.

5.5. Compute the steady–state temperature distribution $T(x, y)$ in the long bar of rectangular cross–section shown in Fig. 5.29.

5.6. The combustion chamber of a jet engine is cooled by water flowing in an annular jacket around it. Fins running spirally in the annular jacket are cast integral with the combustion chamber wall and guide the cooling water. As an approximation we shall consider the fins to be straight and of rectangular cross–section as shown in Fig. 5.30. We shall also assume that the inner surface of the combustion chamber is maintained at the temperature T_g of the combustion gases. The cooling water temperature T_∞ and the heat transfer coefficient h are prescribed constants. Develop the BE Eq and modify the program Be1 to compute the temperature distribution in the fins.

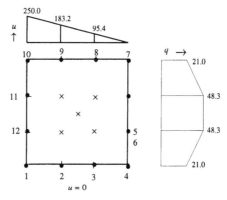

Fig. 5.31.

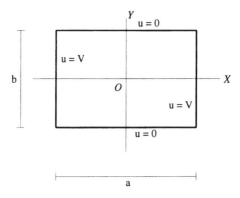

Fig. 5.32.

5.7. Use the programs Be1 and Be2 and verify the results for the problem in Fig. 5.31.

5.8. Find the electrostatic potential $u(x, y)$ inside an elongated box of rectangular cross–section if two opposite sides are at potential $V(= 50)$ volts and the other two sides are grounded (see Fig. 5.32).
[Exact solution: $u(x, y) =$

$$= \frac{4V}{\pi} \sum_{n=0}^{\infty} (-1)^n \frac{\cosh[(2n + 1)\pi x/b]}{\cosh[(2n + 1)\pi a/b]} \frac{\cos[(2n + 1)\pi y/b]}{2n + 1}.]$$

5.9. Find the electrostatic potential $u(x, y)$ inside a semi–infinite rectangular box if the vertical wall is held at potential $V(= 110)$ volts and the horizontal walls are held at potential zero (see Fig. 5.33). [The exact solution is:

$$u(x, y) = \frac{2V}{\pi} \arctan \frac{\sin(\pi y/b)}{\sinh(\pi x/b)}$$

$$= \frac{4V}{\pi} \sum_{n=0}^{\infty} \frac{\sin[(2n + 1)\pi y/b]}{2n + 1} e^{-(2n+1)\pi x/b}, \; x > 0.]$$

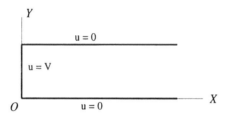

Fig. 5.33.

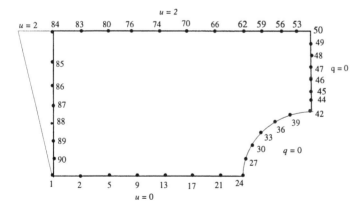

Fig. 5.34.

5.10. Use the programs Be1 and Be2 to solve the potential problem in Fig. 5.34.

5.11. Compute the steady–state temperature distribution and the rate of heat flow at the points A, B, C, D on the surface of the two–dimensional plate with the boundary conditions shown in Fig. 5.35.

5.12. Develop a program for the potential problem for the case of the boundary elements in Fig. 5.4(c) using quadratic elements (call this program Be3).

5.13. Develop numerical formulas for the computation of G_{ii} in the case of cubic elements (rewrite the respective improper integrals by using Gauss quadrature).

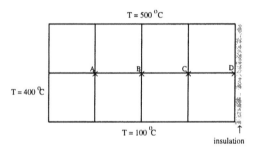

$$T = 500\,^\circ\text{C}$$

$$T = 400\,^\circ\text{C}$$

$$T = 100\,^\circ\text{C}$$

insulation

Fig. 5.35.

5.14. Solve the Laplace equation $\nabla^2 u = 0$ for the square region $R = \{0 \leq x, y \leq \pi\}$, where $u(0, y) = 0 = u(\pi, y) = u(x, \pi)$ and $u(x, 0) = \sin x$. [Exact solution: $u(x, y) = \sin x \sinh(\pi - y)/\sin \pi$.]

5.15. Solve the Laplace equation $\nabla^2 u = 0$ for the square region of Prob. 5.14, with $u(0, y) = 0 = u(\pi, y) = u(x, \pi)$ and $u(x, 0) = 2 \sin^2 x$.

Exact solution: $u(x, y) = \sum_{n=1}^{\infty} b_n \sin nx \sinh n(\pi - y)/\sinh n\pi$;

$b_1 = 16/3\pi, b_2 = 0,$ and

$$b_k = \frac{2}{\pi}\left[\frac{1}{k} - \frac{\cos k\pi}{k} + \frac{\cos(k-2)\pi}{2(k-2)} + \frac{\cos(k+2)\pi}{2(k+2)} - \frac{1}{2(k-2)} - \frac{1}{2(k+2)}\right], \ k = 3, 4, \cdots.]$$

5.16. Solve the Laplace equation $\nabla^2 u = 0$ for the unit disk, given that $u(x, y) = 1 + x^2 - y$ on the unit circle.
[Let the unit circle be parametrized by $x = \cos \theta, y = \sin \theta, -\pi \leq \theta \leq \pi$. Then on the unit circle $u(x, y) = 1 + \cos^2 \theta - \sin \theta = h(\theta)$. The exact

solution inside the unit circle is given by

$$u(x,y) = \frac{1}{2\pi} \int_{-\pi}^{\pi} \frac{h(t)(1-x^2-y^2)}{1+x^2+y^2-2x\cos t - 2y\sin t} \, dt.]$$

5.17. Solve the Laplace equation and compute the two–dimensional electro-static field due to the electrodes shown in Fig. 5.36(a)–(b).

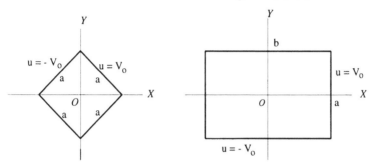

Fig. 5.36.

5.18. Find the distribution of d-c current in a thin rectangular sheet if the current is applied by electrodes at the points $x = -a, y = 0$ and $x = a, y = 0$ (see Fig. 5.37).
[Solve the Poisson equation $-\nabla^2 u = 0$ with the boundary conditions

$$u_x(\pm a, y) = \begin{cases} -j/2\varepsilon\sigma h = A, & \text{for } |y| < \varepsilon \\ 0, & \text{for } |y| > \varepsilon, \end{cases}$$

where j is the current flowing through the sheet by using suitable elec-trodes, σ the conductivity and h the thickness of the sheet. The exact solution for the potential of the current distribution in the sheet is given by

$$u(x,y) = A\left[\frac{x}{b} + \frac{2}{\pi} \sum_{n=1}^{\infty} \frac{\sinh(n\pi x/b)}{n\cosh(n\pi a/b)} \cos\frac{n\pi y}{b}\right] + \text{const.}]$$

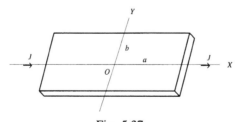

Fig. 5.37.

5.19. Compute the potential flow in an L–shaped region shown in Fig. 5.38.

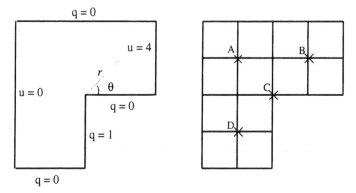

Fig. 5.38.

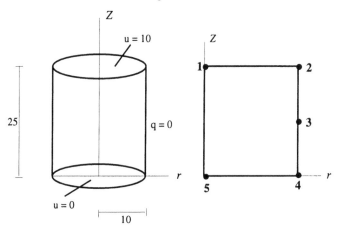

Fig. 5.39.

A series approximation of the exact solution in the vicinity of the re–entrant corner of the L–shaped region is

$$u(r, \theta) = a + br^{2/3} \cos \frac{2\theta}{3} + cr^{4/3} \cos \frac{4\theta}{3} + dr^2 \cos 2\theta + \cdots ,$$

where $a = 0.667$, $b = -0.156$, $c = -0.025$, $d = 0(10^{-8})$.

5.20. Compute the uniform potential flow in a circular pipe of radius 10 units and length 10 units. The boundary conditions, shown in Fig. 5.39, are: Potential on the top surface of the pipe is kept at $u = 10$, while on the bottom surface of the pipe it is kept at zero. The side boundary is assumed to be a solid wall so that the radial velocity must vanish. The analytical solution of the problem is $u = z$.

5.21. Develop a program (Be4) for discontinuous elements discussed in §5.5.

5.22. Modify the program Be5 for the case of mixed boundary conditions (Code= 2, §5.11).

5.23. Determine the temperature distribution in an 8– by 4– cm rectangular plate with a 2–cm circular hole (Fig. 5.9(a)) if the temperature on the outside boundary is 20° C, and the circular hole is filled with a fluid at 200° C. The thermal conductivity of the plate is 1.2, and the convective heat transfer coefficient between the plate and the fluid is 0.15. Use Code= 2 with $q_0 = 0$.

5.24. The fundamental solution u^* for the two–dimensional Laplace equation is given by (4.26). Prove that

$$\frac{1}{2\pi} \int_C \frac{\partial}{\partial n} \left(\frac{1}{|x - x'|} \right) ds(x') = \begin{cases} 1, & \text{for } x \in R \\ 1/2 & \text{for } x \in C \\ 0 & \text{for } x \in R', \end{cases}$$

where R' is the exterior of the region R.

References and Bibliography for Additional Reading

E. Alarcón, A. Martin and F. Paris, *Boundary elements in potential and elasticity theory*, Computers and Structures **10** (1979), 351–362.

P. K. Banerjee and R. Butterfield, *Boundary Element Methods in Engineering Science*, McGraw–Hill, New York, 1981.

———and R. P. Shaw (Eds.), *Developments in Boundary Element Methods*, Vol. 1, 2, Applied Science Publishers, 1982.

C. A. Brebbia and J. Dominguez, *Boundary Elements: An Introductory Course*, 2nd Ed., Computational Mechanics Publications, Southampton/McGraw–Hiil, New York, 1992.

———and S. Walker, *Boundary Element Techniques in Engineering*, Newness–Butterworths, London, 1980.

———, J. Telles and L. Wrobel, *Boundary Element Techniques – Theory and Applications in Engineering*, Springer-Verlag, Berlin, 1984.

C. A. Brebbia (Ed.), *Recent Advances in Boundary Element Methods*, Pentech Press, London, 1978.

———, *The Boundary Element Method for Engineers*, Pentech Press, London, 1980.

———, *New Developments in Boundary Element Methods*, Proc. 2nd Conf. on BEM, Computational Mechanics Publications, Southampton, 1980.

———, *Progress in Boundary Element Methods*, Vols. 1 and 2, Pentech Press, London, 1981, 1983.

———, *Topics in Boundary Element Research*, Springer-Verlag, Berlin, 1984.

G. F. Carrier, M. Krook and C. F. Pearson, *Functions of a Complex Variable: Theory and Technique*, McGraw–Hill, New York, 1966.

G. Chen and J. Zhou, *Boundary Element Methods*, Academic Press, London, 1992.

T. G. B. DeFigueiredo, *A New Boundary Element Formation in Engineering*, Lecture Notes in Engineering, Vol. 68, Springer–Verlag, Berlin, 1991.

L. M. Delves and J. Walsh J. (Eds.), *Numerical Solution of Integral Equations*, Clarendon Press, London, 1974.

J. Fairweather, F. J. Risso, D. J. Shippy and Y. S. Wu, *On the numerical solution of two-dimensional potential problems by an improved boundary integral equation method*, J. Comput. Phys. **31** (1979), 86–112.

R. F. Harrington, K. Pontoppidau, K. Abrahamson and N. C. Albertson, *Computation of Laplacian potential by an equivalent-source method*, Proc. IEE **116** (1969), 1715–1720.

J. L. Hess and A. M. O. Smith, *Calculation of potential flow about arbitrary bodies*, Progress in Aeronautical Sciences (D. Kúchemann, ed.), vol. 8, Pergamon Press, London, 1967.

M. A. Jawson, *Integral equation methods in potential theory, I*, Proc. Royal Soc., Ser. A **275** (1963), 23–32.

M. A. Jawson and G. T. Symm, *Integral Equation Methods in Potential Theory and Elastostatics*, Academic Press, London, 1977.

N. N. Lebedev, I. P. Skalskaya and Y. S. Uflyand, *Worked Problems in Applied Mathematics*, Dover, New York, 1965.

M. A. Mautz and R. F. Harrington, *Computation of rationally symmetric Laplacian potentials*, Proc. IEE **117** (1970), 850–852.

G. F. Roach, *Green's Functions: Introductory Theory with Applications*, Van Nostrand Reinhold, London, 1970.

G. T. Symm, *Integral equation methods in potential theory, II*, Proc. Royal Soc. Ser. A **275** (1963), 33–46.

L. C. Wrobel, *Two-Dimensional Potential Analysis Using Boundary Elements*, Software Series, Computational Mechanics Publications, Southampton, 1986.

6

Linear Elasticity

We will analyze linear elastic continua under the assumption that they undergo small strains. The linear theory of elasticity is based on the following two basic assumptions: (i) The material is subject to an infinitesimal strain and the stress is expressed as a linear function of strain, and (ii) any variation in the orientation of this material due to displacements is negligible. These assumptions lead to small strain and equilibrium equations under an undeformed geometry. The linearity assumption is an attempt to simplify the mathematical aspect of the behavior of solids. Although we assume that the material properties are linear, the deformations in a body may not be completely linear. For example, under certain loads, various materials exhibit plastic deformation while others creep with time, or they may crack in which case the stresses are redistributed.

6.1. Stress and Strain

We will express small strains and related stresses with respect to a right-hand rectangular coordinate system. Thus, in a cartesian system where the coordinates are denoted by $\mathbf{x} = (x_1, x_2, x_3)$, consider an infinitesimal element (Fig. 6.1). The stress vector is defined by

$$\boldsymbol{\sigma} = \begin{bmatrix} \sigma_{11} & \sigma_{12} & \sigma_{13} \\ \sigma_{21} & \sigma_{22} & \sigma_{23} \\ \sigma_{31} & \sigma_{32} & \sigma_{33} \end{bmatrix}. \tag{6.1}$$

Note that if the coordinate system is taken as (x, y, z), instead of (x_1, x_2, x_3), then the normal stresses $\sigma_{11}, \sigma_{22}, \sigma_{33}$ are denoted by $\sigma_x, \sigma_y, \sigma_z$ respectively, and the shearing stresses $\sigma_{12}, \sigma_{13}, \sigma_{21}, \sigma_{23}, \sigma_{31}, \sigma_{32}$ by $\tau_{xy}, \tau_{xz}, \tau_{yx}, \tau_{yz}, \tau_{zx}, \tau_{zy}$ respectively. The equilibrium of the infinitesimal element implies that

$$\sigma_{12} = \sigma_{21}, \quad \sigma_{13} = \sigma_{31}, \quad \sigma_{23} = \sigma_{32}. \tag{6.2}$$

Thus, we need to consider only three independent components of the shearing stress. Corresponding to these stresses, the normal and shearing strains are defined as follows:

Normal strains: $\varepsilon_{ii} = u_{i,i}, \quad i = 1, 2, 3,$

shearing strains: $\varepsilon_{ij} = \dfrac{1}{2} (u_{i,j} + u_{j,i}), \quad i, j = 1, 2, 3 \quad (i \neq j),$ $\qquad (6.3)$

where (u_1, u_2, u_3) are translations along the (x_1, x_2, x_3) directions respectively. As before, only three of the above shearing strains are independent.

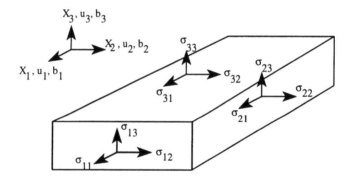

Fig. 6.1. Stresses on an infinitesimal element.

The stress tensor σ satisfies the following three equilibrium equations throughout the element of Fig. 6.1:

$$\frac{\partial \sigma_{ij}}{\partial x_j} + b_i = 0, \quad i, j = 1, 2, 3, \tag{6.4}$$

where b_i are the body forces. The strain-stress relations for an isotropic material are given by

$$\varepsilon_{11} = \frac{\sigma_{11} - \nu \sigma_{22} - \nu \sigma_{33}}{E}, \qquad \varepsilon_{12} = \frac{\sigma_{12}}{G},$$

$$\varepsilon_{22} = \frac{-\nu \sigma_{11} + \sigma_{22} - \nu \sigma_{33}}{E}, \qquad \varepsilon_{23} = \frac{\sigma_{23}}{G}, \tag{6.5}$$

$$\varepsilon_{33} = \frac{-\nu \sigma_{11} - \nu \sigma_{22} + \sigma_{33}}{E}, \qquad \varepsilon_{31} = \frac{\sigma_{31}}{G},$$

where E is the Young's modulus, G the shear modulus, and ν the Poisson's ratio $(0 < \nu < 1/2)$. In matrix form, Eq (6.5) is written as

$$\varepsilon = C\sigma, \tag{6.6}$$

where

$$C = \frac{1}{E} \begin{bmatrix} 1 & -\nu & -\nu & 0 & 0 & 0 \\ -\nu & 1 & -\nu & 0 & 0 & 0 \\ -\nu & -\nu & 1 & 0 & 0 & 0 \\ 0 & 0 & 0 & 2(1+\nu) & 0 & 0 \\ 0 & 0 & 0 & 0 & 2(1+\nu) & 0 \\ 0 & 0 & 0 & 0 & 0 & 2(1+\nu) \end{bmatrix} \tag{6.7}$$

is the matrix that relates the strain vector ε to the stress vector σ. The components c_{ij} of the matrix C

are called elastic compliances. Inversely, the stress-strain relations from (6.6) are given by

$$\sigma = D\varepsilon, \tag{6.8}$$

where the matrix D which relates the stress vector σ to the strain vector ε is

$$D = C^{-1} = \frac{E}{2(1+\nu)(1-2\nu)} \times$$

$$\times \begin{bmatrix} 2(1-\nu) & \nu & \nu & 0 & 0 & 0 \\ 2\nu & (1-\nu) & \nu & 0 & 0 & 0 \\ 2\nu & \nu & (1-\nu) & 0 & 0 & 0 \\ 0 & 0 & 0 & 1-2\nu & 0 & 0 \\ 0 & 0 & 0 & 0 & 1-2\nu & 0 \\ 0 & 0 & 0 & 0 & 0 & 1-2\nu \end{bmatrix}. \tag{6.9}$$

The components d_{ij} of the matrix D are called rigidity coefficients. The relationships (6.6) and (6.8) can also be expressed in terms of the Lame's constants λ, μ which are related to E and ν by

$$\lambda = \frac{E\varepsilon}{(1+\nu)(1-2\nu)}, \quad \mu = \frac{E}{2(1+\nu)} = G.$$

The three types of boundary conditions needed to solve field equations are as follows:

1. Essential conditions (Dirichlet–type): The displacement u is prescribed on the boundary S of a region V, i.e.,

$$\mathbf{u}(\mathbf{x}) = \mathbf{u}^0(\mathbf{x}), \quad \mathbf{x} \in V. \tag{6.10}$$

2. Natural conditions (Neumann–type): The stresses are prescribed on the boundary S, i.e.,

$$\mathbf{p}(\mathbf{x}) = \mathbf{p}^0(\mathbf{x}), \quad \mathbf{x} \in S, \tag{6.11}$$

where $\mathbf{p}$ is the traction vector at a point $\mathbf{x} \in S$, and $\hat{\mathbf{n}}$ is the unit outward normal at that point (Fig. 6.2). Since $\mathbf{p} = \boldsymbol{\sigma} \cdot \hat{\mathbf{n}}$, the following three conditions are satisfied as a result of equilibrium at the boundary:

$$p_i = \sigma_{ij} n_j = p_i^0, \quad i, j = 1, 2, 3. \tag{6.12}$$

where $n_i = \cos(n, x_i)$. Thus, an equilibrium stress field is defined on a set of sufficiently continuous stress functions σ_{ij} which satisfy (6.4),(6.12) and the Hooke's law

$$\sigma_{ij} = \lambda \varepsilon_{kk} \delta_{ij} + 2\mu \varepsilon_{ij}. \tag{6.13}$$

In view of (6.11)–(6.13) and (6.3), we can write the natural conditions in terms of the displacement $\mathbf{u}$ as

$$\lambda \hat{\mathbf{n}} \nabla \cdot \mathbf{u} + 2\mu \frac{\partial \mathbf{u}}{\partial n} + \mu \left(\hat{\mathbf{n}} \times \operatorname{curl} \mathbf{u} \right) = \mathbf{p}^0(\mathbf{x}). \tag{6.14}$$

3. Mixed conditions: They are a combination of the above two types and apply on different disjoint portions of the boundary where displacements are prescribed on one portion and stresses on the other. Another kind of mixed conditions is known as the Robin–type and has the form

$$a\mathbf{u} + b\mathbf{p} = c, \tag{6.15}$$

where a, b, c are constants.

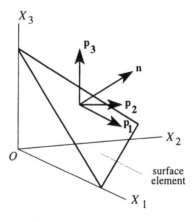

Fig. 6.2. Surface tractions

6.2. Virtual Work

A virtual displacement field is defined as a set of sufficiently continuous

functions δu_i and $\delta \varepsilon_{ij}$ which satisfy

$$\delta u_i = u_i^* \text{ on } S \tag{6.16}$$

and

$$\delta \varepsilon_{ij} = \frac{1}{2}(\delta u_{i,j} + \delta u_{j,i}) \text{ in } V, \tag{6.17}$$

where u_i^* denotes virtual displacements.

It may be noted that the equilibrium stress field as well as the virtual displacement field bears no resemblance to the actual stress and strain field encountered by any real or ideal material under the given circumstances. Also, the two fields bear no relationship relative to each other for any material. However, there exists an integral relationship between these two fields; it is known as the principle of virtual work and is defined as follows:

Let a body be in equilibrium under the action of prescribed body and surfaces forces. The tractions (surfaces forces) p_i may be prescribed only over a portion of the surface S. Suppose that the displacements are known over a portion S_1 of the surface S and over the remaining portion S_2 the tractions are prescribed. If we denote the displacements in the equilibrium state by u_i^*, and consider a class of arbitrary displacements $u_i^* + \delta u_i^*$, consistent with constraints imposed on the body V, then $\delta u_i^* = 0$ over the portion S_1 where the displacements are prescribed, while δu_i^* are arbitrary over the portion S_2 except that they belong to the class C^3 and are of order of magnitude of displacements (under the linearity theory). Such arbitrary displacements are called virtual displacements.

The virtual work δW performed by the external forces $\sigma_{ij,i}+b_j$ and $p_j^0-p_j$ in a virtual displacement δu_j^* is defined by the equation

$$\delta W = \iiint_V (\sigma_{ij,i} + b_j)\delta u_j^* dV - \iint_{S_2} (p_j - p_j^0)\delta u_j^* dS$$
$$= \delta \left[\iiint_V (\sigma ij, i + b_j)u_j^* dV - \iint_{S_2} (p_j - p_j^0)u_j^* dS \right]. \tag{6.18}$$

Since $\delta W = 0$ in the equilibrium state, we obtain from (6.18) the principle of virtual displacements for three dimensional linear elastic problems as

$$\iiint_V (\sigma_{ij,i} + b_j)u_j^* dV - \iint_{S_2} (p_j - p_j^0)u_j^* dS = 0. \qquad (6.19)$$

Note that for two dimensional problems, we should replace V by R, dV by $dx\,dy$, S_2 by C_2, and dS by ds in (6.19). Also, note that if u_j^* does not satisfy the homogeneous boundary conditions on S_1 (i.e., if $u_j^* \neq 0$ on S_1), then (6.19) becomes

$$\iiint_V (\sigma_{ij,i}+b_j)u_j^* dV = \iint_{S_1} (u_j^0 - u_j)p_j^* dS + \iint_{S_2} (p_j - p_j^0)u_j^* dS, \qquad (6.20)$$

where $p_i^* = \sigma_{ij}^* n_j$ are the tractions relative to the u_i^* system.

6.3. Somigliana Identity

The integral relation between the equilibrium stress field and the virtual displacement field is, in general, defined by (6.20) which after applying the divergence theorem (1.5c) yields

$$\iiint_V (\sigma_{jk,j} + b_k)\, u_k^*\, dV = \iiint_V (b_k u_k^* - \sigma_{jk}\varepsilon_{jk}^*)\, dV + \iint_{S_1+S_2} p_k u_k^*\, dS. \qquad (6.21)$$

Thus,

$$\iiint_V b_k u_k^*\, dV - \iiint_V \sigma_{jk}\varepsilon_{jk}^*\, dV = \iint_{S_1} (u_k^0 - u_k)p_k^*\, dS -$$
$$\iint_{S_2} p_k^0 u_k^*\, dS - \iint_{S_1} p_k u_k^*\, dS, \qquad (6.22)$$

where $p_k = n_k\sigma_{jk}$ and $p_k^* = n_k\sigma_{jk}^*$. Since $u_k = u_k^*$ and $\sigma_{jk} = \sigma_{jk}^*$, i.e., $p_k = p_k^*$ on S_1, we find from (6.22) that

$$\iiint_V (\sigma_{jk,j} + b_k)u_k^*\, dV = \iint_{S_1} u_k^0 p_k^*\, dS - \iint_{S_2} p_k^0 u_k^*\, dS -$$
$$- \iint_{S_1} p_k u_k^*\, dS + \iint_{S_2} p_k u_k^*\, dS. \qquad (6.23)$$

We know from Eq (6.4) that the fundamental solution for the stress tensor satisfies the (three) equations

$$\sigma_{jk,j}^* + \delta_l(i) = 0, \tag{6.24}$$

where $\delta_l(i)$ is the Dirac delta function which represents a unit load at a point i in a direction l (which can be x_l, $l = 1, 2, 3$). In view of the translation property (1.55), we get from (6.24)

$$\iiint_V \sigma_{jk,j}^* u_k \, dV = - \iiint_V \delta_l(i) u_k \, dv = -u_l(i).$$

Thus, we find from (6.23) that the fundamental solution u_k^* will satisfy the integral relation

$$u_l(i) + \iint_{S_1} u_k^0 p_k^* \, dS + \iint_{S_2} p_k u_k^* \, dS = \iiint_V b_k u_k^* \, dV + \iint_{S_1} p_k u_k^* \, dS +$$
$$+ \iint_{S_2} p_k^0 u_k^* \, dS, \tag{6.25}$$

which can be written concisely as

$$u_l(i) + \iint_S u_k p_k^* \, dS = \iiint_V b_k u_k^* \, dV + \iint_S p_k u_k^* \, dS, \tag{6.26}$$

where $S = S_1 + S_2$, $u_k = u_k^0$ on S_1, and $p_k = p_k^0$ on S_2.

Since the displacements u_k^* and the tractions $p_k^* (= n_k \sigma_{lk}$ in the direction l) are the fundamental solutions, these quantities represent the displacements and tractions due to a concentrated unit load at a point i in the direction l. The relation (6.26) is valid for a unit force acting in the three directions X_1, X_2, X_3 (see Fig. 6.3), where p_{jk}^* denote the surface forces at the point k generated by the unit load at the point i. Thus, if we consider the unit forces acting in the X_1, X_2, X_3 directions, the relation (6.26) can be written as

$$u_l(i) + \iint_S u_k p_{lk}^* \, dS = \iiint_V b_k u_{lk}^* \, dV + \iint_S p_k u_{lk}^* \, dS, \tag{6.27}$$

where u_{lk}^* and p_{lk}^* denote the displacements and tractions applied at the point

i and acting in the l direction. Eq (6.27) is as the Somigliana identity.

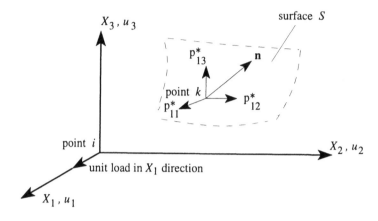

Fig. 6.3. Displacements and tractions.

In a two–dimensional medium, the fundamental solutions for an isotropic plane strain case are

$$u_{lk}^* = \frac{1}{8\pi\mu(1-\nu)r}\left[(3-4\nu)\ln\left(\frac{1}{r}\right)\delta_{lk} + \frac{\partial r}{\partial x_l}\frac{\partial r}{\partial x_k}\right], \qquad (6.28)$$

$$p_{lk}^* = -\frac{1}{4\pi(1-\nu)r}\left[\frac{\partial r}{\partial n}\left\{(1-2\nu)\delta_{lk} + 2\frac{\partial r}{\partial x_l}\frac{\partial r}{\partial x_k}\right\} - \right.$$
$$\left. - (1-2\nu)\left(\frac{\partial r}{\partial x_l}n_k - \frac{\partial r}{\partial x_k}n_l\right)\right]. \quad (6.29)$$

Note that for the spherical coordinate system, defined by

$$\left.\begin{aligned} x_1 &= r\sin\theta\cos\phi, \\ x_2 &= r\sin\theta\sin\phi, \\ x_3 &= r\cos\theta, \end{aligned}\right\} \qquad \begin{aligned} 0 &\le \phi \le 2\pi, \\ 0 &\le \theta \le \pi, \end{aligned} \qquad (6.30)$$

we have $\dfrac{\partial r}{\partial x_k} = \dfrac{r}{x_k}$, and

$$\frac{\partial r}{\partial x_l}n_k - \frac{\partial r}{\partial x_k}n_l = \frac{\partial r}{\partial x_l}\frac{\partial x_k}{\partial r} - \frac{\partial r}{\partial x_k}\frac{\partial x_l}{\partial r} = 0. \qquad (6.31)$$

The fundamental solutions for an isotropic body in a three–dimensional region are given by

$$u^*_{lk} = \frac{1}{16\pi\mu(1-\nu)r}\left[(3-4\nu)\delta_{lk} + \frac{\partial r}{\partial x_l}\frac{\partial r}{\partial x_k}\right], \qquad (6.32)$$

$$p^*_{lk} = -\frac{1}{8\pi(1-\nu)r^2}\left[\frac{\partial r}{\partial n}\left\{(1-2\nu)\delta_{lk} + 3\frac{\partial r}{\partial x_l}\frac{\partial r}{\partial x_k}\right\} - (1-2\nu)\left\{\frac{\partial r}{\partial x_l}n_k - \frac{\partial r}{\partial x_k}n_l\right\}\right], \quad (6.33)$$

where r is the distance from the (boundary) point of application of the load (point i in Fig. 6.3) to the (boundary) point under consideration (point k in Fig. 6.3); $\hat{n}$ is the outward unit normal to the surface S of the body, with n_j as its direction cosines, and δ_{ij} is the Kronecker delta. The fundamental solutions u^*_{lk} have been derived in §4.4.

6.4. Boundary Integral Equation

As we did in the case of potential problems in Chapter 5, we shall start with the integral relation (6.27) and discretize it for the boundary S. Let us assume that the boundary is smooth and it is the part S_2 that contains the point i (the same will hold if the part S_1 contains the point i). If we consider the hemisphere of radius ε on the surface S_2 of a three–dimensional region, as in Fig. 5.1(b), and assume that the point i is at the center of this hemisphere, then in the limiting process the hemisphere reduces to the point i as $\varepsilon \to 0$. This hemisphere divides the surface S_2 into two parts: S_ε and $S_{2-\varepsilon}$. Now, the first integral in Eq (6.27) can be written on S_2 as

$$\iint_{S_2} u_k p^*_{lk}\, dS = \iint_{S_\varepsilon} u_k p^*_{lk}\, dS + \iint_{S_{2-\varepsilon}} u_k p^*_{lk}\, dS. \qquad (6.34)$$

Note that $r = \varepsilon$ and $\partial r/\partial n = 1$ on S_ε. In view of (6.31) and (6.33) we find that

$$\lim_{\varepsilon\to 0}\iint_{S_\varepsilon} u_k p^*_{lk}\, dS$$
$$= \lim_{\varepsilon\to 0}\left\{-\iint_{S_\varepsilon}\frac{u_k}{8\pi(1-\nu)\varepsilon^2}\left[(1-2\nu)\delta_{lk} + 3\frac{\partial\varepsilon}{\partial x_l}\frac{\partial\varepsilon}{\partial x_k}\right]\right\} dS. \qquad (6.35)$$

Since $\partial r/\partial x_l = e_l$, we can take, e.g., $l = 1$ in (6.35), and then using the spherical coordinate system (6.30), we get

$$\lim_{\varepsilon \to 0} \iint_{S_\varepsilon} u_k p_{lk}^* \, dS = -\frac{1}{8\pi(1-\nu)} \int_0^{2\pi} \int_0^{\pi/2} \{ u_1(i)(1-2\nu) +$$

$$+ 3u_1(i) \sin^2 \theta \cos^2 \phi + 3u_2(i) \sin^2 \theta \cos \phi \sin \phi +$$

$$3u_3(i) \sin \theta \cos \theta \cos \phi \} \sin \theta \, d\theta \, d\phi = -\frac{1}{2} u_1(i). \quad (6.36)$$

The same value of the limit of the above integral as $\varepsilon \to 0$ is obtained in the cases when $l = 2$ and 3. Hence, we find that

$$\lim_{\varepsilon \to 0} \iint_{S_\varepsilon} u_k p_{lk}^* \, dS = -\frac{1}{2} u_l(i). \quad (6.37)$$

Now, the last integral in (6.27) on S_2 can be written as

$$\iint_{S_2} p_k u_{lk}^* \, dS = \iint_{S_\varepsilon} p_k u_{lk}^* \, dS + \iint_{S_{2-\varepsilon}} p_k u_{lk}^* \, dS. \quad (6.38)$$

Since from (6.32)

$$\lim_{\varepsilon \to 0} \iint_{S_\varepsilon} p_k u_{lk}^* \, dS = \lim_{\varepsilon \to 0} p_k u_{lk}^* \cdot 2\pi\varepsilon^2 = 0,$$

and since $S_{2-\varepsilon} \to S_2$ as $\varepsilon \to 0$, and recalling that similar results hold for S_1, we find that Eq (6.27) gives the BI Eq for a smooth boundary surface as

$$\frac{1}{2} u_l(i) + \iint_{S_1} u_k^0 p_{lk}^* \, dS + \iint_{S_2} u_k p_{lk}^* \, dS$$

$$= \iiint_V b_k u_{lk}^* \, dV + \iint_{S_1} p_k^0 u_{lk}^* \, dS + \iint_{S_2} p_k u_{lk}^* \, dS. \quad (6.39)$$

In the case of a nonsmooth boundary surface S, the evaluation of the integral of the type (6.36) on S_ε is different from $-\frac{1}{2} u_l(i)$. However, we do not need an exact value in the nonsmooth case. We can take the value of this integral as $c_l(i)u_l(i)$, where $c_l(i)$ is the constant $c(i)$ defined in (5.15) and depends on the geometry of the surface at the point i. Hence, the BI Eq for the nonsmooth body surface can be written, in general, as

$$c_l(i)u_l(i) + \iint_S u_k p_{lk}^* \, dS = \iiint_V b_k u_{lk}^* \, dV + \iint_S p_k u_{lk}^* \, dS. \quad (6.40)$$

The relations (6.39) and (6.40) are the starting point for the boundary element method in linear elastostatics. In the two–dimensional case, the relations (6.39) and (6.40) remain valid if we replace V by R, dV by $dx_1\, dx_2$ (or $dx\, dy$), S by C, and dS by ds.

6.5. Derivation of BE Equation

We will rewrite the BI Eq (6.40) for a two–dimensional isotropic elastic medium. Using the notation $\mathbf{u}^*$ for the displacements u_{lk}^* and $\mathbf{p}^*$ for the tractions p_{lk}^*, we note that both $\mathbf{u}^*$ and $\mathbf{p}^*$ are 2×2 matrices:

$$\mathbf{u}^* = \begin{bmatrix} u_{11}^* & u_{12}^* \\ u_{21}^* & u_{22}^* \end{bmatrix}, \qquad \mathbf{p}^* = \begin{bmatrix} p_{11}^* & p_{12}^* \\ p_{21}^* & p_{22}^* \end{bmatrix}. \tag{6.41}$$

These displacements and tractions are in the k direction due to a unit force applied in the l direction. We will further denote the displacement, traction and body forces acting on the body by $\mathbf{u}$, p, b respectively, each of which is defined as a vector, as follows:

$$\mathbf{u} = [u_1\ u_2]^T, \qquad \mathbf{p} = [p_1\ p_2]^T, \qquad \mathbf{b} = [b_1\ b_2]^T. \tag{6.42}$$

With this notation, Eq (6.40) becomes

$$c(i)\mathbf{u}(i) + \int_C \mathbf{p}^*\mathbf{u}\, ds = \iint_R \mathbf{u}^*\mathbf{b}\, dx_1\, dx_2 + \int_C \mathbf{u}^*\mathbf{p}\, ds, \tag{6.43}$$

where R is a plane region (plate) in the $X_1 X_2$-plane (see Fig. 6.4) which depicts the case of constant boundary elements with mid–nodes (cf. with the case of Fig. 5.4(a), and see Fig. 5.20). The interior cells are used for the evaluation of the domain integral $\iint_R \mathbf{u}^*\mathbf{b}\, dx_1\, dx_2$ in (6.43) which contains the body force terms.

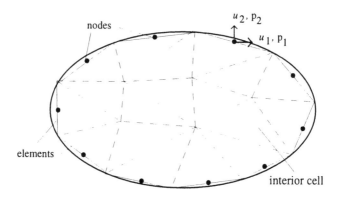

Fig. 6.4. Constant boundary elements and interior cells.

In the case of constant elements, the values of $\mathbf{u}$ and $\mathbf{p}$ are assumed to be constant on each element and equal to their values at its mid–node. Hence, in this case Eq (6.43) reduces to the BE Eq

$$\mathbf{c}(i)\mathbf{u}(i) + \sum_{j=1}^{N}\left\{\int_{\tilde{C}_j}\mathbf{p}^* ds\right\}\mathbf{u}_j = \int\!\!\!\int_{\tilde{R}}\mathbf{u}^*\mathbf{b}\, dx_1\, dx_2 + \sum_{j=1}^{N}\left\{\int_{\tilde{C}_j}\mathbf{u}^* ds\right\}\mathbf{p}_j,$$

$$(6.44)$$

where $\mathbf{u}_j$ and $\mathbf{p}_j$ are the nodal displacement and traction in the element $j = 1, \ldots, N$.

The interior cells are used to numerically integrate the domain integral (body force terms)

$$\mathbf{B}_i = \int\!\!\!\int_{\tilde{R}}\mathbf{u}^*\mathbf{b}\, dx_1\, dx_2,$$

$$(6.45)$$

which appears in (6.44) (for interior cells, see §5.7). If there are M interior cells, then

$$\mathbf{B}_i = \sum_{s=1}^{M}\left\{\sum_{\kappa=1}^{l}(\mathbf{u}^*\mathbf{b})_\kappa w_\kappa\right\}A_s,$$

$$(6.46)$$

where w_κ are the weights used in the Gauss quadrature formula (C.7) and A_s is the area of the interior cell for $s = 1, \ldots, M$.

We obtain a vector $\mathbf{B}_i$ as a result of the numerical integration of the body force terms by (6.46). In BE Eq (6.44) which corresponds to a node i the integral terms within the braces relate to node i with the element segment j over which the integrals are computed. Let us denote these integrals by $\hat{H}_{ij}$

and G_{ij} respectively, i.e.,

$$\hat{H}_{ij} = \int_{C_j} \mathbf{p}^* \, ds, \quad G_{ij} = \int_{C_j} \mathbf{u}^* \, ds, \qquad (6.47)$$

each of which is a 2×2 matrix. Then Eq (6.44) becomes

$$\mathbf{c}(i)\mathbf{u}(i) + \sum_{j=1}^{N} H_{ij}\mathbf{u}_j = \mathbf{B}_i + \sum_{j=1}^{N} G_{ij}\mathbf{p}_j, \qquad (6.48)$$

which relates the value of $\mathbf{u}$ at a mid–node i with the value of $\mathbf{u}$ and $\mathbf{p}$ at all the nodes j, including i. Let us write

$$H_{ij} = \begin{cases} \hat{H}_{ij} & \text{if } i \neq j \\ \hat{H}_{ij} + \mathbf{c}(i) & \text{if } i = j, \end{cases} \qquad (6.49)$$

where $\mathbf{c}(i)$ is a coefficient matrix dependent on the boundary geometry, i.e.,

$$\mathbf{c}(i) = \begin{bmatrix} c(i) & 0 \\ 0 & c(i) \end{bmatrix}, \qquad (6.50)$$

and $c(i)$ is defined in (5.15). Hence, Eq (6.48) can be written as

$$\sum_{j=1}^{N} H_{ij}\mathbf{u}_j = \mathbf{B}_i + \sum_{j=1}^{N} G_{ij}\mathbf{p}_j, \qquad (6.51\text{a})$$

or in matrix form, as

$$HU = B + GP. \qquad (6.51\text{b})$$

Note that in Eq (6.51a) we know N_1 values of the displacements $\mathbf{u}_j$ and N_2 values of the tractions $\mathbf{p}_j$; thus, $2N - (N_1 + N_2)$ values are unknown in this equation. As we did in the case of potential problems (see (5.23)), we will collocate and rearrange Eq (6.51b) in the matrix form

$$AX = B + F, \qquad (6.52)$$

where the unknowns are denoted by the vector X on the left side.

In Eq (6.51a), the integrals H_{ij} and G_{ij} are evaluated numerically by using the 4–point Gauss quadrature formula (C.5) except when $i = j$. The values of H_{ii} are easy to compute using rigid body considerations, but to compute G_{ij} we can use the logarithmically weighted integration formula (C.6), or for the two–dimensional isotropic case, G_{ij} can also be evaluated analytically using (6.28) and Fig. 6.5. Thus,

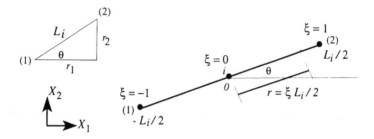

Fig. 6.5. Geometry at the node i.

$$G_{11} = \frac{1}{8\pi\mu(1-\nu)} \left[(3-4\nu) \int_{(1)}^{(2)} \ln\frac{1}{r}\,ds + \int_{(1)}^{(2)} \left(\frac{\partial r}{\partial x_1}\right)^2 ds \right]$$

$$= \frac{1}{8\pi\mu(1-\nu)} \lim_{\varepsilon\to 0} \left[(3-4\nu)2 \int_{\varepsilon}^{L_i/2} \ln\frac{1}{r}\,dr + 2 \int_{\varepsilon}^{L_i/2} \cos^2\theta\,dr \right]$$

$$= \frac{1}{4\pi\mu(1-\nu)} \lim_{\varepsilon\to 0} \left[(3-4\nu)[r - r\ln r]_{\varepsilon}^{L_i/2} + \frac{r_1^2}{L_i^2}\left(\frac{L_i}{2}\right) \right]$$

$$= \frac{L_i}{8\pi\mu(1-\nu)} \left[(3-4\nu)\left(1 + \ln\frac{2}{L_i}\right) + \frac{r_1^2}{L_i^2} \right],$$

$$\tag{6.53}$$

$$G_{12} = \frac{1}{8\pi\mu(1-\nu)} \int_{(1)}^{(2)} \frac{\partial r}{\partial x_1}\frac{\partial r}{\partial x_2}\,ds$$

$$= \frac{1}{8\pi\mu(1-\nu)} \lim_{\varepsilon\to 0} 2 \int_{\varepsilon}^{L_i/2} \sin\theta\cos\theta\,dr$$

$$= \frac{1}{4\pi\mu(1-\nu)} \lim_{\varepsilon\to 0} 2 \int_{\varepsilon}^{L_i/2} \frac{r_1 r_2}{L_i^2}\,,dr = \frac{r_1 r_2}{8\pi\mu(1-\nu)L_i} = G_{21},$$

$$\tag{6.54}$$

$$G_{22} = \frac{1}{8\pi\mu(1-\nu)} \left[(3-4\nu) \int_{(1)}^{(2)} \ln\frac{1}{r}\,ds + \int_{(1)}^{(2)} \left(\frac{\partial r}{\partial x_1}\right)^2 ds \right]$$

$$= \frac{L_i}{8\pi\mu(1-\nu)} \left[(3-4\nu)\left(1 + \ln\frac{2}{L_i}\right) + \frac{r_2^2}{L_i^2} \right],$$

$$\tag{6.55}$$

where L_i, as defined in (5.27), is the length of the element $\tilde{C}_i$. After solving (6.52), we use (6.44) and (6.46) to compute the displacements at an interior point as follows:

$$\mathbf{u}(i) = \sum_{s=1}^{M}\left\{\sum_{\kappa=1}^{l}(\mathbf{u}^*\mathbf{b})_\kappa w_\kappa\right\}A_s + \sum_{j=1}^{N}\left\{\int_{C_j}\mathbf{u}^*\,ds\right\}\mathbf{p}_j - \sum_{j=1}^{N}\left\{\int_{C_j}\mathbf{p}^*\,ds\right\}\mathbf{u}_j.$$
$$(6.56)$$

The stress components at an interior point can be computed from (6.8), i.e.,

$$\sigma_{ij} = \iint_R D_{ij}\mathbf{b}\,dx_1\,dx_2 + \int_C D_{ij}\mathbf{p}\,ds - \int_C S_{ij}\mathbf{u}\,ds$$

$$= \sum_{s=1}^{M}\left\{\iint_{R_s} D_{ij}\,dx_1\,dx_2\right\}\mathbf{b}_s + \sum_{j=1}^{N}\left\{\int_{C_j} D_{ij}\,ds\right\}\mathbf{p}_j - \qquad (6.57)$$

$$- \sum_{j=1}^{N}\left\{\int_{C_j} S_{ij}\,ds\right\}\mathbf{u}_j,$$

where

$$D_{ij} = [D_1\ D_2], \quad S_{ij} = [S_1\ S_2], \quad \mathbf{p} = [p_1\ p_2]^T, \quad \mathbf{u} = [u_1\ u_2]^T,$$
$$(6.58)$$

and for $k = 1, 2,$

$$D_k = \frac{1}{4\pi(1-\nu)r}\left[(1-2\nu)\{\delta_{ki}r_{,j} + \delta_{kj}r_{,i} - \delta_{ij}r_{,k}\} + 2r_{,i}r_{,j}r_{,k}\right], \quad (6.59)$$

$$S_k = \frac{\mu}{2\pi(1-\nu)r^2}\left[2\frac{\partial r}{\partial n}\{(1-2\nu)\delta_{ij}r_{,k} + \nu(\delta_{ik}r_{,j} + \delta_{jk}r_{,i}) - 4r_{,i}r_{,j}r_{,k}\} + \right.$$

$$+ 2\nu(n_i r_{,j}r_{,k} + n_j r_{,i}r_{,k}) + (1-2\nu)(2n_k r_{,i}r_{,j} + n_j\delta_{ik} + n_i\delta_{jk}) -$$
$$\left. - (1-4\nu)n_k\delta_{ij}\right]. \quad (6.60)$$

Program Be11

We shall develop this program for two–dimensional linear elastic boundary value problems for the case of constant elements of Fig. 5.4(a). The input file is created in the following order:

N: Number of boundary elements (same as the number of nodes in this case)

L: Number of interior points where results are to be computed

M: Number of different surfaces 1 through 5.

Last: Number of the last node on each different surface. Enter the last
        node numbers followed by zeros to a total of five entries

mu: shear modulus μ

nu: Poisson's ratio ν

X, Y: Coordinates of extreme points of the elements

Xm, Ym: Coordinates of the mid–nodes

G: Matrix defined in (6.51b)
        After boundary conditions are applied, the matrix A of (6.45)
        is stored in this location.

H: Matrix defined in (6.51b)

Code: = 0 if displacements are prescribed,
        = 1 if tractions are prescribed

Bc: Prescribed boundary conditions

F: Vector defined in (6.52)
        After solution, the values of the unknowns are located here

Xi, Yi: Coordinates of the interior points

The output produces solution values of the displacement components at interior points (2 displacement values each point), and solution values of stresses $\sigma_x, \sigma_y, \tau_{xy}$ at interior points (3 stress values at each point).

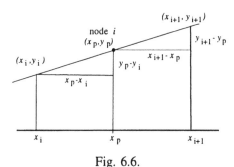

Fig. 6.6.

It can be seen from Fig. 5.8(a) that the equation of the boundary element with mid–nide i is

$$m\left(x_i - x\right) - \left(y_i - y\right) = 0,$$

and thus,

$$\text{Perp} = \begin{cases} \left|m(x_i - x_p) - (y_i - y_p)\right|/\sqrt{1+m^2}, & \text{if } x_i \neq x_{i+1} \\ \left|x_i - x_p\right|, & \text{if } x_i = x_{i+1}. \end{cases}$$

From Fig. 6.6 it is easy to see that

$$\frac{y_{i+1} - y_p}{x_{i+1} - x_p} = \frac{y_p - y_i}{x_p - x_i}$$

which yields

$$(x_p - x_i)(y_{i+1} - y_p) - (x_{i+1} - x_p)(y_p - y_i) \equiv \text{sgn}.$$

(See (3.8) also). Note that sgn has the same sign as slope defined in §5.3, where other variables used in the program are also listed. This program, listed in Appendix D, calls the following functions: Sys11, Quad11, Diag11, Inter11, and Solve.

Example 6.1:
Consider the case of a circular hole under interior pressure embedded in an infinite medium, as shown in Fig. 6.7. The data is: $E = 94500$, $\nu = 0.1$.

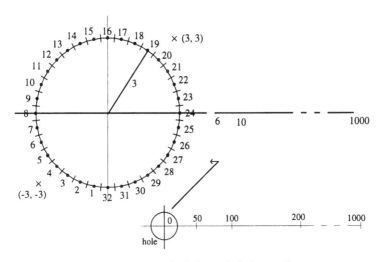

Fig. 6.7. Circular hole in an infinite medium.

Input File:
```
Example 6.1 (Constant elements)
32  10  1 32 0 0 0 0   94500   0.1
-0.294051 -2.98555  -0.870854 -2.87082
-1.41419 -2.64576  -1.90318 -2.31903
-2.31903 -1.90318  -2.64576 -1.41419
-2.87082 -0.870854  -2.98555 -0.294051
-2.98555  0.294051  -2.87082  0.870854
```

```
-2.64576   1.41419   -2.31903   1.90318
-1.90318   2.31903   -1.41419   2.64576
-0.870854  2.87082   -0.294051  2.98555
0.294051   2.98555   0.870854   2.87082
1.41419    2.64576   1.90318    2.31903
2.31903    1.90318   2.64576    1.41419
2.87082    0.870854  2.98555    0.294051
2.98555   -0.294051  2.87082   -0.870854
2.64576   -1.41419   2.31903   -1.90318
1.90318   -2.31903   1.41419   -2.64576
0.870854  -2.87082   0.294051  -2.98555
1  -19.509   1  -98.0785  1  -38.2683  1  -92.388
1  -55.557   1  -83.147   1  -70.7107  1  -70.7107
1  -83.147   1  -55.557   1  -92.388   1  -38.2683
1  -98.0785  1  -19.509   1  -100.     1  0.
1  -98.0785  1   19.509   1  -92.388   1   38.2683
1  -83.147   1   55.557   1  -70.7107  1   70.7107
1  -55.557   1   83.147   1  -38.2683  1   92.388
1  -19.509   1   98.0785  0  0.        1   100.
1   19.509   1   98.0785  1   38.2683  1   92.388
1   55.557   1   83.147   1   70.7107  1   70.7107
1   83.147   1   55.557   1   92.388   1   38.2683
1   98.0785  1   19.509   1   100.     0   0.
1   98.0785  1  -19.509   1   92.388   1  -38.2683
1   83.147   1  -55.557   1   70.7107  1  -70.7107
1   55.557   1  -83.147   1   38.2683  1  -92.388
1   19.509   1  -98.0785  0  0.        1  -100.
4. 0. -4. 0.   2.8284 2.82842 -3. -3.  6. 0.
10. 0. 50. 0. 100. 0. 200. 0. 1000. 0.
```

The output is:

```
Displacements at interior points:
     (Xi,Yi)                    u_x         u_y
(    4.0000,    0.0000)       0.0012       0.0000
(   -4.0000,    0.0000)      -0.0012       0.0000
(    2.8284,    2.8284)       0.0008       0.0008
(   -3.0000,   -3.0000)      -0.0008      -0.0008
(    6.0000,    0.0000)       0.0008       0.0000
(   10.0000,    0.0000)       0.0005       0.0000
(   50.0000,    0.0000)       0.0001       0.0000
(  100.0000,    0.0000)       0.0000       0.0000
(  200.0000,    0.0000)       0.0000       0.0000
```

```
( 1000.0000,    0.0000)      0.0000        0.0000
```

Stresses at interior points:

(Xi,Yi)		sigma X	tau XY	sigma Y
(4.0000,	0.0000)	-56.8348	0.0000	56.8234
(-4.0000,	0.0000)	-56.8348	0.0000	56.8234
(2.8284,	2.8284)	-0.0052	-56.8298	-0.0060
(-3.0000,	-3.0000)	-0.0008	-50.4890	-0.0008
(6.0000,	0.0000)	-25.2413	0.0000	25.2413
(10.0000,	0.0000)	-9.0869	0.0000	9.0869
(50.0000,	0.0000)	-0.3635	0.0000	0.3635
(100.0000,	0.0000)	-0.0909	0.0000	0.0909
(200.0000,	0.0000)	-0.0227	0.0000	0.0227
(1000.0000,	0.0000)	-0.0009	0.0000	0.0009

Note that the displacement (Code= 0) is prescribed zero at the nodes 16, 24, and 32 to keep the plate in equilibrium. Mathematically it means that the value of the constant for the displacement obtained as a result of integration is assumed to be zero. The solution for radial stress obtained from elasticity theory (see Timoshenko and Goodier, pp. 78). is as follows:

Point	Theoretical solution	BEM solution
$(4., 0.)$	-56.25	-56.8348
$(6., 0.)$	-25.0	-25.2413
$(10., 0.)$	-9.0	-9.0869
$(50., 0.)$	-0.36	-0.3635
$(200., 0.)$	-0.0225	-0.0227
$(1000., 0.)$	-0.0009	-0.0009

The extreme points X, Y can be computed in the program, instead of being read, by using the formulas $x_j = 3\cos((j-1)\pi/16 + 17\pi/32)$, $y_j = -3\sin((j-1)\pi/16 + 17\pi/32)$, $j = 0, 1, \cdots, 32$ respectively. The boundary conditions at the nodes $j = 1, \ldots, 32$ are prescribed in order by the formulas

$$\text{Bc in } x\text{–direction:} \quad 100\cos\left(\frac{j\pi}{16} + \frac{\pi}{2}\right), \ j = 0, 1, \ldots, 32,$$

$$\text{Bc in } y\text{–direction:} \quad -100\sin\left(\frac{j\pi}{16} + \frac{\pi}{2}\right), \ j = 0, 1, \ldots, 32. \ \blacksquare$$

Example 6.2:

Consider the problem of a hollow circular pipe of radii $a = 10$ and $b = 15$ units respectively under an internal pressure $p = 100$ (see Fig. 6.8). The other data is: $\mu = 80,000$, and $\nu = 0.25$. Because of axial symmetry, the input file is created with constant elements as in Fig. 5.12(b). For the plane stress case, the displacements is given by

$$u(r) = \frac{pa^2}{E(b^2 - a^2)} \left[(1 - \nu)r + (1 + \nu)\frac{b^2}{r^2} \right], \quad a \leq r \leq b,$$

and the stress by

$$\sigma(r) = \frac{pa^2}{E(b^2 - a^2)} \left(1 - \frac{b^2}{r^2} \right),$$

$$\theta(r) = \frac{pa^2}{E(b^2 - a^2)} \left(1 - \frac{b^2}{r^2} \right),$$

for $a \leq r \leq b$. Both circumferential and radial results for displacements and stresses compares very well with the exact solutions. However, the boundary element results for the stresses in the vicinity of the boundary do not match with the exact solutions; but this was expected. It is found that the boundary element results are, in general, correct for those interior points which lie at a distance more than half an element length away from the boundary.

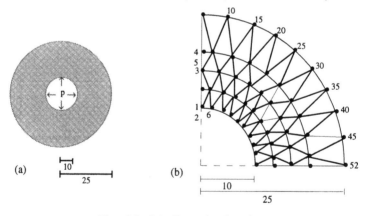

Fig. 6.8. A hollow circular pipe.

If this problem is solved with the finite element method of Fig. 6.8(b), with 52 nodes and 76 triangular elements, the results do not agree with the exact solutions. It means that if constant strains are used in linear elasticity, the resulting finite elements computed at the center of each element will produce poor results. This method should therefore be avoided. ■

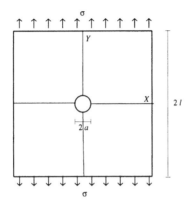

Fig. 6.9. A square plate with a small hole.

Example 6.3:
Consider the problem of stress concentration on a square plate of length $2l$ with a small circular hole of radius a at the center, with $a/l = 0.1$, and $\nu = 0.25$ (see Fig. 6.9). The plate is subject to a uniform tensile stress σ in the y–direction.

It is known that the stress concentration factor $K_t = 3.0$ for an infinite plate. We shall use $K_t = 3.024$ for the finite plate in this example. Because of the axial symmetry we should discretize the quarter region into constant elements as in Fig. 5.12(b). ■

Example 6.4:
Consider a cantilever beam of length 16 and width 2 units, shown in Fig. 6.10(a), which is under plane stress, with $\mu = 80000$ and $\nu = 0.25$. The beam is rigidly supported as the left end so that its displacements along both x and y axes are zero; the right end is free and is subject to deflection which is given by

$$u_A = -\frac{128}{3EI}qa^4 \approx 0.012.$$

Since the length to width ratio of the beam is 5, the shear contribution to the total deflection is very small and should, therefore, be neglected. Because of axial symmetry, a discretization of the half beam has been used as shown in

Fig. 6.10(b).

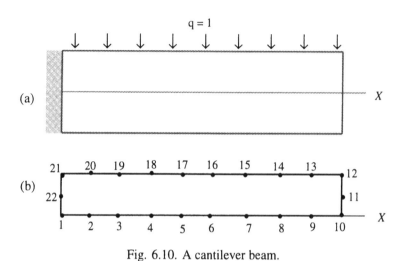

Fig. 6.10. A cantilever beam.

6.6. Torsion of a Cylindrical Bar

Consider the problem of the torsion of a homogeneous isotropic cylindrical bar (see Fig. 6.11). According to the Prandtl theory, the stress function $\phi(x, y)$ in a bar of an arbitrary cross–section R can be reduced to the two–dimensional boundary value problem

$$-\nabla^2 \phi(x, y) = 2G\theta, \quad \text{in } R, \text{ and } u = 0 \text{ on } C, \qquad (6.61)$$

where G is the shear modulus of the material, and θ is the angle of the twist.

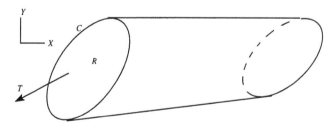

Fig. 6.11. Cross–section of a cylindrical bar.

The exact solution of the boundary value problem (6.61) for an elliptic cross–section ($x^2/a^2 + y^2/b^2 = 1$) is given by

$$\phi(x,y) = \frac{G\theta a^2 b^2}{a^2 + b^2} \left(1 - \frac{x^2}{a^2} - \frac{y^2}{b62}\right), \qquad (6.62)$$

and for an equilateral triangle of altitude a (i.e., of side $2a/\sqrt{3}$) by

$$\phi(x,y) = -G\theta \left[\frac{1}{2}\left(x^2 + y^2\right) - \frac{1}{2a}\left(x^3 - 3xy^2\right) - \frac{2}{27}a^2\right]. \qquad (6.63)$$

(See Examples 5.9, 5.10 and 5.13.) The nonzero stress components are given in terms of the stress function by

$$\tau_{xz} = \frac{\partial \phi}{\partial y}, \quad \tau_{yz} = -\frac{\partial \phi}{\partial x}, \qquad (6.64)$$

and the torque transmitted as

$$T = 2 \iint_R \phi \, dx \, dy. \qquad (6.65)$$

In the case of a rectangular cross–section $R = \{-a \leq x \leq a, -b \leq y \leq b\}$, we use the transformation $x = ax'$, $y = by'$, and $\phi = 2G\theta a^2 u$. Then, if we suppress the primes over x and y, the boundary value problem (6.61) reduces to

$$-\nabla^2 u = 1 \quad \text{in } R' = \{|x| \leq 1, |y| \leq a\}, \quad u = 0 \text{ on } C, \qquad (6.66)$$

where $\alpha = b/a$ is the aspect ratio of the rectangle. Also,

$$\tau_{xz} = 2G\theta a \frac{\partial u}{\partial y}, \quad \tau_{yz} = -2G\theta a \frac{\partial u}{\partial x}, \quad T = 4G\theta a^4 \iint_R u \, dx \, dy. \quad (6.67)$$

The boundary element method of §5.7 (Poisson Eq) is used for the boundary value problem (6.66).

6.7. Three–Dimensional Medium

The boundary surface S of a three–dimensional body V is divided into constant, linear, or quadratic elements (triangles) $\tilde{S}_j$, $j = 1, \ldots, N$, as shown in Fig. 6.12(a)–(c).

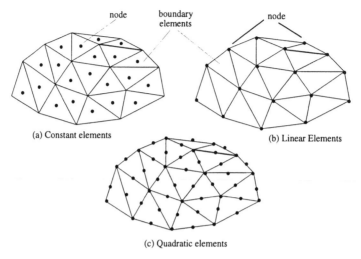

(a) Constant elements   (b) Linear Elements

(c) Quadratic elements

Fig. 6.12. Three–dimensional boundary elements.

the values of $\mathbf{u}$ and $\mathbf{p}$ can be approximated over each boundary element by
using the interpolation functions ϕ defined in (1.27)–(1.29) respectively for
the constant, linear, or quadratic elements. Thus,

$$
\begin{aligned}
\mathbf{u} &= \phi^T \mathbf{u}^n, \\
\mathbf{p} &= \phi^T \mathbf{p}^n,
\end{aligned}
\tag{6.68}
$$

where $\mathbf{u}^n$ and $\mathbf{p}^n$ denote the nodal displacements and tractions respectively
for the elements $n = 1, \ldots, N$. The interpolation functions for $\mathbf{u}$ and $\mathbf{p}$ are
generally different, and usually those for $\mathbf{p}$ are taken one order less than those
for $\mathbf{u}$, obviously because of the order of relationship between $\mathbf{p}$ and $\mathbf{u}$. We
shall henceforth assume that the body forces are absent. Then, just as we did
in §5.7, we shall start with the BI Eq (6.40) and substitute the functions (6.68)
into (6.40). This results in the BE Eq

$$
c\mathbf{u} + \sum_{j=1}^{N} \left(\iint_{\tilde{S}_j} \mathbf{p}^* \phi^T \, dS \right) \mathbf{u}^n = \sum_{j=1}^{N} \left(\iint_{\tilde{S}_j} \mathbf{u}^* \phi^T \, dS \right) \mathbf{p}^n.
\tag{6.69}
$$

We will now integrate (6.69) by using the Gauss quadrature formula; thus,

$$
\begin{aligned}
c\mathbf{u} + \sum_{j=1}^{N} &\left(\sum_{k=1}^{M} |G_k| W_k \left[\mathbf{p}^* \phi^T (\eta_1, \eta_2) \right]_k \right) \mathbf{u}^n \\
&= \sum_{j=1}^{N} \left(\sum_{k=1}^{M} |G_k| W_k \left[\mathbf{u}^* \phi^T (\eta_1, \eta_2) \right]_k \right) \mathbf{p}^n,
\end{aligned}
\tag{6.70}
$$

where $|G|$ is the Jacobian of the three–dimensional transformation and W_k are
the weights. For a node i, Eq (6.70) reduces to a system of algebraic equations

$$
\begin{aligned}
\mathbf{c}(i)\mathbf{u}(i) + \begin{bmatrix} \hat{\mathbf{h}}_{i1} & \hat{\mathbf{h}}_{i2} & \ldots \hat{\mathbf{h}}_{iN} \end{bmatrix} \begin{bmatrix} \mathbf{u}_1 & \mathbf{u}_2 & \ldots \mathbf{u}_N \end{bmatrix} = \\
= \begin{bmatrix} \mathbf{g}_{i1} & \mathbf{g}_{i2} & \ldots \mathbf{g}_{iN} \end{bmatrix} \begin{bmatrix} \mathbf{p}_1 & \mathbf{p}_2 & \ldots \mathbf{p}_N \end{bmatrix},
\end{aligned}
\tag{6.71}
$$

where $\mathbf{u}_j$, $\mathbf{p}_j$ are the unknowns at node j, and $\hat{\mathbf{h}}_{ij}$ and $\mathbf{g}_{ij}$ are the coefficients
from the integration in (6.70) which relate the node i to all the nodes on the
surface elements $\tilde{S}_j$. Also, each one of the matrices $\hat{\mathbf{h}}_{ij}$ and $\mathbf{g}_{ij}$ will contribute
one element only in the case of constant elements, but 2×2 elements in the
case of linear elements and 3×3 elements in the case of quadratic elements.
Let us denote, as in Chapter 5, $\mathbf{h}_{ii} = \hat{\mathbf{h}}_{ii} + \mathbf{c}(i)$. Then, Eq (6.71) reduces to
the system of equations

$$
\begin{aligned}
&\begin{bmatrix} \mathbf{h}_{11} & \mathbf{h}_{12} & \ldots & \mathbf{h}_{1N} \\ \mathbf{h}_{21} & \mathbf{h}_{22} & \ldots & \mathbf{h}_{2N} \\ \vdots & & \ldots & \\ \mathbf{h}_{N1} & \mathbf{h}_{N2} & \ldots & \mathbf{h}_{NN} \end{bmatrix} \begin{bmatrix} \mathbf{u}_1 & \mathbf{u}_2 & \ldots & \mathbf{u}_N \end{bmatrix}^T \\
&= \begin{bmatrix} \mathbf{g}_{11} & \mathbf{g}_{12} & \ldots & \mathbf{g}_{1N} \\ \mathbf{g}_{21} & \mathbf{g}_{22} & \ldots & \mathbf{g}_{2N} \\ \vdots & & \ldots & \\ \mathbf{g}_{N1} & \mathbf{g}_{N2} & \ldots & \mathbf{g}_{NN} \end{bmatrix} \begin{bmatrix} \mathbf{p}_1 & \mathbf{p}_2 & \ldots & \mathbf{p}_N \end{bmatrix}^T, \quad (6.72)
\end{aligned}
$$

or, in matrix form,
$$
HU = GP, \tag{6.73}
$$

which, after using the collocation technique of Chapter 5, can be written as

$$
AX = F. \tag{6.74}
$$

Once the boundary conditions are applied, the system (6.74) is solved for
unknown displacements and tractions over the entire surface, and also for the
interior points where the solution is desired.

6.8. Axisymmetric Problems

In many three–dimensional stress analysis problems, the geometry and loading result in axisymmetry. Such situations involve, e.g., problems with pipes,

discs, and pressure structures. We will use the polar coordinate system (r, θ, z), defined by $x_1 = r \cos \theta$, $x_2 = r \sin \theta$ and $x_3 = z$, determine the fundamental solution for the axisymmetric geometry of Fig. 6.13, and then transform the BI Eq (6.40) for axisymmetric problems in the absence of body forces.

The axisymmetric fundamental solution $U^*(\mathbf{x}, \mathbf{x}')$ is derived from the three–dimensional point source potential u^*, defined by (4.25), by integrating it with respect to θ between the limits 0 to 2π, i.e.,

$$U^*(\mathbf{x}, \mathbf{x}') = \frac{1}{2\pi} \int_0^{2\pi} u^*(\mathbf{x}, \mathbf{x}') \, d\theta(\mathbf{x}'). \qquad (6.75)$$

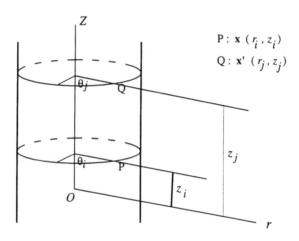

Fig. 6.13.

If we denote the points $P = \mathbf{x}$ and $Q = \mathbf{x}'$ by i and j respectively, then the polar cylindrical coordinates of these points, with $\theta_i = 0$, are given by

For point P: $x_1 = r_i \cos \theta_i = r_i$, $x_2 = r_i \sin \theta_i = 0$, $x_3 = z_i$,
For point Q: $x_1 = r_j \cos \theta_j$, $x_2 = r_j \sin \theta_j$, $x_3 = z_j$.

Thus,

$$u^*(\mathbf{x}, \mathbf{x}') = \frac{1}{4\pi} \left[(r_i + r_j)^2 + (z_i - z_j)^2 - 4r_i r_j \cos^2(\theta_j/2) \right]^{-1/2},$$

where, by substituting $\theta_j = \pi - 2\alpha$ in (6.75), we obtain

$$U^*(\mathbf{x}, \mathbf{x}') \equiv U^*(i, j)$$

$$= \frac{1}{4\pi^2} \int_{-\pi/2}^{\pi/2} \left[(r_i + r_j)^2 + (z_i - z_j)^2 - 4r_i r_j \sin^2 \alpha \right]^{-1/2} d\alpha$$

$$= \frac{1}{2\pi^2 c} \int_0^{2\pi} \frac{1}{\sqrt{1 - m^2 \sin^2 \alpha}} \, d\alpha$$

$$= \frac{1}{2\pi^2 c} K(m, \pi/2),$$

$$(6.76)$$

where

$$m = \frac{2\sqrt{r_i r_j}}{c}, \quad c = \sqrt{(r_i + r_j)^2 + (z_i - z_j)^2},$$

and $K(m, \pi/2)$ is the complete elliptic integral of first kind of modulus m^2 and complementary modulus $m_1 = 1 - m^2$. Note that the differentiation with respect to n is given by

$$\frac{\partial U^*(i, j)}{\partial n} = \frac{\partial U^*(i, j)}{\partial r} \frac{\partial r}{\partial n} + \frac{\partial U^*(i, j)}{\partial z} \frac{\partial z}{\partial n},$$

and the differentials of the elliptic integral $(m, \pi/2)$ are given by

$$\frac{\partial K(m, \pi/2)}{\partial r_j} = \frac{\partial K(m, \pi/2)}{\partial m} \frac{\partial m}{\partial r_j},$$

$$\frac{\partial K(m, \pi/2)}{\partial z_j} = \frac{\partial K(m, \pi/2)}{\partial m} \frac{\partial m}{\partial z_j},$$

$$\frac{\partial K(m, \pi/2)}{\partial m} = \frac{E(m, \pi/2)}{m m_1} - \frac{K(m, \pi/2)}{m},$$

$$\frac{\partial [m K(m, \pi/2)]}{\partial m} = \frac{E(m, \pi/2)}{m_1},$$

$$(6.77)$$

where $E(m, \pi/2)$ is the complete elliptic function of second kind of modulus m^2 and complementary modulus m_1. It is difficult to work with the elliptic functions, but they can be computed by polynomial approximations:

$$K(m, \pi/2) = \left[a_0 + a_1 m_1 + \cdots + a_4 m_1^4 \right] +$$
$$+ \left[b_0 + b_1 m_1 + \cdots + b_4 m_1^4 \right] \ln(1/m_1) + \epsilon(m), \quad (6.78a)$$

$$E(m, \pi/2) = \left[1 + \alpha_1 m_1 + \cdots + \alpha_4 m_1^4 \right] +$$
$$+ \left[\beta_1 m_1 + \cdots + \beta_4 m_1^4 \right] \ln(1/m_1) + \epsilon(m), \quad (6.78b)$$

where $|\epsilon(m)| \leq 2 \times 10^{-8}$, and

$$a_0 = 1.38629, \quad a_1 = 0.09666, \quad a_2 = 0.359,$$
$$a_3 = 0.3742, \quad a_4 = 0.0145,$$
$$b_0 = 0.5, \quad b_1 = 0.12498, \quad b_2 = 0.688,$$
$$b_3 = 0.3328, \quad b_4 = 0.0044,$$
$$\alpha_1 = 0.44325, \quad \alpha_2 = 0.0626,$$
$$\alpha_3 = 0.04757, \quad \alpha_4 = 0.01736,$$
$$\beta_1 = 0.24998, \quad \beta_2 = 0.092,$$
$$\beta_3 = 0.04069, \quad \beta_4 = 0.00526.$$

Since the geometry of an axisymmetric body is independent of the variable θ, only the following displacements, tractions, stresses and strains appear in axisymmetric problems:

u_r, u_z:	Displacements in the radial and axial directions
p_r, p_z:	Tractions in the radial and axial directions
σ_{rr}, σ_{zz}:	Direct stresses in the radial and axial directions
$\varepsilon_{rr}, \varepsilon_{zz}$:	Direct stains in the radial and axial directions
$\sigma_{rz}, \varepsilon_{rz}$:	axisymmetric shearing stress and strain
$\sigma_{\theta\theta}$:	Hoop stress (constant)
$\varepsilon_{\theta\theta}$:	Hoop strain $(= u_r/r)$

There are two methods to discretize the BI Eq (6.40) for axisymmetric problems in the absence of body forces:

Method 1. The displacements and tractions at a node i can be resolved into their radial and axial components by the following relations, with $\theta_i = 0$:

$$u_{x_1}(\mathbf{x}) = u_r(i) \cos \theta_i = u_r(i), \qquad u_{x_1}(\mathbf{x}') = u_r(j) \cos \theta_j,$$
$$u_{x_2}(\mathbf{x}) = u_r(i) \sin \theta_i = 0, \qquad u_{x_2}(\mathbf{x}') = u_r(j) \sin \theta_j,$$
$$u_{x_3}(\mathbf{x}) = u_z(i), \qquad u_{x_3}(\mathbf{x}') = u_z(j),$$
$$p_{x_1}(\mathbf{x}) = p_r(i) \cos \theta_i = p_r(i), \qquad p_{x_1}(\mathbf{x}') = p_r(j) \cos \theta_j,$$
$$p_{x_2}(\mathbf{x}) = p_r(i) \sin \theta_i = 0, \qquad p_{x_2}(\mathbf{x}') = p_r(j) \sin \theta_j,$$
$$p_{x_3}(\mathbf{x}) = p_z(i), \qquad p_{x_3}(\mathbf{x}') = p_z(j). \qquad (6.79)$$

Hence the BI Eq (6.40), in matrix form, reduces to the axisymmetric BE Eq, with $\theta_j = 0$:

$$
\mathbf{c}(i) \left\{ \begin{array}{c} u_r(i) \\ 0 \\ u_z(i) \end{array} \right\} + \sum_{j+1}^{N} \int_{\tilde{C}_j} \int_0^{2\pi} \left[\begin{array}{ccc} P_{rr}^*(i,j) & P_{r\theta}^*(i,j) & P_{rz}^*(i,j) \\ P_{\theta r}^*(i,j) & P_{\theta\theta}^*(i,j) & P_{\theta z}^*(i,j) \\ P_{zr}^*(i,j) & P_{z\theta}^*(i,j) & P_{zz}^*(i,j) \end{array} \right] \times
$$

$$
\times \left\{ \begin{array}{c} u_r(j)\cos\theta_j \\ u_r(j)\sin\theta_j \\ u_z(j) \end{array} \right\} r\, d\theta_j\, ds
$$

$$
= \sum_{j+1}^{N} \int_{\tilde{C}_j} \int_0^{2\pi} \left[\begin{array}{ccc} U_{rr}^*(i,j) & U_{r\theta}^*(i,j) & U_{rz}^*(i,j) \\ U_{\theta r}^*(i,j) & U_{\theta\theta}^*(i,j) & U_{\theta z}^*(i,j) \\ U_{zr}^*(i,j) & U_{z\theta}^*(i,j) & U_{zz}^*(i,j) \end{array} \right] \times
$$

$$
\times \left\{ \begin{array}{c} p_r(j)\cos\theta_j \\ p_r(j)\sin\theta_j \\ p_z(j) \end{array} \right\} r\, d\theta_j\, ds
$$

which simplifies to

$$
\mathbf{c}(i) \left\{ \begin{array}{c} u_r(i) \\ u_z(i) \end{array} \right\} + 2\pi \sum_{j+1}^{N} \int_{\tilde{C}_j} \left[\begin{array}{cc} P_{rr}^*(i,j)) & P_{rz}^*(i,j) \\ P_{zr}^*(i,j) & P_{zz}^*(i,j) \end{array} \right] \left\{ \begin{array}{c} u_r(j) \\ u_z(j) \end{array} \right\} r\, ds
$$

$$
= 2\pi \sum_{j+1}^{N} \int_{\tilde{C}_j} \left[\begin{array}{cc} U_{rr}^*(i,j) & U_{rz}^*(i,j) \\ U_{zr}^*(i,j) & U_{zz}^*(i,j) \end{array} \right] \left\{ \begin{array}{c} p_r(j) \\ p_z(j) \end{array} \right\} r\, ds,
$$

$$
\tag{6.80}
$$

where, with $\theta_j = \pi - 2\alpha$,

$$
U_{rr}^*(i,j) \equiv U_{rr}^*(\mathbf{x}, \mathbf{x}')
$$

$$
= \frac{1}{2\pi} \int_0^{2\pi} \left[U_{x_1 x_1}^*(\mathbf{x}, \mathbf{x}') \cos\theta_j + U_{x_1 x_2}^*(\mathbf{x}, \mathbf{x}') \sin\theta_j \right] d\theta_j
$$

$$
= 2A \int_0^{2\pi} \Big[\frac{(3-4\nu)(1 - 2\cos^2\alpha)}{c\sqrt{1 - m^2 \sin^2\alpha}} +
$$

$$
+ \frac{(r_i - r_j)^2 (1 - 2\cos^2\alpha) - 4 r_i r_j \cos^2\alpha}{c^3 (1 - m^2 \sin^2\alpha)^{3/2}} \Big] d\alpha
$$

$$
= \frac{A}{c r_i r_j} \left[(3-4\nu)(r_i^2 + r_j^2) + 4(1-\nu)(z_i - z_j)^2 \right] K(m, \pi/2) -
$$

$$
- \frac{A}{c r_i r_j} \left[(3-4\nu) c^2 + \frac{a(z_i - z_j)^2}{d} \right] E(m, \pi/2),
$$

$$
\tag{6.81}
$$

$$U_{rz}^*(i,j) \equiv U_{rz}^*(\mathbf{x}, \mathbf{x}') = \frac{1}{2\pi} \int_0^{2\pi} U_{x_1 x_3}^*(\mathbf{x}, \mathbf{x}')\, d\theta_j$$

$$= 2A \int_0^{2\pi} \frac{(r_i - r_j + 2r_j \cos^2 \alpha)(z_i - z_j)}{c^3 (1 - m^2 \sin^2 \alpha)^{3/2}}\, d\alpha$$

$$= \frac{A(z_i - z_j)}{cr_j}\left[K(m, \pi/2) + \frac{r_i^2 - r_j^2 - (z_i - z_j)^2}{d} E(m, \pi/2)\right],$$

$$\text{(6.82)}$$

$$U_{zr}^*(i,j) \equiv U_{rz}^*(\mathbf{x}, \mathbf{x}')$$

$$= \frac{1}{2\pi} \int_0^{2\pi} \left[U_{x_3 x_1}^*(\mathbf{x}, \mathbf{x}') \cos \theta_j + U_{x_3 x_2}^*(\mathbf{x}, \mathbf{x}') \sin \theta_j \right] d\theta_j$$

$$= 2A \int_0^{2\pi} \frac{(r_i - r_j - 2r_i \cos^2 \alpha)(z_i - z_j)}{c^3 (1 - m^2 \sin^2 \alpha)^{3/2}}\, d\alpha$$

$$= \frac{A(z_i - z_j)}{cr_j}\left[-K(m, \pi/2) + \frac{r_i^2 - r_j^2 + (z_i - z_j)^2}{d} E(m, \pi/2)\right],$$

$$\text{(6.83)}$$

$$U_{zz}^*(i,j) \equiv U_{x_3 x_3}^*(\mathbf{x}, \mathbf{x}') = \frac{1}{2\pi} \int_0^{2\pi} U_{zz}^*(i,j)\, d\theta_j$$

$$= 2A \int_0^{2\pi} \left[\frac{3 - 4\nu}{c\sqrt{1 - m^2 \sin^2 \alpha}} + \frac{(z_i - z_j)^2}{c^3 (1 - m^2 \sin^2 \alpha)^{3/2}} \right] d\alpha$$

$$= \frac{2A}{c}\left[(3 - 4\nu) K(m, \pi/2) + \frac{(z_i - z_j)^2}{d} E(m, \pi/2)\right],$$

$$\text{(6.84)}$$

where

$$A = \frac{1}{16\pi^2 \mu(1 - \nu)}, \qquad a = r_i^2 + r_j^2 + (z_i - z_j)^2,$$

$$d = (r_i - r_j)^2 + (z_i - z_j)^2.$$

$$\text{(6.85)}$$

The axisymmetrix virtual tractions P^* can be derived in an analogous manner, but since this involves too much algebra before they can be written in terms of elliptic functions, we shall instead use (6.12) and (6.13) and express them directly in terms of $\mathbf{U}^*$ already evaluated in (6.81)–(6.84). Thus,

$$P_{rr}^* \equiv P_{rr}^*(\mathbf{x}, \mathbf{x}')$$

$$= \mu n_z \left[\frac{\partial U_{rr}^*}{\partial z} + \frac{\partial U_{rz}^*}{\partial r} \right] +$$

$$+ 2\mu n_r \left[\frac{1 - \nu}{1 - 2\nu} \frac{\partial U_{rr}^*}{\partial r} + \frac{\nu}{1 - 2\nu} \left(\frac{U_{rr}^*}{r} + \frac{\partial U_{rr}^*}{\partial z} \right) \right],$$

$$\text{(6.86)}$$

$$P_{rz}^* \equiv P_{rz}^*(\mathbf{x}, \mathbf{x}')$$

$$= \mu n_r \left[\frac{\partial U_{rr}^*}{\partial z} + \frac{\partial U_{rz}^*}{\partial r} \right] +$$

$$+ 2\mu n_z \left[\frac{1-\nu}{1-2\nu} \frac{\partial U_{rz}^*}{\partial z} + \frac{\nu}{1-2\nu} \left(\frac{U_{rr}^*}{r} + \frac{\partial U_{rr}^*}{\partial r} \right) \right], \qquad (6.87)$$

$$P_{zr}^* \equiv P_{zr}^*(\mathbf{x}, \mathbf{x}')$$

$$= \mu n_z \left[\frac{\partial U_{zr}^*}{\partial z} + \frac{\partial U_{zz}^*}{\partial r} \right] +$$

$$+ 2\mu n_r \left[\frac{1-\nu}{1-2\nu} \frac{\partial U_{zr}^*}{\partial r} + \frac{\nu}{1-2\nu} \left(\frac{U_{zr}^*}{r} + \frac{\partial U_{zz}^*}{\partial z} \right) \right], \qquad (6.88)$$

$$P_{zz}^* \equiv P_{zz}^*(\mathbf{x}, \mathbf{x}')$$

$$= \mu n_r \left[\frac{\partial U_{zr}^*}{\partial z} + \frac{\partial U_{zz}^*}{\partial r} \right] +$$

$$+ 2\mu n_z \left[\frac{1-\nu}{1-2\nu} \frac{\partial U_{zz}^*}{\partial z} + \frac{\nu}{1-2\nu} \left(\frac{U_{zr}^*}{r} + \frac{\partial U_{zr}^*}{\partial r} \right) \right], \qquad (6.88)$$

If the point $\mathbf{x}$ lies on the z–axis, both $\mathbf{U}^*$ and $\mathbf{P}^*$ must converge to finite values as $r_i \to 0$. Thus, we expand the expression in (6.81)–(6.88) and put $r_i = 0$; this yields

$$U_{rr}^*\big|_{r_i=0} = U_{rz}^*\big|_{r_i=0} = 0,$$

$$U_{zr}^*\big|_{r_i=0} = -\frac{\pi A r_j (z_i - z_j)}{c^3}, \qquad (6.89a)$$

$$U_{zz}^*\big|_{r_i=0} = \frac{\pi A}{c} \left[(3 - 4\nu) + \frac{(z_i - z_j)^2}{c^2} \right],$$

$$P_{rr}^*\big|_{r_i=0} = P_{rz}^*\big|_{r_i=0} = 0,$$

$$P_{zr}^*\big|_{r_i=0} = \frac{2\pi\mu A(z_i - z_j)}{c^3} \left[2(1+\nu) - \frac{3(z_i - z_j)^2}{c^2} \right] n_r -$$

$$- \frac{2\pi\mu A r_j}{c^3} \left[(1 - 2\nu) + \frac{3(z_i - z_j)^2}{c^2} \right] n_z,$$

$$\qquad (6.89b)$$

$$P_{zz}^*\big|_{r_i=0} = -\frac{2\pi\mu A(z_i - z_j)}{c^3} \left[2(1+\nu) + \frac{3(z_i - z_j)^2}{c^2} \right] n_r +$$

$$+ \frac{2\pi\mu A r_j}{c^3} \left[(1 - 2\nu) + \frac{3(z_i - z_j)^2}{c^2} \right] n_z.$$

For numerical implementation, the BE Eq (6.80), together with (6.81)–(6.89), are written in the matrix form $HU = GP$ which after the collocation technique reduces to the form $AX = F$, as in (6.51) and (6.52) with $B = 0$. Subroutine Bell must be modified accordingly for axisymmetric problems.

Method 2. This is done by using the rotation transformation F (about z–axis)

$$F(\mathbf{x}, \theta_i) = \begin{bmatrix} \cos\theta_i & -\sin\theta_i & 0 \\ \sin\theta_i & \cos\theta_i & 0 \\ 0 & 0 & 1 \end{bmatrix}, F^T(\mathbf{x}', \theta_j) = \begin{bmatrix} \cos\theta_j & \sin\theta_j & 0 \\ -\sin\theta_j & \cos\theta_j & 0 \\ 0 & 0 & 1 \end{bmatrix}$$

(6.90)

for the point $\mathbf{x}$ and $\mathbf{x}'$ respectively. In the absence of body forces, the BI Eq (6.40) reduces to

$$\mathbf{c}(i)\mathbf{u}^{\text{pol}}(i) + \iint_S \left[F^T(\mathbf{x}', \theta_j)\mathbf{P}^* F(\mathbf{x}, \theta_i) \right] \mathbf{u}^{\text{pol}} \, dS$$

$$= \iint_S \left[F^T(\mathbf{x}', \theta_j)\mathbf{u}^* F(\mathbf{x}, \theta_i) \right] \mathbf{p}^{\text{pol}} \, dS$$

(6.91)

where the superscript 'pol' refers to the value of the quantity in the polar coordinate system. If we choose the source point i such that $\theta_j = 0$, then we find that

$$F^T(\mathbf{x}', \theta_j)\mathbf{p}F(\mathbf{x}, \theta_i) =$$
$$= \begin{bmatrix} p_{11}\cos\theta_i + p_{12}\sin\theta_i & -p_{11}\sin\theta_i + p_{12}\cos\theta_i & p_{13} \\ p_{21}\cos\theta_i + p_{22}\sin\theta_i & -p_{21}\sin\theta_i + p_{22}\cos\theta_i & p_{23} \\ p_{31}\cos\theta_i + p_{32}\sin\theta_i & -p_{31}\sin\theta_i + p_{32}\cos\theta_i & p_{33} \end{bmatrix}. \quad (6.92)$$

Note that p_{ij} in (6.92) are still in their cartesian form. The choice of the point i such that θ_j vanishes is therefore important in solving axisymmetric problems. This method also leads to (6.80)–(6.89), but the details are too cumbersome to include here.

Example 6.5:
Consider the problem of stress concentration due to a spherical cavity of radius a in an infinite medium. The assumption of axisymmetry will be valid if we replace the infinite medium by a solid circular cylinder of large radius R and height H. The upper and lower surfaces of the cylinder parallel to the plane $z = 0$ are subjected to a tensile stress σ_0. The exact solution for the axial stress through the midplane of a large cylinder is given by

$$\sigma_{zz} = \sigma_0 \left[1 + \frac{4 - 5\nu}{2(7 - 5\nu)} \left(\frac{R}{r}\right)^3 + \frac{9}{2(7 - 5\nu)} \left(\frac{R}{r}\right)^5 \right].$$

Use the data $a = 0.25, R = 1.0, H = 4.0, \sigma_0 = 1.0, \nu = 0.3$ and the mesh of 32 linear elements shown in Fig. 6.14.

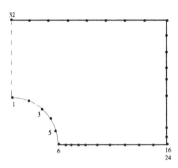

Fig. 6.14.

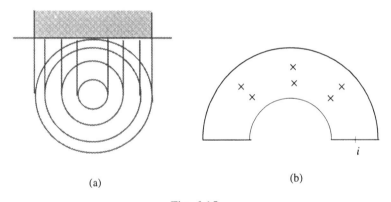

(a)   (b)

Fig. 6.15.

Example 6.6:

In order to investigate the impedance of a rigid foundation in a concrete circular structure (Fig. 6.15), we are interested in computing the stresses when subjected to torsional, vertical, horizontal, or rocking displacements.

The boundary conditions are:

For torsional displacements: $u_r = 0 = u_z, u_\theta = u_\theta^0(r); p_r = 0 = p_z, p_\theta = p_\theta^0(r)$.

For vertical displacements: $u_r = u_r^0(r), u_\theta = 0, u_z = u_z^0; p - r = p_r^0(r), p_\theta = 0, p_z = p_z^0(r)$.

For horizontal and rocking displacements: $u_r = u_r^0 \cos\theta$, $u_\theta = -u_\theta^0 \sin\theta$, $u_z = u_z^0 \cos\theta$; $p_r = p_r^0 \sin\theta$, $p_\theta = -p_\theta^0 \sin\theta$, $p_z = -p_z^0 \cos\theta$.

One–half of the symmetric regions between any two concentric rings can be discretized (Fig. 6.15(b)), using constant, linear or quadratic elements $\tilde{S}_j$, $j = 1, \ldots, N$. If we choose constant elements, then the values of $\mathbf{u}$ and $\mathbf{p}$ are constant, say, $\mathbf{u}^0$ and $\mathbf{p}^0$ respectively, on each boundary element $\tilde{S}_j$. If we further take the point i such that $\theta = 0$, then from (6.91) and (6.92) we obtain the following system of algebraic equations for the case of Fig. 6.15(b) when $0 \le \theta \le \pi$:

$$\frac{1}{2} \begin{Bmatrix} u_r^0 \\ u_\theta^0 \\ u_z^0 \end{Bmatrix} + 2 \iint_S [A] \begin{bmatrix} u_r^0(\mathbf{x}') & u_\theta^0(\mathbf{x}') & u_z^0(\mathbf{x}') \end{bmatrix}^T dS =$$

$$= 2 \iint_S [B] \begin{bmatrix} p_r^0(\mathbf{x}') & p_\theta^0(\mathbf{x}') & p_z^0(\mathbf{x}') \end{bmatrix}^T dS$$

(6.93)

where $[A]$ and $[B]$ are 3×3 matrices given by

$$[A] = \begin{bmatrix} p_{11}\cos\theta + p_{12}\sin\theta & 0 & p_{13} \\ 0 & -p_{21}\sin\theta + p_{22}\cos\theta & 0 \\ p_{31}\cos\theta + p_{32}\sin\theta & 0 & p_{33} \end{bmatrix}, \quad (6.94)$$

$$[B] = \begin{bmatrix} u_{11}\cos\theta + u_{12}\sin\theta & 0 & u_{13} \\ 0 & -u_{21}\sin\theta + u_{22}\cos\theta & 0 \\ u_{31}\cos\theta + u_{32}\sin\theta & 0 & u_{33} \end{bmatrix}, \quad (6.95)$$

integration is performed over the half–ring $0 \le \theta \le \pi$ (hence the factor 2), and $dS = r \, dr \, d\theta$.

Example 6.7:
A discretized model of a thick hollow cylinderical structure (quarter symmetric region) is shown in Fig. 6.16(a) with 94 boundary elements. For the axisymmetric case, this problem can be solved as in Example 6.2. If the cylinder is subject to an internal pressure P, the in the plane stress case the displacements and stresses are given by

$$u(r) = \frac{Pa^2}{E(b^2 - a^2)} \left[(1 - \nu)r + (1 + \nu)\frac{b^2}{r} \right],$$

$$\sigma(r) = \frac{Pa^2}{b^2 - a^2} \left(1 - \frac{b^2}{r^2} \right), \quad \sigma(\theta) = \frac{Pa^2}{b^2 - a^2} \left(1 + \frac{b^2}{r^2} \right), \quad a \le r \le b.$$

Linear elements as in Fig. 5.13(a) are used to create the input file, where the nodal points given below in the polar coordinate system (r, θ) must be changed to the cartesian system for the input file:

10.0	0.0	10.0	$\pi/8$	10.0	$\pi/4$	10.0	$\pi/6$
15.0	0.0	15.0	$\pi/8$	15.0	$\pi/4$	15.0	$\pi/6$
20.0	0.0	20.0	$\pi/8$	20.0	$\pi/4$	20.0	$\pi/6$
30.0	0.0	30.0	$\pi/8$	30.0	$\pi/4$	30.0	$\pi/6$ ∎

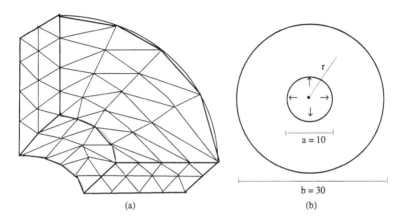

(a)   (b)

Fig. 6.16.

6.9. Problems

6.1. For a two–dimensional linear elastic media, show that

$$\lim_{\varepsilon \to 0} \int_{C_\varepsilon} u_k p_{lk}^* \, ds = -\frac{1}{2} u_l(i), \quad \lim_{\varepsilon \to 0} \int_{C_\varepsilon} p_k u_{lk}^* \, ds = 0,$$

where $C_2 = C_\varepsilon \cup C_{2-\varepsilon}$ represents a semicircle of radius ε and center at the point i. (It is obvious that the boundary $C = C_1 + C_2$ is smooth.)

6.2. Compute the two–dimensional stress field around the circular hole in an infinite plate (Example 6.3).

6.3. Compute the two–dimensional stress field around an elliptic hole in an infinite plate.

6.4. Compute the displacements in a thin uniform rectangular plate loaded by a uniform tensile load on one face as shown in Fig. 6.17. Assume that the plate is made of an isotropic steel with the value of Young's modulus of $E = 3 \times 10^7$ psi and a Poisson's ratio of $\nu = 0.3$.

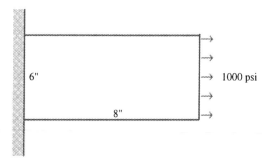

Fig. 6.17. Displacements and Tractions.

6.5. Develop the BE Eq (6.43) for linear boundary elements and write a subroutine for this case.

6.6. Using the plane stress assumption, compute all nodal displacements in a plate made of an isotropic material (Fig. 6.18).

6.7. Consider a thin elastic plate held between two smooth rigid walls and subjected to uniform compression c as shown in Fig. 6.19. Compute the displacements and stresses in the plate.

6.8. Compute the stress field for the torsion problem $-\nabla^2 u = 1$ for the cross–section of the rectangular region in Fig. 6.20.

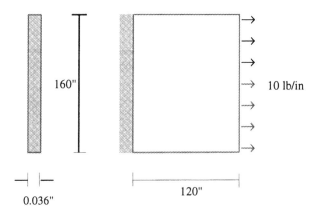

Fig. 6.18.

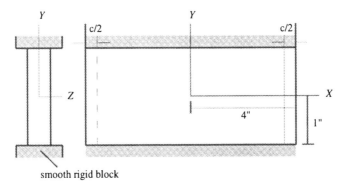

Fig. 6.19.

6.9. Use the subroutine Be2 or Be5 to evaluate the stress field for the torsion problem for the elliptic cross–section of Fig. 5.17.

6.10. Use the boundary element method to compute the stress field $\phi(x, y)$ for the torsion problem for a circular cross–section of radius 5 cm.

6.11. Compute the stress field $\phi(x, y)$ for the torsion problem of the equilateral triangular cross–section of side 30 inches (Fig. 6.21).

6.12. Solve the torsion problem for the square prism, shown in Fig. 6.22. Because of symmetry, discretize the octant of the section since warping u is zero along the coordinate and diagonal axes.

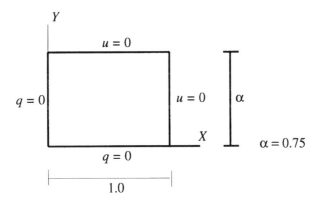

Fig. 6.20. Displacements and Tractions.

6.13.. Solve the torsion problem for a rectangle shown in Fig. 6.23. The

quarter region is analyzed (because of symmetry) and the corner at node
19 has been discretized in a finer manner.

6.14. Solve the torsion problem for an infinite plate with a central hole
(notched plate) as shown in Fig. 6.24. The mixed boundary conditions
are:

$t_1 = 0, t_2 = 0$ at nodes 28, 1, 2 and 11 – 17;
$t_1 = 0, u_2 = 0$ at nodes 3 – 10; $u_1 = 0, t_2 = 0$ at nodes 18 – 24;
$t_1 = 0, t_2 = 1$ at nodes 25 – 27.

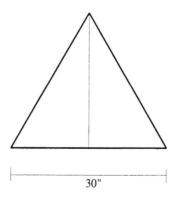

Fig. 6.21. Displacements and Tractions.

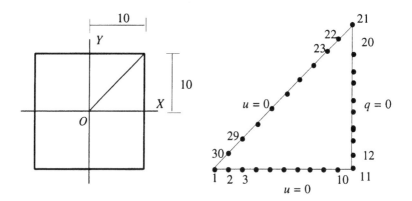

Fig. 6.22. A square prism, and discretization.

6.15. Determine the displacements for the reinforced concrete syphon structure (Fig. 6.25) which is designed to withstand an internal pressure of 6

feet of water and work under an external pressure of 15 feet of water.

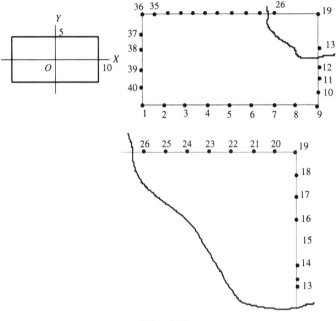

Fig. 6.23.

6.16. Let a thick walled circular cylinder of inner radius R_1 and outer radius R_2, and length sufficiently large so that the linear strain conditions are applicable, be subjected to internal pressure P, with $\nu = 0.3$ (see Fig. 6.26) for the quarter axisymmetric region). Compute the displacements and stresses with the data $R_1 = 3.0$, $R_2 = 6.0$, $P = 1.0$, $\nu = 0.3$, and compare results with the plane strain Lame solution which is

$$u_r = C_1 r + \frac{C_2}{r}, \quad \sigma_{rr} = C_3 - \frac{C_4}{r^2}, \quad \sigma_{\theta\theta} = C_3 + \frac{C_4}{r^2},$$

where $C_1 = 0.1733, C_2 = 15.6, C_3 = 0.3333, C_4 = 12.0$.

6.17. Solve the problem of deformation of a ring (torus) with inner radius R_1 and outer radius R_2, subjected to two diametrically opposite forces of equal magnitude F (see Fig. 6.26 for the quarter region). Assume that the ring is sufficiently thick for linear strain conditions to be applicable. Compute the stress along the x–axis. This problem is of practical interest for computation of stresses in close–coiled helical springs where, because of axisymmetry, we consider the quarter ring sector.

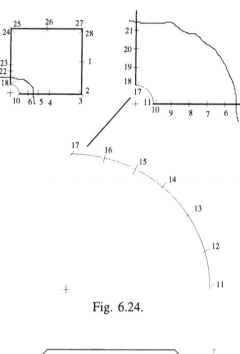

Fig. 6.24.

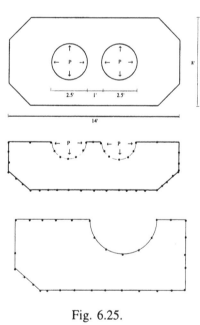

Fig. 6.25.

6.18. Consider a hollow sphere of inner radius R_1 and outer radius R_2, subjected to an internal pressure P_1 and an external pressure P_2. Since the problem is axisymmetric, the BE mesh shown in Fig. 6.27 can be used to determine displacements and stresses. Use the data $R_1 = 1.0$, $R_2 = 2.0$, $P_1 = 3.0$, $P_2 = 5.0$, and $\nu = 0.3$. The exact solution is

$$u_r = \frac{C_1}{r^2} + C_2 r, \quad \sigma_{rr} = \frac{C_3}{r^3} + C_4, \quad \sigma_{\theta\theta} = \frac{C_5}{r^3} + C_6,$$

where $C_1 = 1.4857, C_2 = -1.0857, C_3 = -2.2857, C_4 = -2.7143$, $C_5 = 1.1429, C_6 = -2.7143$.

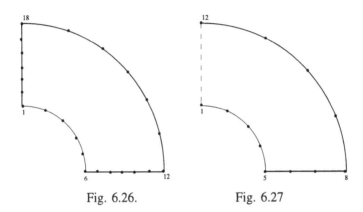

Fig. 6.26.   Fig. 6.27

References and Bibliography for Additional Reading

E. Alarcón, C. A. Brebbia and J. Dominguez, *The boundary element method in elasticity*, Int. J. Mech. Sci. **20** (1978), 625–639.

P. K. Banerjee, *Boundary Element Methods in Engineering*, 2nd Ed., McGraw–Hill, New York, 1993.

A. A. Becker, *The Boundary Element Method in Engineering*, McGraw–Hill, London, 1992.

C. A. Brebbia, *The Boundary Element Method for Engineers*, Pentech Press, London, 1980.

C. A. Brebbia, J. C. F. Telles and L. C. Wrobel, *Boundary Element Techniques – Theory and Applications*, Springer–Verlag, Berlin, 1984.

C. A. Brebbia (Ed.), *Recent Advances in Boundary Elements*, Proc. 1st Conf. on Boundary Element Method, Pentech Press, London, 1978.

C. A. Brebbia and J. Dominguez, *Boundary Elements – An Introductory Course*, Computational Mechanics Publishers, Southampton and McGraw–Hill, New York, 1989.

G. Chen and J. Zhou, *Boundary Element Methods*, Academic Press, London, 1992.

T. A. Cruise and G. J. Meyers, *Two dimensional fracture mechanics analysis*, J. Struct. Div. Proc. ASCE **103** (1977), 309–320.

T. A. Cruse T. A., D. W. Snow and R. B. Wilson, *Numerical solutions in axisymmetric elasticity*, Comput. Structures **7** (1977), 445–451.

M. S. Gómez–Lera and E. Alarcón, *Elastostatics*, Ch. 3 in Boundary Element Methods in Mechanics (D. E. Beskos, Ed.), North–Holland, Amsterdam, 1987.

P. G. Hodge, Jr., *Continuum Mechanics — In Introductory Text for Engineers*, McGraw–Hill, New York, 1970.

M. A. Jaswon and G. T. Symm, *Integral Equation Methods in Potential Theory and Elastostatics*, Academic Press, London, 1977.

———— and A. R. Ponter, *An integral equation solution of the torsion problem*, Proc. Roy. Soc., Ser. A **273** (1963), 237–245.

J. H. Kane, *Boundary Element Analysis in Engineering*, Prentice–Hall, Englewood Cliffs, NJ, 1994.

T. Kermanidis T., *A numerical solution for axially symmetrical elasticity problems*, Int. J. Solids Structures **11** (1975), 493–500.

S. G. Lekhnitski, *Theory of Elasticity of an Anisotropic Elastic Body*, Holden–Day, San Francisco, 1963.

A. E. H. Love, *A Treatise on the Mathematical Theory of Elasticity*, 4th Ed., Cambridge Univ. Press, 1944.

R. D. Mindlin, *Force at a point in the interior of a semi–infinite solid*, Physics **7** (1936), 145ff.

N. I. Muskhelishvilli, *Some Basic Problems in the Mathematical Theory of Elasticity*, Noordhoff, Groningen, 1963.

Y. C. Pan and T. W. Chou, *Point force solution for an infinite transversely isotropic solid*, Trans. ASME, J. Appl. Mech. **43** (1976), 608–612.

————, *Green's function solutions for semi–infinite transversely isotropic materials*, Int. J. Engng. Sci. **17** (1979), 545–551.

H. G. Poulos and E. H. Davis, *Elastic Solutions for Soil and Rock Mechanics*, Wiley, New York, 1974.

J. N. Reddy, *An Introduction to the Finite Element Method*, McGraw–Hill, New York, 1984.

F. J. Rizzo and D. J. Shippy, *A method for stress determination in plane anisotropic bodies*, J. Composite Materials **4** (1970), 36–61.

R. J. Roark and W. C. Young, *Formulas for Stress and Strain*, McGraw–Hill, London, 1975.

I. S. Sokolnikoff, *Mathematical Theory of Elasticity*, McGraw–Hill, 1956.

S. Timoshenko and J. N. Goodier, *Theory of Elasticity*, McGraw–Hill, New York, 1951.

S. K. Vogel and F. J. Rizzo, *An integral equation formulation of three dimensional anisotropic elastostatic boundary value problems*, J. Elasticity **3** (1973), 203–216.

O. C. Zienkiewicz and G. S. Holister G. S. (Eds.), *Stress Analysis,*, Wiley, London, 1966.

7

Ocean Acoustics and Helmholtz Equation

Due to rapid advancement in offshore technology to design and construct offshore structures safely and economically, it is important to understand the interaction of water waves with such structures. Since the offshore drilling platforms in the Gulf of Mexico in the late 1940's, various types of offshore structures, e.g., mobile drilling rigs, submersible platforms, jackup platforms, drill ships or drill barges, and semisubmersible drilling platforms have been constructed. The development and production at offshore sites are primarily carried out with fixed platforms. The most commonly used platforms are the template or jacket structures and their extensions which include space frames that use skirt piles or pile clusters, or enlarged legs to provide for self–buoyancy during installation. Another class of fixed structures are the gravity platforms which depend on their excessive weight rather than piles. The problem of underwater acoustic radiation and scattering by these and other submerged structures is also important for their sonar detection.

7.1. Water Waves

The Navier–Stokes equations for an incompressible Newtonian fluid with constant viscosity are

$$\frac{\partial \mathbf{v}}{\partial t} + (\mathbf{v} \cdot \nabla)\mathbf{v} = -\frac{1}{\rho}\nabla(p + \rho g h) + \nu\nabla^2 \mathbf{v}, \qquad (7.1)$$

where $\mathbf{v} = (u, v, w)$ is the velocity vector in the x, y, z directions; ρ is the density and ν the kinematic viscosity of the fluid, and gh is the body–force potential. The conservation of mass for an incompressible fluid requires that the volumetric dilatation be zero, i.e,

$$\nabla \cdot \mathbf{v} = \frac{\partial u}{\partial x} + \frac{\partial v}{\partial y} + \frac{\partial w}{\partial z} = 0. \tag{7.2}$$

This is also known as the equation of continuity. The solution of the four unknowns (u, v, w, p) from Eqs (7.1) and (7.2) are fully determinate if the initial and boundary conditions are prescribed.

The rates of rotation of a fluid particle about the x, y, z axes are given by

$$\mathbf{\Omega} = \frac{1}{2} \nabla \times \mathbf{v}, \tag{7.3a}$$

or in component form, by

$$\Omega_x = \frac{1}{2}\left(\frac{\partial w}{\partial y} - \frac{\partial v}{\partial z}\right), \quad \Omega_y = \frac{1}{2}\left(\frac{\partial u}{\partial z} - \frac{\partial w}{\partial x}\right), \quad \Omega_z = \frac{1}{2}\left(\frac{\partial v}{\partial x} - \frac{\partial u}{\partial y}\right). \tag{7.3b}$$

Rotation is related to two other concepts which are circulation and vorticity (see §8.2). Circulation Γ in a volume V of the fluid with surface area S is defined by

$$\Gamma = \iiint_V \mathbf{v} \cdot dV = \iiint_V (u\,dx + v\,dy + w\,dz) = \iint_S \nabla \times \mathbf{v} \cdot dS =$$

$$= 2\iint_S \mathbf{\Omega} \cdot \hat{n}\,dS = \iint (\omega_x\,dy\,dz + \omega_y\,dz\,dx + \omega_z\,dx\,dy). \tag{7.4}$$

where the components of the vorticity vector ω are $\omega_x = 2\Omega_x, \omega_y = 2\Omega_y, \omega_z = 2\Omega_z$. Thus, we find that $\Delta\Gamma = \omega_n\,\Delta S$ is the component of the vorticity vector normal to the surface element ΔS, i.e., the flux of vorticity through the surface is equal to the circulation along the boundary of the surface. An irrotational motion exists only when all components of the rotation vector ω are zero, i.e.,

$$\frac{\partial w}{\partial y} - \frac{\partial v}{\partial z} = 0, \quad \frac{\partial u}{\partial z} - \frac{\partial w}{\partial x} = 0, \quad \frac{\partial v}{\partial x} - \frac{\partial u}{\partial y} = 0. \tag{7.5}$$

In this case we can define a continuous, differentiable, scalar function $\phi = \phi(x, y, z, t)$ such that its gradients satisfy Eq (7.5), i.e., $\nabla \times \mathbf{v} = \mathbf{0}$ is the n.a.s.c. for the existence of such a function. The variation of the potential ϕ

along a portion of a streamline of length δs is $\delta \phi = \dfrac{\partial \phi}{\partial s} \delta s$. Thus, the velocity is given by the gradient of the potential which is known as the velocity potential. Since

$$\mathbf{v} = \nabla \cdot \phi, \qquad (7.6)$$

the velocity components are given in cartesian coordinates by

$$u = \frac{\partial \phi}{\partial x}, \quad v = \frac{\partial \phi}{\partial y}, \quad w = \frac{\partial \phi}{\partial z}, \qquad (7.7a)$$

and in cylindrical polar coordinates by

$$v_r = \frac{\partial \phi}{\partial r}, \quad v_\theta = \frac{1}{r}\frac{\partial \phi}{\partial \theta}, \quad w = \frac{\partial \phi}{\partial z}. \qquad (7.7b)$$

Since (7.7a, b) satisfy (7.5) automatically, the potential flow is irrotational.

As a consequence of the continuity equation (7.2) the velocity potential ϕ satisfies the Laplace equation

$$\nabla^2 \phi = \frac{\partial^2 \phi}{\partial x^2} + \frac{\partial^2 \phi}{\partial y^2} + \frac{\partial^2 \phi}{\partial z^2} = 0. \qquad (7.8)$$

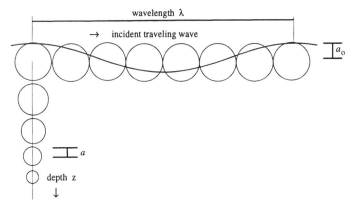

Fig. 7.1. Surface wave travelling from left to right.

Internal waves, generated inside the ocean medium, are similar to surface waves, except that they are gravity waves with varying vertical displacement depending on variations in density with depth and temperature or between two water masses of different densities. The motion of internal waves can be noticed both below and above this interface. The material velocity variations in these waves have, in general, a lesser acoustic effect than temperature variations. The vertical and horizontal components of the material velocity for a long–crested wave in deep water are usually totally orbital near the water surface (Fig. 7.1). The depth dependence of the displacement amplitude a is given by $a = a_0\, e^{-2\pi z/\lambda}$, where a_0 is the incident wave amplitude at the surface, z the depth, and λ the surface wavelength. Besides this type of well defined underwater motion, there always is the turbulent motion which is non-isotropic (independent of direction). There also exists an isotropic turbulent motion which depends on depth and surface conditions. We shall, however, not study turbulent motion at all.

7.2. Simple Amplitude Wave Theory

We shall first consider a two–dimensional wave interaction problem. Let the cartesian coordinate system be defined as follows: The x–axis is taken in the direction of wave propagation, z is measured upwards from the still water level and y is perpendicular to both x and z axes. We will assume that there are two–dimensional waves in the (x, z)–plane, which are progressive in the positive x–direction and propagate over a smooth water bed of constant undisturbed depth d; also that the waves preserve their form, that there are no underlying currents, and that the free water surface is free from any surface tension (i.e., it is uncontaminated). Under these assumptions the general form of a wave train is shown in Fig. 7.2.

Note that h is the wave height (vertical distance from crest to trough), λ the wavelength (distance between successive crests or troughs), T the wave period (time interval between successive crests, or troughs, passing a particular point), and $c = \lambda/T$ the wave speed. Then the wave angular frequency is given by $\omega = 2\pi/T$ and the wave number by $k = 2\pi/\lambda$, thus $c = \omega/k$. Dimensionless quantities are defined as follows: Wave height is expressed as h/gT^2, or h/λ (wave steepness), or h/d (relative height); the water depth is expressed as d/gt^2, or kd, or d/λ; the Ursell number U is defined by $U = h\lambda^2/d^3$ for steeper waves in shallow water.

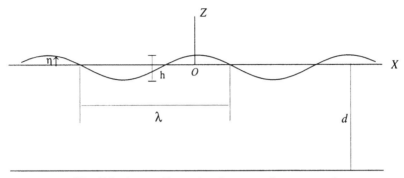

Fig. 7.2. A two–dimensional progressive wave train.

The existence of the velocity potential ϕ and the validity of the Laplace equation follow from the assumptions of an irrotational flow in an incompressible fluid. Thus, the velocity potential ϕ satisfies the Laplace equation (7.8), i.e.,

$$\frac{\partial^2 \phi}{\partial x^2} + \frac{\partial^2 \phi}{\partial z^2} = 0, \tag{7.9}$$

with the boundary conditions

$$\frac{\partial \phi}{\partial z} = 0 \quad \text{at } z = -d, \tag{7.10a}$$

$$\frac{\partial \eta}{\partial t} + \frac{\partial \phi}{\partial x}\frac{\partial \eta}{\partial x} - \frac{\partial \phi}{\partial z} = 0 \quad \text{at } z = \eta, \tag{7.10b}$$

$$\frac{\partial \phi}{\partial t} + \frac{1}{2}\left[\left(\frac{\partial \phi}{\partial x}\right)^2 + \left(\frac{\partial \phi}{\partial z}\right)^2\right] + g\eta = f(t) \quad \text{at } z = \eta, \tag{7.10c}$$

$$\phi(x, z, t) = \phi(x - ct, z), \tag{7.10d}$$

where $\eta(x, t)$ is the free surface elevation measured above the still water level at $z = 0$. It should be noted that the boundary condition (7.10a) corresponds to the condition at the seabed which imposes a zero vertical component on the fluid particle velocity at the seabed; the boundary conditions (7.10b) and (7.10c) denote the kinematic and dynamic free surface boundary conditions respectively; thus (7.10b) means that the fluid particle velocity normal to the free surface is equal to the velocity of the free surface in that direction, whereas (7.10c) says that the pressure at the free surface, expressed in terms of the Bernoulli equation, is constant (this follows from the assumption that the atmospheric pressure immediately above the fluid surface is itself constant and the free surface has zero surface tension); the boundary condition (7.10d) expresses the periodic nature of the wave train. It should be noted that the above conditions are based on assumptions that are seldom justified in real situations. Some of the most objectionable assumptions are: (i) there is no underlying current; (ii) the depth is constant, and (iii) the wave train is two–dimensional and of permanent form. However, these assumptions are necessary to develop various wave theories, the simplest of which is the small amplitude wave theory which we shall examine further.

For small amplitude waves, the free surface boundary conditions (7.10b, c) reduce to

$$\frac{\partial \eta}{\partial t} - \frac{\partial \phi}{\partial z} = 0 \quad \text{at } z = 0, \tag{7.11b}$$

$$\frac{\partial \phi}{\partial t} + g\eta = 0 \quad \text{at } z = 0. \tag{7.11c}$$

Combining (7.11b) and (7.11c) we get

$$\frac{\partial^2 \phi}{\partial t^2} + g\frac{\partial \phi}{\partial z} = 0 \quad \text{at } z = 0, \tag{7.12}$$

and

$$\eta = -\frac{1}{g} \frac{\partial \phi}{\partial t}\bigg|_{z=0}. \tag{7.13}$$

Since the boundary condition (7.10d) expresses a periodicity condition, the solution to the problem (7.9)–(7.13) can be found by using the separation of variables method. Thus, we assume that

$$\phi(x, z, t) = Z(z)X(x - ct), \tag{7.14}$$

which after substitution into Eq (7.9) gives the following two equations:

$$\frac{\partial^2 Z}{\partial z^2} - k^2 Z = 0, \tag{7.15}$$

$$\frac{\partial^2 X}{\partial x^2} + k^2 X = 0, \tag{7.16}$$

where the sign of k^2 is chosen such that it gives a periodic (rather than a hyperbolic) solution in $(x - ct)$. The general solution Z and X of (7.15) and (7.16) are found to be

$$Z = A_1 \cosh kz + A_2 \sinh kz,$$
$$X = A_3 \cos k(x - ct) + A_4 \sin k(x - ct). \tag{7.17}$$

The constants A_1, A_2, A_3, A_4 are determined from the boundary conditions. Let us define the time $t = 0$ when a wave crest crosses the plane $x = 0$. Then, in view of (7.13), we must have $A_3 = 0$. The boundary condition (7.10a) at seabed implies that $A_2 = A_1 \tanh kd$. Substituting the expressions for Z and X into (7.14) we obtain

$$\phi = A\frac{\cosh k(z+d)}{\cosh kd} \sin k(x-ct), \qquad (7.18)$$

where the constant $A = A_1 A_4$. This constant A can be determined by using Eq (7.13) to give η by relating the range in η to the wave height h, thus yielding $A = gh/2kc$. It is clear from the solution (7.18) that the velocity potential is periodic in the x direction with wavelength $\lambda = 2\pi/k$, wave period $T = 2\pi/ck$, and wave number $c = \omega/k = \lambda/T$. The remaining boundary condition (7.12) can now be used to express c in terms of k, which on substitution of (7.18) into (7.12) gives

$$\omega^2 = gk\tanh kd, \quad \text{or} \quad c^2 = \frac{g}{k}\tanh kd. \qquad (7.19)$$

This is the so called linear dispersion relation. It describes the increase in wave speed with the wavelength. Hence, in view of (7.19), the velocity potential defined by (7.18) can also be written as

$$\begin{aligned}\phi &= \frac{gh}{2\omega}\frac{\cosh k(z+d)}{\cosh kd} \sin\theta, \\ &= \frac{\pi h}{kT}\frac{\cosh k(z+d)}{\cosh kd} \sin\theta,\end{aligned} \qquad (7.20)$$

where $\theta = k(x-ct) = kx-\omega t$. If we take the solution of (7.16) in the complex form as $X = Be^{-ik(x-ct)}$, then the velocity potential (7.20) is written as

$$\phi = \Re\frac{gh}{2\omega}\frac{\cosh k(z+d)}{\cosh kd} e^{-ik(x-ct)} =$$

$$\Re\frac{gh}{2\omega}\frac{\cosh k(z+d)}{\cosh kd} e^{-i\omega t+ikx}. \qquad (7.21)$$

Let us now consider a three–dimensional wave interaction problem as described in Fig. 7.3.

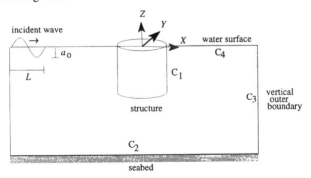

Fig. 7.3. A three–dimensional wave interaction problem.

Again, the velocity potential ϕ is defined by (7.6) and the continuity equation satisfies the Laplace equation (7.8). As in the boundary condition (7.12), the water surface condition for small amplitude waves is given by

$$\frac{\partial^2 \phi}{\partial t^2} + g\frac{\partial \phi}{\partial n} = 0. \tag{7.22}$$

As in (7.11b), a moving–body surface will produce a normal velocity which is expressed as

$$\hat{n} \cdot \mathbf{v} - \frac{\partial \phi}{\partial n} = 0, \tag{7.23}$$

where $\mathbf{v}$ is the velocity vector of the structure. As in the derivation of (7.21), the potential ϕ_0 in the fluid for the incident wave can be found as

$$\phi_0 = \Re\frac{ga_0}{\omega}\frac{\cosh k(z+d)}{\cosh kd}\, e^{-i\omega t + i(k_1 x + k_2 y)}, \tag{7.24}$$

where a_0 is the surface incident wave amplitude, k_1 and k_2 the wave number components in the x and y directions with $k_1 = k\cos\alpha$, $k_2 = k\sin\alpha$, α being the angle of the incident wave with the x–axis, and k satisfies the dispersion expression (7.19).

The presence of the structure in Fig. 7.2 will disturb the incident waves and as a result the total potential of the fluid can be written in terms of the incident and diffracted wave potentials as

$$\phi = \phi_0 + U, \tag{7.25}$$

where

$$U = \Re\left(u\, e^{-i\omega t}\right). \tag{7.26}$$

Here u is the diffracted wave potential which is governed by the equation of continuity

$$\nabla^2 u = 0, \tag{7.27}$$

and satisfies the following boundary conditions:

1. On the surface C_1 of the structure:

$$q + \frac{\partial \phi_0}{\partial n} = \hat{n} \cdot \mathbf{v}. \tag{7.28}$$

2. On the seabed C_2:

$$q = 0. \tag{7.29}$$

3. On the vertical outer boundary C_3 which is sufficiently far from the structure, the diffracted outward wave satisfies the radiation condition:

$$q - iku = 0. \qquad (7.30)$$

4. On the undisturbed water surface C_4:

$$q - \frac{\omega^2}{g} u = 0, \qquad (7.31)$$

where $q = \partial u / \partial n$. The boundary value problem (7.27)–(7.31) can be solved by the boundary element method described in Chapter 5 for the Laplace equation.

In order to compute the wave forces, we proceed as follows: After calculating the potential u around the structure, we determine the pressure p by using the linearized Bernoulli equation

$$p = -\rho \frac{\partial}{\partial t} \phi, \qquad (7.32)$$

where ρ is the density of the fluid, and ϕ is given by (7.25). After the pressures at various points on the structure are known, the wave forces F acting on the structure can be evaluated by integration over the surface of the structure; thus,

$$F = \Re\{f e^{-i\omega t}\}, \qquad (7.33)$$

where

$$\{f\} = \left\{ \begin{matrix} f_x \\ f_y \end{matrix} \right\} = i\rho\omega \frac{\tanh kd}{k} \int_{C_1} \left\{ \begin{matrix} n_x \\ n_y \end{matrix} \right\} \phi \, ds. \qquad (7.34)$$

Finally, the water surface elevation D at a point is given by

$$D = \Re\left\{ \frac{i\omega}{g}(\phi_0 + u)e^{-i\omega t} \right\}. \qquad (7.35)$$

7.3. Helmholtz Equation

Consider the wave equation

$$v_{tt} = c^2 \nabla^2 v. \qquad (7.36)$$

If we set $v(x, y, z) = u(x, y, z) e^{-i\omega t}$, then the wave Eq (7.36) reduces to the Helmholtz equation

$$\left(\nabla^2 + k^2\right)u = 0, \qquad (7.37)$$

where $k = \omega/c$ is the wave number. As derived in Chapter 4, the fundamental solution u^* of the Helmholtz equation in a two–dimensional region is given by (4.42*), and in a three–dimensional region by (4.43*). Different solutions of the Helmholtz equation and the related wave phenomena will not be discussed here.

For a three–dimensional Helmholtz equation in a region V, the weak variational formulation with the fundamental solution u^* leads to

$$\begin{aligned}
0 &= \iiint_V \left[\left(\nabla^2 + k^2\right) u\right] u^* \, dV \\
&= \iiint_V u \left(\nabla^2 + k^2\right) u^* \, dV - \iint_S uq^* \, dS + \iint_S u^* q \, dS,
\end{aligned} \qquad (7.38)$$

where S is the boundary surface of the volume V. Since u^* is the solution of the equation $(\nabla^2 + k^2)u^* + \delta(i) = 0$, we find that

$$\iiint_V u \left(\nabla^2 + k^2\right) u^* \, dV = - \iiint_V u\delta(i) \, dV = -u(i), \qquad (7.39)$$

and thus (7.38) leads to

$$c(i)u(i) + \iint_S uq^* \, dS = \iint_S u^* q \, dS. \qquad (7.40)$$

Again, by considering a hemisphere S_ε of radius ε (Fig. 5.1(b)) and assuming that the surface S_2 (say) is smooth, we can take the point i on S_2 at the center of the hemisphere S_ε, divide S_2 into two parts $S_{2-\varepsilon}$ and S_ε, and let $\varepsilon \to 0$. Then, using (4.43*),

$$q^* = \frac{d}{dr} \frac{1}{4\pi r} e^{-ikr} = -\frac{1}{4\pi}\left[\frac{ik}{r} + \frac{1}{r^2}\right]e^{-ikr},$$

and we find that

$$\lim_{\varepsilon \to 0} \int_{S_\varepsilon} uq^* \, dS = -\frac{u}{4\pi} \lim_{\varepsilon \to 0}\left(\frac{ik}{\varepsilon} + \frac{1}{\varepsilon^2}\right) e^{-ik\varepsilon} \int_{S_\varepsilon} dS = -\frac{u}{2}, \qquad (7.41)$$

$$\lim_{\varepsilon \to 0} \int_{S_\varepsilon} qu^* \, dS = -\frac{q}{4\pi} \lim_{\varepsilon \to 0} \frac{1}{\varepsilon} e^{-ik\varepsilon} \int_{S_\varepsilon} dS = 0. \qquad (7.42)$$

Hence in the three–dimensional case the BI Eq is given by

$$\frac{u(i)}{2} + \iint_S uq^* \, dS = \iint_S u^* q \, dS, \tag{7.43}$$

or, in general, by

$$c(i)u(i) + \iint_S uq^* \, dS = \iint_S u^* q \, dS, \tag{7.44}$$

where $c(i)$ has the same values as in (5.15).

If we consider the general three–dimensional Helmholtz equation

$$\rho\nabla \cdot \left(\rho^{-1}\nabla u\right) + k^2 u = 0, \quad \text{on } V, \tag{7.45}$$

then, writing this equation as $\nabla \cdot \left(\rho^{-1}\nabla u\right) + \rho^{-1}k^2 u = 0$, the weak variational formulation with the function u^* gives

$$
\begin{aligned}
0 &= \iiint_V \left[\nabla \cdot \left(\rho^{-1}\nabla u\right) + \rho^{-1}k^2 u\right] u^* \, dV \\
&= -\iiint_V \rho^{-1}\left[\nabla u^* \cdot \nabla u - k^2 u u^*\right] dV + \iint_S \rho^{-1} u^* \frac{\partial u}{\partial n} \, dS \\
&= \iiint_V \rho^{-1} u (\nabla^2 + k^2) u^* \, dV - \iint_S \rho^{-1} u \frac{\partial u^*}{\partial n} \, dS + \iint_S \rho^{-1} u^* \frac{\partial u}{\partial n} \, dS \\
&= \iiint_V u\left[\nabla \cdot \left(\rho^{-1}\nabla u^* + \rho^{-1}k^2 u^*\right)\right] dV - \iint_S u\rho^{-1} q^* \, dS+ \\
&\quad + \iint_S u^* \rho^{-1} q \, dS.
\end{aligned}
\tag{7.46}
$$

If u^* is the fundamental solution of

$$\nabla \cdot \left(\rho^{-1}\nabla u^*\right) + \rho^{-1}k^2 u^* + \delta(i) = 0,$$

then, as in (7.39),

$$\iint_V u\left[\nabla \cdot \left(\rho^{-1}\nabla u^*\right) + \rho^{-1}k^2 u^*\right] = -u(i).$$

Further, it is easy to see that $\lim_{\varepsilon \to 0} u\rho^{-1} q^* \, dS = -u(i)/2$, and $\lim_{\varepsilon \to 0} u^* \rho^{-1} q \, dS = 0$. This leads to the general three–dimensional BI Eq

$$c(i)u(i) + \iint_S u\rho^{-1} q^* \, dS = \iint_S u^* \rho^{-1} q \, dS. \tag{7.47}$$

For a two–dimensional region R with boundary C, the weak variational formulation is

$$
\begin{aligned}
0 &= \iint_R \left[(\nabla^2 + k^2)u \right] u^* \, dx \, dy \\
&= \iint_R u \left(\nabla^2 + k^2 \right) u^* \, dx \, dy - \int_C uq^* \, ds + \int_C u^* q \, ds,
\end{aligned}
\tag{7.48}
$$

which, after using the translation property (1.55), yields the BI Eq

$$
u(i) + \int_C uq^* \, ds - \int_C u^* q \, ds = 0,
\tag{7.49}
$$

which is similar to (5.5) for the case of two–dimensional Laplace equation, with the only difference that u^* in (7.49) is now defined by (4.42*). Using the same technique as in deriving the result (5.14), we consider a semicircle of radius ε on the boundary of R as in Fig. 5.1(a). Assuming that the boundary C is smooth and the point i is on the C_2 portion of C and at the center of this semicircle, we divide C_2 into two parts C_ε and $C_{2-\varepsilon}$, and let $\varepsilon \to 0$. Then, since

$$
\begin{aligned}
q^* &= -\frac{i}{4} \frac{d}{dr} H_0^{(2)}(kr) \\
&= -\frac{i}{4} \frac{d}{dr} \left[J_0(kr) - iY_0(kr) \right] = \frac{ik}{4} \left[J_1(kr) - iY_1(kr) \right],
\end{aligned}
\tag{7.50}
$$

we find that

$$
\begin{aligned}
\lim_{\varepsilon \to 0} \int_{C_\varepsilon} uq^* \, ds &= -\lim_{\varepsilon \to 0} \frac{uik}{4} \left[J_1(k\varepsilon) - iY_1(k\varepsilon) \right] \int_{C_\varepsilon} ds \\
&= -\frac{ui\pi}{4} \left[\lim_{\varepsilon \to 0} k\varepsilon J_1(k\varepsilon) - i \lim_{\varepsilon \to 0} k\varepsilon Y_1(k\varepsilon) \right] \\
&= \left(-\frac{ui\pi}{4} \right) \left(-\frac{2i}{\pi} \right) = -\frac{u}{2}.
\end{aligned}
\tag{7.51}
$$

The details for finding the limits in (7.51) are as follows: Since $J_1(0) = 0$, and $Y_1(0) = -\infty$, the first limit is obviously $\lim_{z \to 0} zJ_1(z) = 0$. For the second limit in (7.51), we first note that

$$
\begin{aligned}
Y_1(z) = &-\frac{2}{\pi z} \sum_{k=0}^{1} \frac{(-k)!}{k!} (z^2/4)^k + \frac{2}{\pi} \ln(z/2) J_1(z) - \\
&\frac{z}{2\pi} \sum_{k=0}^{\infty} \{ \psi(k+1) + \psi(k+2) \} \frac{(-z^2/4)^k}{k!(k+1)!},
\end{aligned}
\tag{7.52}
$$

where $\psi(z) = \dfrac{d}{dz} \ln \Gamma(z) = \Gamma'(z)/\Gamma(z)$ is the Digamma function. Now using (7.52) we find that

$$\lim_{z \to 0} z Y_1(z) = -\frac{2}{\pi}. \qquad (7.53)$$

In the same manner, we find that

$$\begin{aligned}
\lim_{\varepsilon \to 0} \int_{C_\varepsilon} u^* q \, ds &= - \lim_{\varepsilon \to 0} \frac{iq}{4} H_0^{(2)}(k\varepsilon) \int_{C_\varepsilon} ds \\
&= -\frac{iq\pi}{4} \lim_{\varepsilon \to 0} \varepsilon H_0^{(2)}(k\varepsilon) \qquad (7.54) \\
&= -\frac{iq\pi}{4} \left[\lim_{\varepsilon \to 0} \varepsilon J_0(k\varepsilon) - i \lim_{\varepsilon \to 0} \varepsilon Y_0(k\varepsilon) \right] = 0.
\end{aligned}$$

The first limit in (7.54) is obvious. For the second limit, we use

$$\begin{aligned}
Y_0(z) = \frac{2}{\pi} \left\{ \ln(z/2) + \gamma \right\} J_0(z) &+ \frac{2}{\pi} \left\{ \frac{z^2/4}{(1!)^2} - \right. \\
- \left(1 + \frac{1}{2}\right) \frac{(z^2/4)^2}{(2!)^2} &\left. + \left(1 + \frac{1}{2} + \frac{1}{3}\right) \frac{(z^2/4)^3}{(3!)^2} + \cdots \right\}. \qquad (7.55)
\end{aligned}$$

Then it is evident that $\lim_{z \to 0} z Y_0(z) = 0$.

Taking the values of these limits into account, Eq (7.43) leads to the two–dimensional BI Eq for the Helmholtz equation as

$$\frac{u(i)}{2} + \int_C u q^* \, ds = \int_C u^* q \, ds, \qquad (7.56)$$

which is similar to (5.13), with the difference that u^* is defined by (4.42*). As in §5.1, we can, in general, write the two–dimensional BI Eq (7.56) as

$$c(i) u(i) + \int_C u q^* \, ds = \int_C u^* q \, ds, \qquad (7.57)$$

where $c(i)$ is defined by (5.15).

7.4. BEM: Helmholtz Equation

We will discretize the boundary C into n elements of which n_1 belong to C_1 and n_2 to C_2, and consider the case of constant elements (see Fig. 5.4(a)).

This means that the values of u are constant on each boundary element and equal to the prescribed value at the mid–node. As in Chapter 5, the discretized form of Eq (7.57) at a node i is

$$c(i)u(i) + \sum_{j=1}^{n} \hat{H}_{ij} u_j = \sum_{j=1}^{n} G_{ij} q_j, \tag{7.58}$$

where

$$\hat{H}_{ij} = \int_{C_j} q^* \, ds, \quad G_{ij} = \int_{C_j} u^* \, ds. \tag{7.59}$$

Note that the integral over C_j relates the node i to the element with node j over which the integral is evaluated. If we set H_{ij} as in Chapter 5, then we can write Eq (7.58) in matrix form as

$$HU = GQ. \tag{7.60}$$

Note that n_1 values of u and n_2 values of q are known on C, where $n = n_1 + n_2$. We can reorder the above matrix equation such that all the unknown quantities (viz., n_2 of u and n_1 of q) are on the left side. This leads to the matrix equation

$$AX = F, \tag{7.61}$$

where X is the vector of unknown u and q. Once this equation is solved, we can find the value of u at an interior point by using the discretized form of Eq (7.56) which is

$$u(i) = \sum_{j=1}^{n} \left[q_j G_{ij} - u_j \hat{H}_{ij} \right]. \tag{7.62}$$

The internal fluxes $q_x = \partial u / \partial x$ and $q_y = \partial u / \partial y$ can be computed by differentiating Eq (7.49), i.e.,

$$\begin{aligned} q_x(i) &= \int_{\tilde{C}} q \frac{\partial u}{\partial x} \, ds - \int_{\tilde{C}} q \frac{\partial q^*}{\partial x} \, ds \\ q_y(i) &= \int_{\tilde{C}} q \frac{\partial u^*}{\partial y} \, ds - \int_{\tilde{C}} u \frac{\partial q^*}{\partial x} \, ds. \end{aligned} \tag{7.63}$$

To evaluate the integrals, note that $\hat{H}_{ii} = 0$ for node i, since $\hat{n}$ and r are orthogonal, i.e.,

$$H_{ii} = \int_{C_i} q^* \, ds = \int_{C_i} \frac{\partial u^*}{\partial n} \, ds = \int_{C_i} \frac{\partial u^*}{\partial r} \frac{\partial r}{\partial r} \, ds = 0.$$

To compute G_{ii}, we use Fig. 5.3, and set $r = L_i\xi/2$. Then,

$$
\begin{aligned}
G_{ii} &= \int_{C_i} u^* \, ds = 2 \int_0^1 u^* \, d\xi \\
&= -\frac{i}{2} \int_0^1 H_0^{(2)}(kL_i\xi/2)\,(L_i/2)\,d\xi \\
&= \frac{iL_i}{4} \int_0^1 \left[-J_0(kL_i\xi/2) + iY_0(kL_i\xi/2) \right] d\xi.
\end{aligned}
\tag{7.64}
$$

Now using polynomial approximations, we obtain

$$
\begin{aligned}
-J_0(x) + iY_0(x) &= -J_0(x) + \frac{2i}{\pi}\left\{ \ln\!\left(\frac{x}{2}\right) + \gamma \right\} J_0(x) + \frac{2i}{\pi}\frac{x^2}{2^2} - \\
&\quad - \frac{3}{2}\frac{x^4}{2^2 4^2} + \frac{11}{6}\frac{x^6}{2^2 4^2 6^2} + \cdots \bigg\} \\
&\equiv -J_0(x) + \frac{2i}{\pi}\left\{ \gamma J_0(x) + \frac{x^2}{4} - \frac{3x^4}{128} + \frac{11x^6}{13824} \right\} + \\
&\quad + \frac{2i}{\pi} \ln\!\left(\frac{x}{2} J_0(x)\right).
\end{aligned}
\tag{7.65}
$$

Notice that the last term in (7.65) has a logarithmic singularity, while the remaining part is regular. This regular part can be evaluated by using the Gauss quadrature formula. For the singular part, we set $x/2 = az$, so that the singular part becomes

$$
\begin{aligned}
\frac{2i}{\pi} \ln\!\left(\frac{x}{2} J_0(x)\right) &\equiv \frac{2i}{\pi} \ln\!\left(\frac{x}{2}\left\{ 1 - \frac{x^2}{4} + \frac{x^4}{64} - \frac{x^6}{2304} \right\} \right) \\
&= \frac{2i}{\pi}\left\{ 1 - a^2 z^2 + \frac{a^4 z^4}{4} - \frac{a^6 z^6}{6^2} \right\} \ln az.
\end{aligned}
\tag{7.66}
$$

The contribution to G_{ii} from this singular part is then given by

$$
\begin{aligned}
G_{ii}^{\text{sing}} &= \frac{2i}{\pi} \int_0^1 J_0(kL_i\xi/2) \ln(kL_i\xi/4)\,d\xi \\
&= \frac{2i}{\pi}\left[\ln\!\left(\frac{kL_i}{4}\right) - 1 - \frac{(kL_i)^2}{2^4}\left\{ \frac{\ln(kL_i/4)}{3} - \frac{1}{9} \right\} + \right.\\
&\quad \left. + \frac{(kL_i)^4}{2^{10}}\left\{ \frac{\ln(kL_i/4)}{5} - \frac{1}{25} \right\} - \frac{(kl_i)^6}{2^{14}3^2}\left\{ \frac{\ln(kL_i/4)}{7} - \frac{1}{49} \right\} \right].
\end{aligned}
\tag{7.67}
$$

Hence,

$$
G_{ii} = G_{ii}^{\text{reg}} + G_{ii}^{\text{sing}}.
\tag{7.68}
$$

Note that if the field point i is far away from the source point, i.e., when $\mathbf{r} = |\mathbf{x} - \xi|$ is large, then, instead of the polynomial approximations for J_0 and Y_0, we can use the asymptotic expansion for $H_0^{(2)}(x)$ for large x, i.e.,

$$H_0^{(2)}(x) = \sqrt{\frac{2}{\pi x}} e^{-i(x-\pi/4)} \qquad (7.69)$$

in (7.64) and evaluate G_{ii} in this special case.

7.5. Underwater Acoustic Scattering

Sound waves in a fluid medium are produced by the vibration of a solid body in contact with the fluid, or by vibrating forces — natural or mechanical — acting directly on the fluid, or by the violent motion of the fluid imposed by external forces, or by oscillatory thermal effects, or by an obstacle. In each case energy is transferred from the source to the fluid, and a motion of one part of the fluid medium is transmitted to its other parts. But the frontal region of a sound wave at one instant may be regarded as the source for subsequent wave propagation, although no new energy is being transmitted in this case. In most situations the energy that is originally nonacoustic is changed into acoustic energy at the source which, in the acoustic sense, is a region of space in the fluid medium or in contact with it.

We shall assume that the fluid medium exterior to the source region is infinite in extent, initially uniform and at rest, and the acoustic pressure generated outside the source is small enough to produce linear equations. Further, the motion generated by the source is assumed to be isotropic, i.e., it has no preferred direction. It will produce a wave which propagates spherically outward. When a sound wave encounters an obstacle, a part of the wave is absorbed by it and another part is deflected from its original course, thus causing a scattering. For example, when a plane or spherical wave strikes an obstacle in its path, in addition to the undisturbed wave there is a scattered wave which spreads out from the obstacle in all directions, distorting and interfering with the original wave. In such a situation the obstacle is called a scatterer. We note in passing that if the scatterer is very large compared to the wavelength, as is the case of light waves, half of the scattered wave spreads out more or less uniformly in all directions and the remaining half is concentrated behind the scatterer, creating a sharp–edged shadow there. This happens in geometric

optics, and the half scattered wave is called the reflected wave while the one
creating the shadow is called the interfering wave. Our interest lies in the
other case when the scatterer is very small compared to the wavelength, as
in the case of sound waves; then all scattered waves are propagated out in all
directions, and there is no interfering wave at all. In all intermediate cases
where the scatterer is about the size of the wavelength, there appears a variety
of interference phenomena.

Let us consider a closed surface S which incloses a volume. We can
distinguish between the interior and exterior regions relative to the surface S
(scatterer) and denote them by V_{int} and V_{ext} respectively. It is obvious that
$S = V_{int} \cap V_{ext}$, where V_{ext} is unbounded, and $V_{int} \cup S \cup V_{ext}$ form the entire
space. The governing equation for the acoustic pressure amplitude p in V_{int} or
V_{ext} (or both) is given by the Helmholtz equation (with time dependence of
the form $e^{-i\omega t}$)

$$\left(\nabla^2 + k^2\right) p = 0, \tag{7.70}$$

which is Eq (7.37) with u replaced by p. There are two problems: (i) The
interior problem, where Eq (7.70) is valid in V_{int} and we seek the values of
p at interior points and relate them to the normal derivative along the surface
S; (ii) The exterior problem, where Eq (7.70) holds throughout V_{ext} and we
seek the values of p at exterior points and relate them to the normal derivative
along the surface S. The geometry of both of these problems is represented
in Fig. 7.4.

For the exterior problem, the law of causality requires that p should satisfy
the Sommerfeld radiation condition at a large distance r from the source in
V_{ext}, i.e.,

$$\lim_{r\to\infty} r\left(\frac{\partial p}{\partial r} - ikp\right) = 0. \tag{7.71}$$

We shall apply the result of Problem 1.2 to Eq (7.70) and obtain

$$\nabla \cdot (G\nabla p) - \nabla G \cdot \nabla p = G\nabla \cdot \nabla p,$$
$$\nabla \cdot (p\nabla G) - \nabla p \cdot \nabla G = p\nabla \cdot \nabla G,$$

which, on subtracting one from the other, yields

$$\nabla \cdot (G\nabla p) - \nabla \cdot (p\nabla G) = G\nabla \cdot \nabla p - p\nabla \cdot \nabla G$$
$$= G\nabla^2 p - p\nabla^2 G$$
$$= G\left(\nabla^2 + k^2\right)p - p\left(\nabla^2 + k^2\right)G, \tag{7.72}$$

where G is a function of space variables.

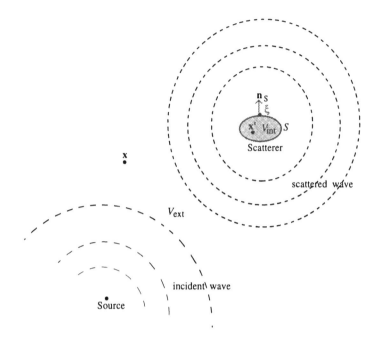

Fig. 7.4. Underwater Acoustic Scattering.

Let $\mathbf{n}_S$ be an outward normal to the surface S. Then, for the interior problem, since $\left(\nabla^2 + k^2\right) p_{\text{int}} = 0$ in V_{int}, we find from (7.72) that

$$\nabla \cdot (G\nabla p) - \nabla \cdot (p\nabla G) = -p \left(\nabla^2 + k^2\right) G. \qquad (7.73)$$

We apply the divergence theorem (1.5c) to (7.73) and obtain

$$-\iiint_{V_{\text{int}}} p_{\text{int}} \left(\nabla^2 + k^2\right) G \, dV_{\text{int}} = \iint_S (G\nabla p_{\text{int}} - p_{\text{int}}\nabla G) \cdot \mathbf{n}_S \, dS. \qquad (7.74)$$

For the exterior problem, where the normal $\mathbf{n}_S$ still points outward into V_{ext}, we notice that $\left(\nabla^2 + k^2\right) p_{\text{ext}} = 0$ in V_{ext}. Then, applying (1.5c) to (7.73), we get

$$-\iiint_{V_{\text{ext}}} p_{\text{ext}} \left(\nabla^2 + k^2\right) G \, dV_{\text{ext}}$$

$$= -\iint_S (G\nabla p_{\text{ext}} - p_{\text{ext}}\nabla G) \cdot \mathbf{n}_S \, dS + I_R, \qquad (7.75)$$

where

$$I_R = r^2 \int_0^{2\pi} \int_0^{\pi} \left[G\frac{\partial p_{\text{ext}}}{\partial r} - p_{\text{ext}}\frac{\partial G}{\partial r} \right] \sin\theta \, d\theta \, d\phi$$

is the surface integral over the outer sphere obtained by using the spherical coordinate system (6.30). Note that the minus sign in front of the double integral on the right side of (7.75) appears because $\mathbf{n}_S$ is pointing inward into V_{ext}. The function G must satisfy the Sommerfeld radiation condition (7.71); thus,

$$G \frac{\partial p_{\text{ext}}}{\partial r} - p_{\text{ext}} \frac{\partial G}{\partial r} = O(1/r^3). \tag{7.76}$$

Hence, Eq (7.75) reduces to

$$\iiint_{V_{\text{ext}}} p_{\text{ext}} \left(\nabla^2 + k^2 \right) G \, dV_{\text{ext}} = \iint_S (G \nabla p_{\text{ext}} - p_{\text{ext}} \nabla G) \cdot \mathbf{n}_S \, dS. \tag{7.77}$$

The simplest choice for the function G is the fundamental solution

$$p^*(\mathbf{x}, \mathbf{x}_0) = \frac{1}{4\pi r} e^{-ikr}, \quad r = |\mathbf{x} - \mathbf{x}_0|, \tag{7.78}$$

defined by (4.43*), where we have written p^* instead of u^* such that

$$\left(\nabla^2 + k^2 \right) p^* = -\delta(\mathbf{x} - \mathbf{x}_0), \tag{7.79}$$

where $\mathbf{x}_0$ represents an arbitrary point which may be replaced respectively by $\mathbf{x} \in V_{\text{ext}}$, or $\mathbf{x}' \in V_{\text{int}}$, or $\xi \in S$. In view of Eq (7.79) and the translation property of the Dirac delta function, we find that (7.74) and (7.77) reduce to

$$\iint_S (p^* \nabla p_{\text{int}} - p_{\text{int}} \nabla p^*) \cdot \mathbf{n}_S \, dS = \begin{cases} p_{\text{int}}(\mathbf{x}_0), & \text{if } \mathbf{x}_0 \in V_{\text{int}} \\ 0, & \text{if } \mathbf{x}_0 \in V_{\text{ext}} \cup S, \end{cases} \tag{7.80a}$$

and

$$\iint_S (p^* \nabla p_{\text{ext}} - p_{\text{ext}} \nabla p^*) \cdot \mathbf{n}_S \, dS = \begin{cases} p_{\text{ext}}(\mathbf{x}_0), & \text{if } \mathbf{x}_0 \in V_{\text{ext}} \\ 0, & \text{if } \mathbf{x}_0 \in V_{\text{int}} \cup S, \end{cases} \tag{7.80b}$$

Both (7.80a) and (7.80b) are collectively known as the Helmholtz–Kirchhoff integral corollary.

   Some methods for solving the underwater scattering problems involving large aspect ratio, prolate spheroidal, and finite cylindrical scatterers are: (i) Transition matrix (T–matrix) or null–field method (Waterman 1969) for arbitrarily shaped bodies; (ii) spherical coordinate–based null–field method (Bates and Wall 1977, Hackman 1984) and its modification (Boström 1980) for prolate spheroidal geometries; (iii) iterative orthogonalization method (Waterman 1971, Werby and Green 1983) for superspheroidal geometries; (iv) the CHIEF (combined Helmholtz integral equation formulation) algorithm

(Schenck 1968, Wu and Barach 1987, Wu and Seybert 1991); and (v) the CONDOR (compact outward normal derivative overlap relation) algorithm (Burton and Miller 1971). Of these methods, the first four are applied to elastic targets, and we shall not consider them here. Of the last two methods, the CHIEF algorithm has been used very effectively and we shall study it in the next section.

7.6. The CHIEF Algorithm

For scattering problems we shall consider three formulations: (i) Simple source formulation; (ii) Surface Helmholtz integral formulation; and (iii) Interior Helmholtz integral formulation. In the sequel, we shall use the expression (7.78) for p^*, and denote a point in V_{ext} by $\mathbf{x}$, a point in V_{int} by $\mathbf{x}'$, and a point on the surface S by ξ (Fig. 7.4).

(i) In the simple source formulation the BI Eq is

$$p(\mathbf{x}) = \frac{i\omega\rho}{4\pi} \iint_S \sigma(\xi) \frac{e^{-ik|\mathbf{x}-\xi|}}{|\mathbf{x}-\xi|}\, dS(\xi), \quad \mathbf{x} \in V_{ext}, \tag{7.81}$$

where $\sigma(\xi)$ is the source density which is initially unknown. Eq (7.81) satisfies the Helmholtz equation (7.70), the radiation condition (7.71), and the boundary condition

$$\left.\frac{\partial p}{\partial n}\right|_\xi = -\frac{i\omega\rho}{4\pi} v(\xi), \tag{7.82}$$

where $v(\xi)$ is the prescribed normal velocity at the point $\xi \in S$. The condition (7.82) determines the source density $\sigma(\xi)$. The integrand in (7.81) becomes singular as $\mathbf{x} \to \xi$, i.e., as $|\mathbf{x} - \xi| \to 0$, but the integral exists in the limit and is continuous. Substituting the boundary condition (7.82) into (7.81) we obtain

$$v(\xi) = \sigma(\xi) - \iint_S \sigma(\xi) \frac{\partial}{\partial n_\xi} \left[\frac{e^{-ik|\mathbf{x}-\xi|}}{|\mathbf{x}-\xi|}\right] dS(\xi), \quad \xi \in S. \tag{7.83}$$

In order to evaluate the normal velocity $v(\xi)$ on the surface S, we shall carry out the BE formulation of Eq (7.83) by discretizing the surface S into N triangular or rectangular boundary elements $\tilde{S}_1, \ldots, \tilde{S}_N$, depending on the geometry of the scatterer, such that $\tilde{S} = \cup_{j=1}^N \tilde{S}_j$, and assume that the source density is constant on each boundary element (see Fig. 6.12 for triangular boundary elements). Then the point ξ takes N different values, each of which becomes the ξ–point for its corresponding boundary element. This reduces Eq (7.83) to a system of algebraic equations

$$v_i = A_{ij}\sigma_j, \quad i, j = 1, 2, \ldots, N, \tag{7.84}$$

where

$$A_{ij} = \delta_{ij} - \iint_{\tilde{S}_j} \frac{\partial}{\partial n_\xi} \left[\frac{e^{-ik|\xi_i - \xi|}}{|\xi_i - \xi|} \right] dS(\xi), \tag{7.85}$$

$$v_i = v(\xi_i), \quad \sigma_j = \sigma(\xi_j).$$

The solution (7.81) does not remain valid for certain characteristic wave numbers (Schenck, 1986).

(ii) In the surface Helmholtz integral formulation, the BI Eq for the pressure p in the region V_{ext} is given by

$$p(\mathbf{x}) = \frac{1}{4\pi} \iint_S \left\{ p(\xi) \frac{\partial}{\partial n_\xi} \left[\frac{e^{-ik|\mathbf{x} - \xi|}}{|\mathbf{x} - \xi|} \right] + i\omega\rho \frac{e^{-ik|\mathbf{x} - \xi|}}{|\mathbf{x} - \xi|} \right\} dS(\xi), \xi \in S. \tag{7.86}$$

In order to determine the pressure p in the region V_{ext} from (7.86), we must know both the normal velocity $v(\xi)$ and the pressure $p(\xi)$ on the surface S. But since these two quantities are linearly dependent, we can determine one if the other is known. In the limiting process as the point $\mathbf{x} \in V_{\text{ext}}$ approaches a point $\zeta \in S$ in Eq (7.86), we find that

$$2\pi p(\zeta) - \iint_S p(\xi) \frac{\partial}{\partial n_\xi} \left[\frac{e^{-ik|\zeta - \xi|}}{|\zeta - \xi|} \right] dS(\xi)$$

$$= i\omega\rho \iint_S v(\xi) \frac{e^{-ik|\zeta - \xi|}}{|\zeta - \xi|} dS(\xi), \xi \in S. \tag{7.87}$$

Notice that the normal derivative in (7.83) is taken with respect to the free variable ξ, but in (7.87) it is taken with respect to the dummy variable ξ. As in the simple source case, Eq (7.87) can be discretized into its BE formulation and transformed into a system of algebraic equations like (7.84). This formulation also fails at certain Dirichlet eigen–frequencies, by nonuniqueness.

(iii) For the interior Helmholtz integral formulation, the integral representation (7.86) is always valid for $\mathbf{x} \in V_{\text{ext}}$. But we know that $p(\mathbf{x}') = 0$ for $\mathbf{x}' \in V_{\text{int}}$, and then (7.86) becomes

$$\iint_S \left\{ p(\xi) \frac{\partial}{\partial n_\xi} \left[\frac{e^{-ik|\mathbf{x}' - \xi|}}{|\mathbf{x}' - \xi|} \right] dS(\xi) + v(\xi) \frac{e^{-ik|\mathbf{x}' - \xi|}}{|\mathbf{x}' - \xi|} \right\} dS(\xi) = 0,$$

$$\mathbf{x}' \in V_{\text{int}}. \tag{7.88}$$

If we know the normal velocity $v(\xi)$, we can determine the surface pressure from (7.88). This means that we use (7.88) to determine p on S and then use (7.86) to determine p in V_{ext}.

Of these three formulations, the formulation (i) is not suitable for numerical computation because the solution (7.81) is not valid for all velocity distributions and wave numbers. A combination of formulations (ii) and (iii) provides a basis for the CHIEF algorithm whose six–point details are as follows:

1. Discretize the surface S into N triangular or rectangular boundary elements $\tilde{S}_j$, $j = 1, \ldots, N$ (or Fig. 6.12 for triangular elements). Each $\tilde{S}_j$ must be small so that we can remove the mean value of the pressure out from the integral sign. Eqs (7.87) and (7.88) reduce to the corresponding BE Eqs

$$2\pi p(\zeta) - \sum_{j=1}^{N} p(\xi_j) \iint_{\tilde{S}_j} \frac{\partial}{\partial n_\xi} \left[\frac{e^{-ik|\zeta-\xi|}}{|\zeta - \xi|} \right] dS(\xi)$$

$$= i\omega\rho \sum_{j=1}^{N} v(\xi_j) \iint_{\tilde{S}_j} \frac{e^{-ik|\zeta-\xi|}}{|\zeta - \xi|} dS(\xi), \xi \in S. \quad (7.89)$$

$$\sum_{j=1}^{N} p(\xi_j) \iint_{\tilde{S}_j} \frac{\partial}{\partial n_\xi} \left[\frac{e^{-ik|\mathbf{x}'-\xi|}}{|\mathbf{x}' - \xi|} \right] dS(\xi)$$

$$= -i\omega\rho \sum_{j=1}^{N} v(\xi_j) \iint_{\tilde{S}_j} \frac{e^{-ik|\mathbf{x}'-\xi|}}{|\mathbf{x}' - \xi|} dS(\xi), \mathbf{x}' \in V_{\text{int}}. \quad (7.90)$$

2. Require that Eq (7.89) holds for all N surface points (boundary nodes) $\zeta = \xi_i$, $i = 1, \ldots, N$, where $\xi_i = \xi_j$ for $i = j$; and that Eq (7.90) holds for all M interior points $\mathbf{x}' \in V_{\text{int}}$, $i = 1, \ldots, M$.

3. For $j = 1, \ldots, N$, define

$$H_{ij} \equiv \begin{cases} 2\pi\delta_{ij} - \iint_{\tilde{S}_j} \frac{\partial}{\partial n_\xi} \left[\frac{e^{-ik|\xi_i-\xi|}}{|\xi_i - \xi|} \right] dS(\xi), & i = 1, \ldots, N, \\ -\iint_{\tilde{S}_j} \frac{\partial}{\partial n_\xi} \left[\frac{e^{-ik|\mathbf{x}'_i-\xi|}}{|\mathbf{x}'_i - \xi|} \right] dS(\xi), & i = N+1, \ldots, M. \end{cases}$$

$$(7.91)$$

4. For $m = 1, \ldots, N$, define

$$
G_{im} \equiv
\begin{cases}
i\omega\rho \iint_{\tilde{S}_m} \dfrac{e^{-ik|\xi_i - \xi|}}{|\xi_i - \xi|}\, dS(\xi), & i = 1, \ldots, N, \\[2ex]
i\omega\rho \iint_{\tilde{S}_m} \dfrac{e^{-ik|\mathbf{x}'_i - \xi|}}{|\mathbf{x}'_i - \xi|}\, dS(\xi), & i = N+1, \ldots, M.
\end{cases}
\tag{7.92}
$$

Eqs (7.91) and (7.92) can be combined into the following system of equations:

$$
H_{ij}p_j(\xi_j) = G_{im}v_m(\xi_m) \quad \text{for } i = 1, \ldots, N+M,
$$
$$
\text{and } j, m = 1, \ldots, N, \tag{7.93a}
$$

or in matrix form

$$
HP = GV, \tag{7.93b}
$$

which is an overdetermined system.

5. Solve the system (7.93b) for P by a least–square orthonormalizing technique which is as follows: Use the collocation process of §5.3 and transform Eq (7.93b) into the form

$$
AX = F. \tag{7.93c}
$$

Then the method of least–squares gives the following set of equations:

$$
A^T A X = A^T F, \tag{7.93d}
$$

which is consistent and determinate.

6. Define an error vector E with the components

$$
E_i \equiv G_{im}v_m - H_{ij}\tilde{p}_j, \tag{7.94}
$$

where $\tilde{p}_j$ is a vector determined from (7.93d) for a prescribed set of velocities and represents the least–square approximation of the surface pressure $p(\xi)$. Once $\tilde{p}_j$ is determined, use the BE approximation of Eq (7.86) which is

$$
p(\mathbf{x}) \approx \tilde{p}_j \tilde{H}_j(\mathbf{x}) + v_m \tilde{G}_m, \quad \mathbf{x} \in V_{\text{ext}}, \tag{7.95}
$$

where

$$
\tilde{H}_j(\mathbf{x}) = \frac{1}{4\pi} \iint_{\tilde{S}_j} \frac{\partial}{\partial n_\xi} \left[\frac{e^{-ik|\mathbf{x} - \xi|}}{|\mathbf{x} - \xi|} \right] dS(\xi), \tag{7.96}
$$

$$
\tilde{G}_m(\mathbf{x}) = \frac{i\omega\rho}{4\pi} \iint_{\tilde{S}_j} \frac{e^{-ik|\mathbf{x} - \xi|}}{|\mathbf{x} - \xi|}\, dS(\xi), \tag{7.97}
$$

Gauss quadrature formula (C.7) for triangular boundary elements, the computer code CHIEF and the CHIEF Users Manual (see bibliography) are used for solving scattering problems. This algorithm can be applied to the examples in the next section.

7.7. Some Structure Types

We shall consider the following two two–dimensional structure types which are used in the boundary element method to study the interaction of water waves on some offshore structures:

1. **Vertically integrated structure.** This type of structure has a constant horizontal section throughout the water depth as shown in Fig. 7.5. The incidental wave potential ϕ_0 is defined by

$$\phi_0 = \hat{\phi}_0(x, y)\frac{\cosh k(z + h)}{\cosh kh},$$

   where

$$\hat{\phi}_0(x, y) = -\frac{iga_0}{\omega} e^{i(k_1 x + k_2 y)}.$$

   In fact, ϕ_0 is the velocity potential defined in the (x, y)–plane by (7.24). Thus, the diffracted potential is given by

$$u(x, y) = \hat{u}(x, y)\frac{\cosh k(z + h)}{\cosh kh},$$

   where $\hat{u}$ satisfies the Helmholtz equation $(\nabla^2 + k^2)\hat{u} = 0$ in R with the boundary conditions (7.28)–(7.31). This leads to the BE Eq

$$c(i)\hat{u}(i) + \int_{\tilde{C}} \hat{u}q^* \, ds = \int_{\tilde{C}} \hat{u}^* q \, ds,$$

   as in (7.57). After solving this equation for $\hat{u}$, the potential for the problem is computed as

$$u = (\phi_0 + \hat{u})\frac{\cosh k(z + h)}{\cosh kh}. \qquad (7.98)$$

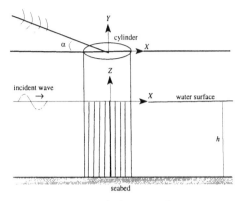

Fig. 7.5. Vertically integrated structure.

The total three–dimensional force on the vertical structure is obtained by integrating (7.98) in the vertical direction and is given by

$$\left\{ \begin{array}{c} f_x \\ f_y \\ f_z \end{array} \right\} = i\rho\omega \frac{\tanh kh}{k} \int_{\tilde{C}} \left\{ \begin{array}{c} n_x \\ n_y \\ n_z \end{array} \right\} \hat{u}\,ds,$$

and

$$\left\{ \begin{array}{c} f_{yz} \\ f_{zx} \end{array} \right\} = \left(z_0 + \frac{1}{k}\tanh\frac{kh}{2} \right) \left\{ \begin{array}{c} f_y \\ -f_x \end{array} \right\},$$

where z_0 is the center of the moments f_{yz} and f_{zx}.

2. **Structure with a vertical axis or plane of symmetry.** The structure in Fig. 7.6 has a vertical axis of symmetry with two symmetrical parts C_1 and C_2.

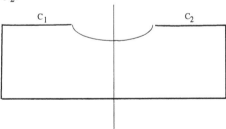

Fig. 7.6. Structure with a vertical axis of symmetry.

If the boundary conditions on the z–axis are not known, a full discretization of the boundary C_1 and C_2 leads to the matrix BE Eq

$$\begin{bmatrix} H_{11} & H_{12} \\ H_{21} & H_{22} \end{bmatrix} \left\{ \begin{array}{c} u_1 \\ u_2 \end{array} \right\} = \begin{bmatrix} G_{11} & G_{12} \\ G_{21} & G_{22} \end{bmatrix} \left\{ \begin{array}{c} q_1 \\ q_2 \end{array} \right\}, \tag{7.99}$$

where the subscript 1 and 2 in $\{u\}$ and $\{q\}$ refer to the values on C_1 and C_2 respectively. Note the $H_{11} = H_{22}$ and $H_{12} = H_{21}$; also, $G_{11} = G_{22}$ and $G_{12} = G_{21}$. We shall use the following transformation to replace each variable by a symmetric (indexed 's') and an antisymmetric (indexed 'a') component:

$$u_s = \frac{u_1 + u_2}{2}, \quad u_a = \frac{u_1 - u_2}{2},$$

$$q_s = \frac{q_1 + q_2}{2}, \quad q_a = \frac{q_1 - q_2}{2}. \tag{7.100}$$

Substituting (7.100) into (7.99) and rearranging, we get

$$\begin{bmatrix} H_{11} + H_{12} & 0 \\ 0 & H_{11} - H_{12} \end{bmatrix} \begin{Bmatrix} u_s \\ u_a \end{Bmatrix} = \begin{bmatrix} G_{11} + G_{12} & 0 \\ 0 & G_{11} - G_{12} \end{bmatrix} \begin{Bmatrix} q_s \\ q_a \end{Bmatrix}.$$

This matrix equation, being in the diagonal form, yields directly the solution for the symmetric and antisymmetric parts.

The wave forces are given by

$$\begin{Bmatrix} f_x \\ f_y \\ f_z \end{Bmatrix} = 2i\rho\omega \int_{\tilde{C}} \begin{bmatrix} 0 & n_x \\ n_z & 0 \\ 0 & n_{zx} \end{bmatrix} \begin{Bmatrix} u_s \\ u_a \end{Bmatrix} ds.$$

Equation of motion for the structures determines the movement of the offshore structure and investigates the changes in forces on the structure during motion. Consider the following system

$$[M]\{\ddot{X}\} + [C]\{\dot{X}\} + [K]\{X\} = \{f\}e^{-i\omega t}, \quad \{X\} = \{X_0\}e^{-i\omega t}. \tag{7.101}$$

Here the matrix $[M]$ is given by the structural mass matrix $[M_0]$ and the added mass $[M_a]$, i.e., $[M] = [M_0] + [M_a]$. The damping matrix $[C]$ can be obtained together with $[M_a]$ by solving

$$[M_0]\{\ddot{X}\} + [C]\{\dot{X}\} = -i\rho\omega \int_{\tilde{C}} \{n\}[H]^{-1}[Q]\, ds\, \{\dot{X}\}.$$

Here $[K]$ is the total stiffness matrix due to the contribution of the hydraulic force and due to the mooring system. The transient form of Eq (7.101) is

$$\left(-\omega^2[M] - i\omega[C] + [K]\right)\{X_0\} = \{f_0\},$$

whose solution is given by

$$\{X_0\} = \left[K - \omega^2 M - i\omega C\right]^{-1}\{f\}.$$

We shall now consider some numerical examples.

Example 7.1:
Submerged half–cylinder resting on the seabed is shown in Fig. 7.7a. The
half BE mesh with 30 boundary elements is depicted in Fig. 7.7b.

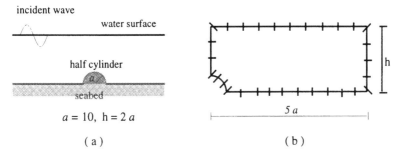

Fig. 7.7. (a) Submerged half–cylinder; (b) BE Mesh.

Example 7.2:
Fully submerged cylinder (Fig. 7.8a) with the half BE mesh consisting of 42
elements is shown in Fig. 7.8b.

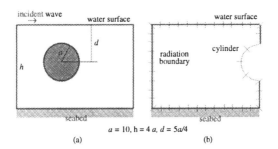

Fig. 7.8. (a) Fully submerged cylinder; (b) BE mesh.

Example 7.3:
Half–floating cylinder of Fig. 7.7a is discretized into the half BE mesh as in
Fig. 7.7b. The added mass and damping coefficients for this half–cylinder are
computed by using the formulas

$$[C_m]_{3\times3} = -\frac{[M_a]}{\rho\pi a^2/2}, \quad [C_d]_{3\times3} = -\frac{[C]}{\rho\pi(a^2/2)\omega}.$$

Example 7.4:
Moored floating cylinder of Fig. 7.9 is investigated under the assumption

that the mooring lines remain taut and straight. This type of structure is used
as a breakwater and is normally moored as shown in the figure. The data is
$a = 16$, $h = 35$, $d = 20$, $w = 80$, $k/lg = 2,632$, $m/(\rho g) = 0.4782$, and
$I/(\rho g) = 69.832$.

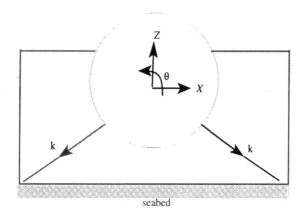

Fig. 7.9. Moored floating cylinder.

Example 7.5:
Surface wave on a sloping bottom as shown in Fig. 7.10 depicts the bottom
sloping toward the land. Two distinct parts are considered, one sloping and
the other straight. The horizontal seabed is taken as the boundary C_2 and the
sloping part as C_1.

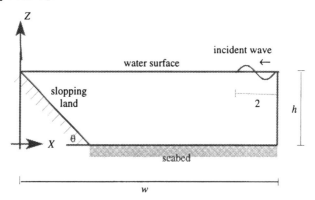

Fig. 7.10. A sloping beach.

Example 7.6:
The square caisson (a water–tight chamber of sheet iron or wood) is depicted
in Fig. 7.11. The BE mesh consists of elements on the quarter of the structure.

Example 7.7:
The elliptic cylinder $x^2/a^2 + y^2/b^2 = 1$ of Fig. 7.12 takes the BE mesh on the quarter structure as in Example 5.6.

Example 7.8:
A floating hemisphere at the free water surface is shown in Fig. 7.13a, with its BE mesh of the half region depicted in Fig. 7.13b.

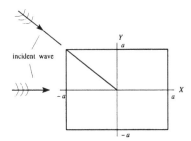

Fig. 7.11. A square cassion.

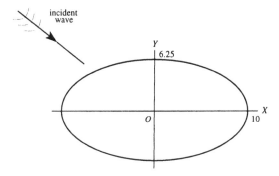

Fig. 7.12. An elliptic cylinder.

Example 7.9:
The surface elevation around a vertical cylinder is depicted in Fig. 7.14. The potential on the water surface is computed by using the formulas

$$\{f\}_{6\times1} = i\rho\omega \int_{C_1} \{n\}_{6\times1}\phi\,ds, \quad \{F\} = \Re[\{f\}e^{-i\omega t}].$$

where ϕ is given by (7.25). The real and imaginary parts of $\{f\}$ are used to compute the surface elevation by the formula

$$\frac{\xi}{a_0} = \text{Real} \times \cos\omega t + \text{Imaginary} \times \sin\omega t.$$

Note that the above examples represent the two–dimensional analysis of some useful offshore structures.

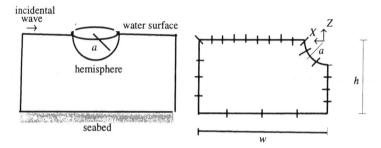

Fig. 7.13. (a) A floating hemisphere, (b) BE mesh.

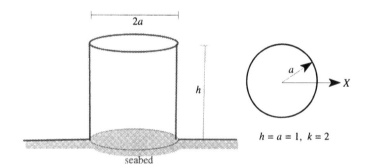

Fig. 7.14. Vertical cylinder.

7.8. Problems

7.1. Develop a program for the two–dimensional boundary element method of §7.4.

7.2. Formulate the boundary element method for the three–dimensional BE Eq (7.47) for constant elements.

7.3. Develop a program for Prob. 7.2.

7.4. Develop a program for the BE Eq (7.60).

7.5. Develop a two–dimensional BEM for the Helmholtz equation $(\nabla^2 + k^2)u = 0$ in R, with the boundary conditions $u = 0$ on C_1 and $q + hu = 0$ on C_2, where $C = C_1 \cup C_2$ is the boundary of R and $k = \omega/c > 0$. [Hint: From the weak variational formulation

$$\iint_R \left(\frac{\partial u^*}{\partial x} \frac{\partial u}{\partial x} + \frac{\partial u^*}{\partial y} \frac{\partial u}{\partial y} - \lambda u u^* \right) dA + \int_{C_2} h u u^* \, ds = 0$$

develop the BI Eq, like (7.49), for this problem.]

7.6. The so called Rayleigh quotient for the Helmholtz problem is the functional

$$k^2 = \frac{\iint_R \nabla^2 u \, dA + \int_{C_2} h u^2 \, ds}{\iint_R u^2 \, dA}.$$

Show that $\delta(k^2) = 0$ leads to the problem $(\nabla^2 + k^2)u = 0$, with the boundary conditions $u = 0$ on C_1 and $q = 0$ on C_2.

7.7. The equation of motion governing the classical problem of a square vibrating membrane D with all edges fixed against transverse displacement is

$$T\nabla^2 w = \rho \frac{\partial^2 w}{\partial t^2},$$

where T is the initial tension in the membrane and ρ the density. The boundary condition is taken such that w vanishes on all edges of the membrane. Set $w(x, y, t) = u(x, y)e^{i\omega t}$, and we get $(\nabla^2 + k^2)u = 0$ in D where $u = 0$ on C (which is the boundary of D), and $k^2 = \rho\omega^2/T$. Solve this square membrane problem using the method of §7.4.

7.8. Setup and solve the Helmholtz problem for the square using a quarter symmetry model.

7.9. Consider the problem of the vibration of a circular membrane. Solve the corresponding Helmholtz equation by using the method of §7.4.

7.10. Setup and solve the Helmholtz problem for the circle using the half–quarter symmetry region shown in Fig. 7.15.

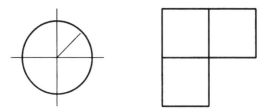

Fig. 7.15.   Fig. 7.16.

7.11. Setup and solve the Helmholtz problem for the L–shaped region in Fig. 7.16, assuming that $u = 0$ everywhere on the boundary.

7.12. Formulate the boundary element method for the cases of the linear elements of Fig. 5.2(a)–(c), to solve the two and three–dimensional Helmholtz BE Eqs.

7.13. Write a computer program for the Problem 7.5.

7.14. Show that the fundamental solution for the undamped wave equation in a two–dimensional region

$$c^2 \left(\frac{\partial^2 u^*}{\partial x^2} + \frac{\partial^2 u^*}{\partial y^2} \right) - \frac{\partial^2 u^*}{\partial t^2} + \delta(i) = 0$$

is

$$u^*(r) = -\frac{H(ct - r)}{2\pi c(c^2 t^2 - r^2)}, \quad r = \sqrt{x^2 + y^2}.$$

7.15. Formulate the boundary element method for the cases of the linear elements of Fig. 5.2(a)–(c), to solve the undamped wave equation of Problem 7.14.

7.16. Use boundary element method to completely solve the following two–dimensional second order model equation

$$-\frac{\partial}{\partial x} \left(a_{11} \frac{\partial u}{\partial x} + a_{12} \frac{\partial u}{\partial y} \right) - \frac{\partial}{\partial y} \left(a_{21} \frac{\partial u}{\partial x} + a_{22} \frac{\partial u}{\partial y} \right) + cu - f = 0,$$

with prescribed boundary conditions for the cases of the linear elements of Fig. 5.2(a)–(c); here a_{ij} $(i, j = 1, 2)$, c, and f are known functions of x and y. (Note that if $a_{11} = a_{22} = 1$ and $a_{12} = a_{21} = 0 = c$, then the above equation becomes the Poisson equation in R. Further, if $f = 0$, then it becomes the Laplace equation.)

7.17. Solve the interior and the exterior problem for the Helmholtz equation.

[The 'interior' problem implies that we should find the acoustic potential u which satisfies the Helmholtz equation with first and second order partial derivatives of u being continuous within and on the closed boundary C of a two–dimensional region R. All (mathematical) sources of disturbance, i.e., singularities at which u or its first or second order partial derivatives are discontinuous, lie outside C.

The 'exterior' problem implies that we find the acoustic potential u which satisfies the Helmholtz equation with first and second order derivatives of u being continuous outside and on the boundary C. In addition, u must satisfy both the magnitude and the Sommerfeld's radiation conditions, viz., $|ru| < A$ as $r \to \infty$, and

$$\sqrt{r}\left[\frac{\partial u}{\partial r} + i\kappa u\right] \to 0 \quad \text{as } r \to \infty.$$

Sommerfeld's radiation condition is derived as follows: Since $u^* = p^*$ satisfies Eq (7.79), for any point outside the boundary C we have $c = 0$ and Eq (7.49) gives $\int_C u^* q \, ds = \int_C u q^* \, ds$, or in terms of the fundamental solution

$$\int_C H_0^{(2)}(kr) q \, ds = \int_C u \frac{\partial H_0^{(2)}(kr)}{\partial n} \, ds.$$

We can simplify it if the field point is assumed to be far away from our region of interest. Thus, we can use an asymptotic expansion for the Hankel function and its derivative. If C is taken as a circle of large radius r, then the above equation gives (note that $\mathbf{n} = \mathbf{r}$ over a circle)

$$\sqrt{\frac{2}{k\pi}} e^{i\pi/4} \int_C \frac{1}{\sqrt{r}}\left(\frac{\partial u}{\partial r} + iku\right) e^{-ikr} r \, d\theta = 0,$$

where $ds = r \, d\theta$; that is,

$$\int_0^{2\pi} \sqrt{r}\left(\frac{\partial u}{\partial r} + iku\right) d\theta = 0,$$

which gives the radiation condition.]

7.18. Solve the problem of the water waves incident upon a cylinder with porous (impervious) wall. (This problem is related to a porous wall type offshore structure used in the North Sea, and is known as the Ekofisk–type structure.) The incident waves impinge on the exterior protective porous wall before they hit the solid cylindrical core with some reduced energy (see Fig. 7.17a). Compute the height of the incident waves and the force on the cylinder and the protective wall.

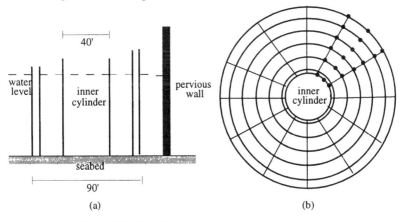

Fig. 7.17. The Ekofisk–type offshore structure.

[The method of solution involves dividing the region into three parts: (i) the region between the inner solid cylinder and the exterior porous wall (the discretization is shown in Fig. 7.17b); (ii) the region outside the exterior wall to some arbitrary circular boundary; and (iii) the remaining semi–infinite region outside the arbitrary circular boundary (use the exterior BEM for this part).]

7.19. Use the CHIEF algorithm to determine the pressure p on a rigid box–type structure (rectangular parallelopiped of dimension $2a \times 2a \times a$) which is sitting on a rigid, infinite plane. Because of symmetry only the one–half of the box of dimensions $2a \times a \times a$ is shown in Fig. 7.18. There are 12 rectangular boundary elements with 30 quadratic nodes. Note that there are no elements and nodes on the bottom surface. Consider an incident plane wave coming from the x–axis direction.

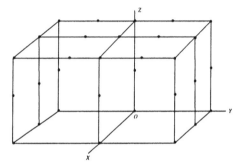

Fig. 7.18. Box–like rigid structure.

References and Bibliography for Additional Reading

G. K. Batchelor, *An Introduction to Fluid Dynamics*, Cambridge Univ. Press, 1967.

R. H. T. Bates and D. J. N. Wall, *Null field approach to scalar diffraction: I. General method; II. Approximate methods; III. Inverse methods*, Phil. Trans. Royal Soc., London, Ser. A **287** (1977), 45–117.

R. P. Banaugh and W. Goldsmith, *Diffraction of steady acoustic waves by surfaces of arbitrary shape*, J. Acoust. Soc. Am. **35** (1963), 1590–1601.

J. E. Black, *Wave forces on vertical axisymmetric bodies*, J. Fluid Mech. **67** (1975), 369–376.

A. Boström, *Scattering by an elastic obstacle in a fluid and a smooth elastic obstacle*, in Acoustic, Electromagnetic and Elastic Wave Scattering (V. V. Varadan and V. K. Varadan, Eds.), Pergamon Press, New York, 1980, pp. 91–100.

C. A. Brebbia (Ed.), *Progress in Boundary Element Methods, Vols. 1 and 2*, Pentech Press, London, 1981, 1983.

———, *Boundary Element Methods*, Proc. 5th Conf. on BEM, Springer-Verlag,

Berlin, 1983.

———— and S. Walker, *The Boundary Element Techniques in Engineering*, Newnes–Butterworths, London, 1979.

C. Bretschneider (Ed.), *Topics in Ocean Engineering*, vol. II, Gulf Publishing Co., Houston, 1970.

A. J. Burton and G. F. Miller, *The application of integral equation methods to the numerical solution of some exterior boundary value problems*, Proc. Royal Soc. London **A 323** (1971), 201–210.

R. D. Ciskowski and C. A. Brebbia (Eds.), *Boundary Element Methods in Acoustics*, Computational Mechanics Publications, Southampton and Elsevier Applied Science, London, 1991.

D. Colton and R. Kress, *Integral Equation Methods in Scattering Theory*, Wiley, New York, 1983.

L. G. Copley, *Fundamental results concerning integral representations in acoustic radiation*, J. Acoust. Soc. Am. **44** (1968), 28–32.

P. J. Davis, *Orthonormalizing Codes in Numerical Analysis*, in Survey of Numerical Analysis (J. Todd, Ed.), McGraw–Hill, New York, 1962.

J. D. Fenton, *Wave forces on vertical bodies of revolution*, J. Fluid Mech. **85** (1978), 241-255.

M. B. Friedman and R. P. Shaw, *Diffraction of a plane shock wave by an arbitrary rigid cylindrical obstacle*, Trans. ASME, J. Appl. Mech. **29** (1962), 40–46.

C. J. Garrison and P. Y. Chow, *Wave forces on submerged bodies*, J. Waterways Harbours Coastal Eng. Div. ASCE **98** (1972), 375-392.

R. H. Hackman, *The transition matrix for acoustic and elastic wave scattering in prolate spheroidal coordinates*, J. Acoust. Soc. Am. **75** (1984), 35–45.

V. W. Harms, *Diffraction of water waves by isolated structures*, J. Waterways Port Coastal Ocean Div. ASCE **105** (1979), 131–147.

L. S. Hwang and E. D. Tuck, *On the oscillations of harbours of arbitrary shape*, J. Fluid Mech. **42** (1970), 447–464.

M. de St. Isaacson, *Vertical cylinders of arbitrary section in waves*, J. Waterways Port Coastal Ocean Div. ASCE **104** (1978), 309–324.

F. John, *On the motion of floating bodies II*, Commun. Pure Appl. Math. **3** (1950), 45–101.

J. J. Lee, *Wave-induced oscillations in harbours of arbitrary geometry*, J. Fluid Mech. **45** (1971), 375–394.

W. J. Mansur and C. A. Brebbia, *Formulation of the boundary element method for transient problems governed by the scalar wave equation*, Appl. Math. Modelling **6** (1982), 307–312.

F. Mattioli, *Wave-induced oscillations in harbours of variable depth*, Comput. Fluids **6** (1978), 161–172.

P. M. Morse and K. U. Ingard, *Theoretical Acoustics*, Princeton Univ. Press, Princeton, 1986.

J. E. Murphy, Characteristic wave numbers for simple source formulations of acoustic radiation and diffraction problems, J. Acoust. Soc. Am. **63** (1978), 272–273.

T. F. Ogilvie, *First and second-order forces on a cylinder submerged under a free surface*, J. Fluid Mech. **16** (1963), 451–472.

M. Rahman, *Numerical response of an arbitrarily shaped harbour*, Appl. Math. Modelling **5** (1981), 109–121.

T. Sarpkaya and M. Isaacson, *Mechanics of Wave Forces on Off-Shore Structures*, Van Nostrand Reinhold, New York, 1981.

H. A. Schenck, *Improved integral formulations for acoustic radiation problems*, J.

Acoust. Soc. Am. **44** (1969), 41–58.

H. Schlichting and E. Truckenbrodt, *Aerodynamics of the Airplane*, Translated by
  H. J. Ramm, McGraw–Hill, New York, 1979.

R. P. Shaw, *An outer boundary integral equation applied to transient wave scattering
  in an inhomogeneous medium*, Trans. ASME, J. Appl. Mech. **42** (1975), 147–
  152.

A. Sommerfeld, *Partial Differential Equations in Physics*, Lectures on Theoretical
  Physics, Vol. VI, Academic Press, New York, 1964.

P. C. Waterman, *New formulation of acoustic scattering*, J. Acoust. Soc. Am. **45**
  (1969), 1417–1429.

_____ Symmetry, unitarity and geometry in electromagnetic scattering, Phys. Rev.
  D3 (1971), 825–839.

T. W. Wu and A. F. Seybert, *Acoustic Radiation and Scattering*, in Boundary Element Methods in Acoustics (R. D. Ciskowski and C. A. Brebbia, Eds.), Computational Mechanics Publications, Southampton and Elsevier Applied Science,
  London, 1991.

_____ and D. Barach, *CHIEF Users Manual*, NOSC TD 970, Naval Ocean Systems
  Center, San Diego, CA, A-2-1-A3-4, 1987; Revision 1, Sept 1988.

8

Some Fluid Flows and D'Arcy Equation

In many boundary value problems, the equilibrium and constitutive constraints lead to a differential equation formulation in the space and time variables, which often contains all compatibility requirements. However, in general, some equations must be supplemented by additional constraints to satisfy additional compatibility requirements. One of the examples of this situation is the problem of groundwater flows or seepage flows through the soil which, besides the Navier–Stokes equations, requires D'Arcy's law for a complete formulation of the problem. As we shall see, these and some aerodynamic problem reduce to potential flow problems.

8.1. Aerodynamic Flows

In Chapter 5 we studied some types of potential flow problems. Such problems also arise in flows around lifting bodies where the location of the trailing vortex sheets is important. For aerodynamic applications, we shall consider the velocity potential form, i.e., if $\mathbf{v} = (u, v)$ is the two–dimensional velocity field, then $\dfrac{\partial \phi}{\partial x} = u$, $\dfrac{\partial \phi}{\partial y} = v$. This leads to the Poisson equation

$$\nabla^2 \phi = b, \tag{8.1}$$

which, as in §5.7, leads to the BI Eq

$$c(i)\phi(i) = \iint_R \phi^* b \, dx \, dy + \int_C \left(\phi q^* - \phi^* q \right) ds, \tag{8.2}$$

where $c(i)$ is defined in (5.15). The fundamental solution ϕ^* in (8.2) is given by

$$\phi^* = -\frac{1}{2\pi} \ln r, \; r = |\mathbf{x} - \mathbf{x}'|, \tag{8.3}$$

as defined in (C.3) or (4.26). The function b on the right side of (8.1) represents the sources that may be present in any particular problem, or nonlinear terms that are to be approximated in a compressible flow problem. If $b \neq 0$, then the surface integral in (8.2) leads to a domain integral of the form

$$B_i = \iint_{\tilde{R}} \phi^* b \, dx \, dy, \tag{8.4}$$

which has been discussed in §5.9 and will be further studied in Chapter 9. For incompressible flow problems, $b = 0$ in general, and they are governed by potential flows which have been studied in Chapter 5.

In aircraft aerodynamics, the theory of lift of airfoils produces potential flows. However, in a compressible inviscid flow, there are elastic forces, besides the inertia forces, which act on the homogeneous fluid. The effects of compressibility become significant for the Mach numbers $M \geq 0.3$. This means that the validity of the potential flow dynamics holds only for very small Mach numbers with the limiting case $M \to 0$. In aerodynamics, the velocity and body dimensions are relatively large. The Reynolds number $Re = vL/\nu$, where ν is the kinematic viscosity, is therefore relatively very large, which implies that the friction forces are much smaller than the inertia forces. The kinematic viscosity $\nu \to 0$ for inviscid flows (this corresponds to $Re \to \infty$). Thus, the fluid flows with very small viscosity correspond very well, for practical purposes, with those without viscosity. However, we cannot neglect even a very small viscosity in the boundary–layer theory.

With altitude the Mach number M increases while the speed of sound decreases. Also, the kinematic viscosity increases rapidly with increase in altitude and thus the Reynolds number decreases considerably. The potential flow theory does not remain valid in this case.

A purely subsonic or purely supersonic flow is governed by the potential equation

$$\left(1 - M_\infty^2\right) \frac{\partial^2 \phi}{\partial x^2} + \frac{\partial^2 \phi}{\partial y^2} + \frac{\partial^2 \phi}{\partial z^2} = 0, \quad M_\infty \neq 1, \tag{8.5}$$

where M_∞ is the Mach number of the incident flow, and the velocity $\mathbf{v} = (u, v, w)$ is defined in terms of the velocity potential ϕ, viz., $u = \partial\phi/\partial x$, $v =$

$\partial\phi/\partial y$, and $w = \partial\phi/\partial z$. Eq (8.5) is of the elliptic type for purely subsonic flows, and of the hyperbolic type for purely supersonic flows. A transonic flow results when the Mach number $M = 1$; in this case the undisturbed flow velocity is equal to the speed of sound. The governing equation for a transonic flow is

$$-\frac{\gamma+1}{U_\infty}\frac{\partial\phi}{\partial x}\frac{\partial^2\phi}{\partial x^2} + \frac{\partial^2\phi}{\partial y^2} + \frac{\partial^2\phi}{\partial z^2} = 0, \tag{8.6}$$

where U_∞ is the incident flow velocity along x–axis, and $\gamma = c_p/c_v$ is the isentropic exponent (in compressible flows). Eq (8.6) is nonlinear in ϕ, and numerical computation is required in transonic flows.

Example 8.1:
Consider the wing profile in Fig. 8.1(a) which shows a profile teardrop (NACA 0010) and in Fig. 8.1(b)which shows an elliptic wing profile, where the linear nodes are marked. For potential/heat transfer flows, the methods of Chapter 5 apply.

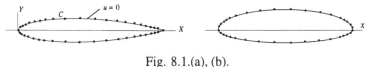

Fig. 8.1.(a), (b).

Note that the profile teardrop which defines the thickness distribution is the ratio of the wing profile thickness t and the chord of length c along the span. Various shapes of wing profiles which have the same lift distribution can be obtained by varying the profile or the angle of attack, or the chord. A structurally secure wing profile can be constructed simply by making geometrically similar profiles and angle of attack equal along the span, such that the chord c at each point is proportional to the lift. Fig. 8.1(b) shows the elliptic lift distribution by making the wing consisting of two semi–ellipses. In this case the centers of pressure of each elliptic shape are all on a straight line and, thus, this wing is approximated very well by a straight vortex filament. ∎

8.2. Navier–Stokes Equations

As mentioned in §7.1, the momentum equation for the incompressible Navier–Stokes flow is

$$\frac{\partial\mathbf{v}}{\partial t} + (\mathbf{v}\cdot\nabla)\mathbf{v} = -\frac{1}{\rho}\nabla p + \nu\nabla^2\mathbf{v}, \tag{8.7}$$

and the equation of continuity is

$$\nabla \cdot \mathbf{v} = 0. \tag{8.8}$$

If we divide the flow into its kinematic and dynamic components, the former is described by Eq (8.8) whereas the latter component is described by the vorticity vector $\boldsymbol{\omega} = \nabla \times \mathbf{v}$. If we take the curl of Eq (8.7), the pressure p is eliminated and then the dynamic component of the problem is governed by

$$\frac{\partial \boldsymbol{\omega}}{\partial t} = \boldsymbol{\omega} \cdot \nabla \mathbf{v} - \mathbf{v} \cdot \nabla \boldsymbol{\omega} + \nu \nabla^2 \boldsymbol{\omega}. \tag{8.9}$$

The BI Eq related to (8.9) in a region V is given by

$$c(i)\mathbf{v}(i) = - \iiint_V (\boldsymbol{\omega} \times \nabla u^*) \, dV +$$
$$+ \int_0^{t'} dt' \iint_S \left[\nabla u^* \mathbf{v} \cdot \hat{\mathbf{n}} + (\mathbf{v} \times \hat{\mathbf{n}} \times \nabla u^*) \right] dS. \tag{8.10}$$

Here u^* is the fundamental solution for the diffusion equation

$$\frac{\partial u}{\partial t} = k \nabla^2 u, \tag{8.11}$$

which is given by

$$u^* = \frac{1}{[4\pi k(t - t_0)]^{m/2}} e^{-r^2/4k(t-t_0)}, \tag{8.12}$$

where m is the dimension of the region V ($m = 2$ for a two–dimensional region, see (4.47)). The BI Eq corresponding to (8.9) for vorticity is given by

$$\boldsymbol{\omega} = \int_V [\boldsymbol{\omega} u^*]_{t=0} \, dV + \int_0^t dt' \int_S (\boldsymbol{\omega} \cdot \nabla \mathbf{v} - \mathbf{v} \cdot \nabla \boldsymbol{\omega}) u^* \, dS +$$
$$+ \nu \int_0^t dt' \int_S (\boldsymbol{\omega} \cdot \nabla u^* - u^* \nabla \boldsymbol{\omega} \cdot \hat{n}) \, dS, \tag{8.13}$$

which expresses the entire kinematics of a viscous flow in the form of an integral representation for the vorticity vector $\boldsymbol{\omega}$. The first integral in (8.13) represents the contribution of the initial vorticity distribution — through the process of diffusion because of Eq (8.11) — to the distribution of vorticity at a subsequent time. The second integral in (8.13) which arises because of the diffusion equation (8.11), satisfied by the vorticity field, is inhomogeneous. The inhomogeneous term represents the effects of vorticity stretching and of convection. The first term of this second integral gives the effect of vorticity stretching which is absent in two–dimensional flows. The third integral in (8.13) gives the effects of the presence of the boundary S (of the region V) on the vorticity. The integration of the third integral is performed only on the boundary surface S, on which the vorticity is being generated continually and is being diffused from the boundary into the fluid region. Since Eq (8.13) is time–dependent, these consideration will not be discussed any further in this chapter.

Note that for the velocity potential the appropriate boundary condition is $q = \partial\phi/\partial n = 0$ because the velocity in this formulation must be kept tangent to the boundary, whereas for the stream function the boundary condition on a fixed boundary is $\psi = \text{const}$ (chosen equal to 0). Since the functions ϕ and ψ are conjugate harmonic function, the curves $\phi = \text{const}$ are orthogonal to all the curves $\psi = \text{const}$ that intersect them. In either case, the derivatives of ϕ and ψ are important and ϕ or ψ can be assigned arbitrarily at some point in the region or on a part of the boundary. The fluid flow problems in Examples 8.2 and 8.3 explain this difference in the two formulations.

Consider the flow in the region shown in Fig. 8.2. The fluid is assumed to enter with a constant velocity at the left end and leave with the same constant velocity at the right end. The circular region is a solid obstruction. We shall investigate the effect of this obstruction on the flow.

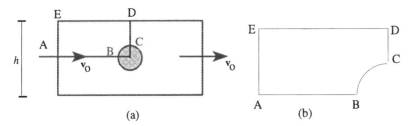

Fig. 8.2. Flow with a solid circular obstruction.

We shall use symmetry and consider only the quadrant as the model for analysis in the two formulations:

(a) **Velocity Potential Formulation.** In this case there is no flow orthogonal to the portions ABC and DE, i.e., $\hat{n}\cdot\nabla\phi = \partial\phi/\partial n = 0$. On the side CD we take $\phi = 0$ which gives $u_y = \partial\phi/\partial y = 0$, as expected. On the portion EA, we have $u_x = \partial\phi/\partial x = v_0$ prescribed.

(b) **Stream Function Formulation.** Since the portion ABC of the boundary is a stream line and since the derivatives are needed to establish the flow, we shall take $\psi = 0$ on the portion ABC. Using the prescribed boundary condition, we have $u_x = \partial\psi/\partial y = v_0$ on the portion EA of the boundary. Integrating this condition we get $\psi = v_0 y + f(x)$, which, on using the condition that $\psi = 0$ at $y = 0$ yields $f(x) = 0$. Thus, $\psi = v_0 y$. Now, on the portion DE, we have $\psi = v_0 h$. But since ψ depends linearly on y between E and A (recall that $\psi = v_0 y$), the y–component of the velocity is zero by symmetry, i.e., $u_y = -\partial\psi/\partial x = 0$. The stream function ψ for the constant and linear boundary elements on the quadrant in Fig. 8.2 can be computed by the methods of Chapter 5 (see Problem 8.17).

Example 8.2 (Velocity potential formulation):
From (8.7) we find that the equation of motion of a uniform incompressible viscous fluid in a steady one–dimensional flow in the z–direction is given by

$$-\frac{\partial p}{\partial z} + \mu\left(\frac{\partial^2 w}{\partial x^2} + \frac{\partial^2 w}{\partial y^2}\right) = 0,$$

where $\mu = \nu/\rho$ is the viscosity of the fluid and w is the velocity component in the z direction. Let $\partial p/\partial z = -P$ be a constant pressure gradient. Then the above equation can be written as

$$\nabla^2 w = -\frac{P}{\mu}.$$

For a pipe of elliptic cross–section $\dfrac{x^2}{a^2} + \dfrac{y^2}{b^2} = 1$ (see Fig. 5.16), the velocity field is given by

$$w = \frac{P}{2\mu\left(a^{-2} + b^{-2}\right)}\left(1 - \frac{x^2}{a^2} - \frac{y^2}{b^2}\right),$$

where a, b are the semi–axes of the ellipse. Let $P/\mu = 2$, $a = 2$, and $b = 1$. Then the problem reduced to the Poisson equation

$$\nabla^2 w = -2$$

with the boundary condition $w = 0$ on the boundary C. For the solution of this problem, see Examples 5.6, 5.11 and 5.13. ∎

Example 8.3 (Stream function formulation):
Consider a two–dimensional motion in the polar cylindrical coordinate system

(r, θ) with velocity components (u_r, u_θ) such that the vorticity vector $\boldsymbol{\omega} = \nabla \times \mathbf{v}$ has only one component

$$\zeta = \frac{1}{r} \frac{\partial (r u_\theta)}{\partial r} - \frac{1}{r} \frac{\partial u_r}{\partial \theta} \tag{8.14}$$

normal to the (r, θ)–plane. If the stream function ψ is defined by

$$u_r = \frac{1}{r} \frac{\partial \psi}{\partial \theta}, \quad u_\theta = -\frac{\partial \psi}{\partial r}, \tag{8.15}$$

then equation of continuity (8.8) becomes

$$\frac{1}{r} \frac{\partial (r u_r)}{\partial \theta} + \frac{1}{r} \frac{\partial u_\theta}{\partial \theta} = 0$$

and is satisfied automatically. If we substitute (8.14) and (8.15) into the Navier–Stokes equations (8.7), then the equations of motion become

$$\nabla^2 \psi = -\zeta,$$

$$\nabla^2 \zeta = \rho \left[\frac{1}{r} \frac{\partial \psi}{\partial \theta} \frac{\partial \zeta}{\partial r} - \frac{1}{r} \frac{\partial \psi}{\partial r} \frac{\partial \zeta}{\partial \theta} \right],$$

which, being a coupled system of second order elliptic equations, can be combined into a single fourth order nonlinear equation

$$\nabla^4 \psi = -\frac{\rho}{r} \frac{\partial (\psi, \nabla^2 \psi)}{\partial (r, \theta)} = -\rho J(\psi, \nabla^2 \psi), \tag{8.16}$$

where $\nabla^4 \equiv \nabla^2(\nabla^2)$ is the biharmonic operator, and J is the Jacobian. Since $\rho \to 0$ for a slow viscous flow, Eq (8.16) reduces to the biharmonic equation

$$\nabla^4 \psi = 0 \tag{8.17}$$

for slow or creeping flows. Eq (8.17), by using the Green's formula (1.7), leads to the BI Eq (as in (8.2))

$$c(i)\psi(i) = -\iint_R \left[\frac{\psi}{r^2} \frac{\partial r}{\partial n} + \frac{1}{r} \frac{\partial \psi}{\partial n} + r \frac{\partial}{\partial n}(\nabla^2 \psi) - (\nabla^2 \psi) \frac{\partial r}{\partial n} \right] dA, \tag{8.18}$$

where, unlike (8.2), $c(i) = 2$ if i is inside R, and $c(i) = 1$ if i is on the smooth boundary C of R.

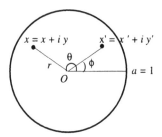

Fig. 8.3. Slow circular flow.

For example, consider the slow viscous motion inside a circular cylinder of radius a and center at the origin (see Fig. 8.3). A Fourier series solution of Eq (8.17) is given by

$$\psi = a_0 + b_0 r^2 + \sum_{n=1}^{\infty} \left[\left(a_n r^n + b_n r^{n+2}\right) \cos n\theta + \left(c_n r^n + d_n r^{n+2}\right) \sin n\theta \right],$$

$$(8.19)$$

where the coefficients a_i, b_i, c_i, d_i are to be determined by satisfying the boundary conditions

$$\psi = f(\theta), \quad \frac{\partial \psi}{\partial r} = -u_\theta = g(\theta) \quad \text{on } r = 1. \qquad (8.20)$$

An integral representation of the series solution (8.19) is given in Problem 8.14. ∎

8.3. Porous Media Flows

The flow through a granular region (anisotropic porous medium) is usually governed by the D'Arcy's law

$$\mathbf{v} = -\boldsymbol{\kappa} \cdot \nabla \phi, \qquad (8.21)$$

where $\mathbf{v}$ is the seepage velocity, $\boldsymbol{\kappa}$ is the permeability tensor and ϕ the velocity potential (piezometric head), i.e., $\phi = p/\rho g + z$, p being the pore-water pressure intensity and ρ the fluid density. Then the equation of continuity (8.8) gives the equation of conservation of mass

$$\nabla \cdot \left(\boldsymbol{\kappa} \nabla \phi\right) = 0. \qquad (8.22)$$

In a three–dimensional cartesian coordinate system, Eq (8.21) becomes

$$\left\{\begin{array}{c} u \\ v \\ w \end{array}\right\} = -\begin{bmatrix} \kappa_{xx} & \kappa_{xy} & \kappa_{xz} \\ \kappa_{yx} & \kappa_{yy} & \kappa_{yz} \\ \kappa_{zx} & \kappa_{zy} & \kappa_{zz} \end{bmatrix} \left\{\begin{array}{c} \partial\phi/\partial x \\ \partial\phi/\partial y \\ \partial\phi/\partial z \end{array}\right\}, \tag{8.23}$$

where $\mathbf{v} = \{u, v, w\}$ are velocity components in the x, y, z directions. In porous media, since the soil is frequently deposited in layers and is more compact in one direction, the soil grains have a preferred direction (see §8.6). They may also have seams which allow water to flow more easily in one direction than another. Since κ is symmetric, i.e., $\kappa_{xy} = \kappa_{yx}$, $\kappa_{yz} = \kappa_{zy}$, $\kappa_{zx} = \kappa_{xz}$, there are only six different components of κ out of the nine components. This means that the flow in soils in one direction is equal and opposite under equal and opposite pressure gradients. For the matrix κ with only six different components we can find a coordinate system in which the coordinate axes are in the direction of the eigenvectors of the matrix κ, thus making the off–diagonal elements zero. To do so, we write

$$\kappa \cdot A = \lambda\mathbf{A}, \tag{8.24}$$

where λ is a scalar and $\mathbf{A} = [A_1\ A_2\ A_3]^T$ is an unknown vector. Eq (8.24) is resolved into three simultaneous algebraic equations

$$\begin{align} (\kappa_{xx} - \lambda)A_1 + \kappa_{xy}A_2 + \kappa_{xz}A_3 &= 0 \\ \kappa_{yx}A_1 + (\kappa_{yy} - \lambda)A_2 + \kappa_{yz}A_3 &= 0 \\ \kappa_{zx}A_1 + \kappa_{zy}A_2 + (\kappa_{zz} - \lambda)A_3 &= 0. \end{align} \tag{8.25}$$

For a nontrivial solution, the determinant of the coefficient matrix must be zero, i.e.,

$$\det\begin{vmatrix} \kappa_{xx} - \lambda & \kappa_{xy} & \kappa_{xz} \\ \kappa_{yx} & \kappa_{yy} - \lambda & \kappa_{yz} \\ \kappa_{zx} & \kappa_{zy} & \kappa_{zz} - \lambda \end{vmatrix} = 0. \tag{8.26}$$

This cubic equation in λ has three roots which are the eigenvalues of the permeability tensor κ. Hence these three values of λ must be distinct, and they will determine the corresponding directions of the three eigenvectors A_1, A_2, A_3, which will form the coordinate system, with principal coordinate axes x', y', z', say, in which the tensor κ is diagonal. Thus, in the new coordinate system (x', y', z')

$$\kappa_{i'j'} = \sum_{i=1}^{3}\sum_{j=1}^{3} a_{i'i}a_{j'j}\kappa_{ij}, \quad i', j' = 1, 2, 3, \tag{8.27}$$

where i', j' refer to the directions x', y', z', and $a_{i'i}$ and $a_{j'j}$ are the cosines of the angles between two coordinate systems, viz., $a_{1'2} \equiv a_{x'y} = \cos\theta_{x'y}$, where $\theta_{x'y}$ is the angle between x' and y axes.

With this transformation (which makes the off–diagonal elements zero), Eq (8.23) becomes in the (x', y', z') system

$$
\left\{\begin{array}{c} u' \\ v' \\ w' \end{array}\right\} = - \left[\begin{array}{ccc} \kappa_{x'x'} & 0 & 0 \\ 0 & \kappa_{y'y'} & 0 \\ 0 & 0 & \kappa_{z'z'} \end{array}\right] \left\{\begin{array}{c} \partial\phi/\partial x' \\ \partial\phi/\partial y' \\ \partial\phi/\partial z' \end{array}\right\}, \tag{8.28}
$$

Then Eq (8.8) gives

$$
\frac{\partial}{\partial x'}\left(\kappa_{x'x'}\frac{\partial\phi}{\partial x'}\right) + \frac{\partial}{\partial y'}\left(\kappa_{y'y'}\frac{\partial\phi}{\partial y'}\right) + \frac{\partial}{\partial z'}\left(\kappa_{z'z'}\frac{\partial\phi}{\partial z'}\right) = 0, \tag{8.29}
$$

which, in the case of a homogeneous porous medium, becomes

$$
\kappa_{x'x'}\frac{\partial^2\phi}{\partial x'^2} + \kappa_{y'y'}\frac{\partial^2\phi}{\partial y'^2} + \kappa_{z'z'}\frac{\partial^2\phi}{\partial z'^2} = 0. \tag{8.30}
$$

The transformation $X = x'$, $Y = y'\sqrt{\kappa_{x'x'}/\kappa_{y'y'}}$, $Z = z'\sqrt{\kappa_{x'x'}/\kappa_{z'z'}}$, reduces Eq (8.30) to the Laplace equation

$$
\nabla^2\phi \equiv \frac{\partial^2\phi}{\partial X^2} + \frac{\partial^2\phi}{\partial Y^2} + \frac{\partial^2\phi}{\partial Z^2} = 0, \tag{8.31}
$$

which shows how a non–isotropic (anisotropic) problem can be solved as an isotropic problem by using a rotation and stretching of coordinates.

In groundwater applications, κ is usually known as a function of space coordinates only in a local sense, i.e., there are 'zones' in the granular region where κ is taken as a constant and has a value which may differ from its value in a vertical adjacent 'zone'. This local concept is justified by the vertical structure of the soil in which beds or layers usually possess well–defined boundaries.

If the coefficients $\kappa_{x'x'} = \kappa_{y'y'} = \kappa_{z'z'}$ in (8.30), then the porous medium is isotropic; if these quantities are different, then it is anisotropic (see §8.6). The values of the permeability κ, in units of cm/sec, for some porous media are: $\ln\kappa = -8$ to -10 for noncracked rock; $\ln\kappa = -7$ to -11 for clay; $\ln\kappa = -5$ to -8 for peat; $\ln\kappa = -3$ to -7 for silt and loess; $\ln\kappa = -1$ to -3 for sand; and $\ln\kappa = 2$ to -1 for gravel.

8.4. Steady–state Groundwater Flows

We shall consider large scale groundwater flow problems where the horizontal length scale is much larger than the vertical thickness of the aquifer. In such cases we can simplify the governing equations by assuming that the vertical flow is small as compared to the horizontal flow.

Example 8.4:
The boundary conditions for steady–state saturated groundwater flows are of the Dirichlet or Neumann type. We shall consider the case of steady–state seepage flow through an earth dam (Fig. 8.4), and solve $\nabla^2 \phi = 0$ under the assumption that the dam is homogeneous.

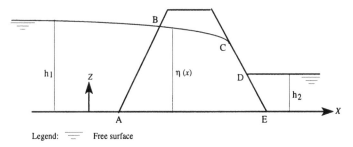

Fig. 8.4. An earth dam.

Note that along the seepage surface CD, the potential is the same as the vertical elevation because the atmospheric pressure is assumed zero; along the impervious boundary AE, the normal flux is zero; and finally, along the free surface BC which is a stream line for steady flow, the normal flux is also zero. Thus, the boundary conditions are:

$$\phi = \begin{cases} h_1 & \text{along } AB \text{ (upstream surface)} \\ h_2 & \text{along } DE \text{ (downstream surface)}, \end{cases}$$
$$\phi = z \quad \text{along } CD,$$
$$q = \frac{\partial \phi}{\partial n} = 0 \quad \text{along } AE,$$
$$\phi = \eta, \quad q = 0 \quad \text{along } BC,$$

$$(8.32)$$

where $z = \eta(x)$ is the free surface elevation. ∎

Example 8.5:

Consider a confined aquifer bounded below and above by strata $z = h_1$ and $z = h_2$ respectively (Fig. 8.5). Integrating the equation of continuity (8.8) between these two strata we find that

$$\int_{h_1}^{h_2} \frac{\partial u}{\partial x}\, dz + \int_{h_1}^{h_2} \frac{\partial v}{\partial y}\, dz + \int_{h_1}^{h_2} \frac{\partial w}{\partial z}\, dz = 0, \qquad (8.33)$$

or, using the formula (1.3) we get

$$\frac{\partial}{\partial x} \int_{h_1}^{h_2} u\, dz - u(x,y,h_2)\frac{\partial h_2}{\partial x} + u(x,y,h_1)\frac{\partial h_1}{\partial x} + \frac{\partial}{\partial y} \int_{h_1}^{h_2} v\, dz -$$

$$-v(x,y,h_2)\frac{\partial h_2}{\partial y} + v(x,y,h_1)\frac{\partial h_1}{\partial y} + w(x,y,h_2) - w(x,y,h_1) = 0. \qquad (8.34)$$

Since there is no flow across these two strata, we have

$$u(x,y,h_1)\frac{\partial h_1}{\partial x} + v(x,y,h_1)\frac{\partial h_1}{\partial y} = w(x,y,h_1),$$

$$u(x,y,h_2)\frac{\partial h_2}{\partial x} + v(x,y,h_2)\frac{\partial h_2}{\partial y} = w(x,y,h_2). \qquad (8.35)$$

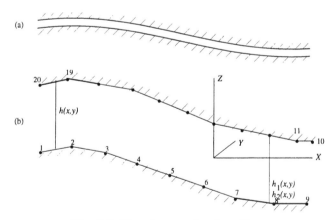

Fig. 8.5. Flow in a confined aquifer.

Thus, Eq (8.34) can be written as

$$\frac{\partial \tilde{u}}{\partial x} + \frac{\partial \tilde{v}}{\partial y} = 0, \qquad (8.36)$$

where

$$\tilde{u} = \int_{h_1}^{h_2} u\,dz, \quad \tilde{v} = \int_{h_1}^{h_2} v\,dz \qquad (8.37)$$

are the mean vertical velocities multiplied by the aquifer thickness $(h_2 - h_1) \equiv h$. We can now integrate the D'Arcy's law (8.21) in a similar manner and get

$$\int_{h_1}^{h_2} u\,dz = -\kappa \int_{h_1}^{h_2} \frac{\partial \phi}{\partial x}\,dz, \qquad (8.38)$$

$$\int_{h_1}^{h_2} v\,dz = -\kappa \int_{h_1}^{h_2} \frac{\partial \phi}{\partial y}\,dz. \qquad (8.39)$$

Since the horizontal potential gradient is not a function of the vertical coordinate z, we obtain the approximate values of $\tilde{u}$ and $\tilde{v}$ from (8.38) and (8.39) as

$$\tilde{u} = -\kappa(h_2 - h_1)\frac{\partial \phi}{\partial x} = -\kappa h\frac{\partial \phi}{\partial x}, \qquad (8.40a)$$

$$\tilde{v} = -\kappa(h_2 - h_1)\frac{\partial \phi}{\partial y} = -\kappa h\frac{\partial \phi}{\partial y}. \qquad (8.40b)$$

Denoting κh by T, which is called transmissivity, we find from (8.37) and (8.40) that

$$\frac{\partial}{\partial x}\left(T\frac{\partial \phi}{\partial x}\right) + \frac{\partial}{\partial y}\left(T\frac{\partial \phi}{\partial y}\right) = 0, \qquad (8.41)$$

which is analogous to (8.22). If $T = \text{const}$, then Eq (8.41) reduces to the Laplace equation. ∎

Example 8.6:
Consider an unconfined aquifer as shown in Fig. 8.6.

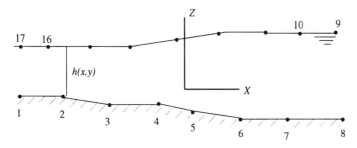

Fig. 8.6. Flow in an unconfined aquifer.

Note that the coordinate system is not rectangular cartesian. We assume that the vertical variation in the lower confining aquifer is small, and that ϕ does not change vertically, which implies that ϕ is equal to the depth of the fluid, i.e.,

$$\phi(x,y) = h(x,y). \tag{8.42}$$

From (8.40) we get $T = \kappa h = \kappa \phi$, and then (8.41) gives

$$\frac{\partial}{\partial x}\left(\kappa\phi\frac{\partial\phi}{\partial x}\right) + \frac{\partial}{\partial y}\left(\kappa\phi\frac{\partial\phi}{\partial y}\right) = 0. \tag{8.43}$$

In the case when $\kappa = $ const, we get

$$\nabla^2\phi^2 = 0, \tag{8.44}$$

which is Laplace equation with the square of the potential ϕ. ∎

If there are sources and sinks in the form of wells in an aquifer, the fluid is injected into or pumped out from a point in such a medium. The D'Arcy's equation (8.21) for conservation of mass is then modified to

$$\nabla \cdot (\kappa\nabla\phi) = -\sum_{i=1}^{N} Q_i\delta(x - x_i)\delta(y - y_i)\delta(z - z_i), \tag{8.45}$$

where N is the number of sources and sinks, Q_i is the volume rate of flow (which is positive for a source and negative for a sink) of the i-th source or sink whose coordinates are (x_i, y_i, z_i), and $\delta(p)$ is the Dirac delta function which is zero for $p \neq 0$ and unity for $p = 0$. For a homogeneous medium, Eq (8.45) becomes

$$\nabla^2\phi = -\sum_{i=1}^{N} \frac{Q_i}{\kappa}\delta(x - x_i)\delta(y - y_i)\delta(z - z_i). \tag{8.46}$$

Example 8.7:
The case of a leaky aquifer is shown in Fig. 8.7. In this case the confining layers (aquitards) are not entirely permeable; instead they leak, i.e., allow some transmission of fluid into or out of the aquifer. Let us assume that the leakage rate v_1 is proportional to the potential (head) difference across the confining layer, i.e.,

$$v_1 = \frac{\kappa_1}{h_1}(\phi - \phi_1), \tag{8.47}$$

where κ_1 is the permeability of the aquitard, h_1 is the thickness of the confining layer, and ϕ_1 is the potential (head) outside of the confining layer. Then the equation of continuity (8.8) becomes

$$\frac{\partial u}{\partial x} + \frac{\partial v}{\partial y} = \frac{\kappa_1}{h_1}(\phi_1 - \phi), \tag{8.48}$$

and (8.41) yields

$$\frac{\partial}{\partial x}\left(T\frac{\partial \phi}{\partial x}\right) + \frac{\partial}{\partial y}\left(T\frac{\partial \phi}{\partial y}\right) + \frac{\kappa_1}{h_1}(\phi_1 - \phi) = 0. \tag{8.49}$$

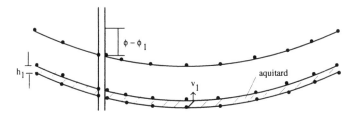

Fig. 8.7. A leaky aquifer.

Let us define $\hat{\phi} = \phi - \phi_1$, and assume that ϕ_1 and T are constant. Then (8.49) gives

$$\frac{\partial^2 \hat{\phi}}{\partial x^2} + \frac{\partial^2 \hat{\phi}}{\partial y^2} - \frac{\kappa_1}{h_1}\hat{\phi} = 0. \tag{4.50}$$

If we take $\alpha = \kappa_1/h_1 T$, Eq (8.50) leads to a Helmholtz–type equation

$$(\nabla^2 - \alpha)\hat{\phi} = 0, \tag{8.51}$$

which can be analyzed as in §7.3, with the obvious difference that α is positive and real, with the corresponding fundamental solution (in a two–dimensional region) as $\hat{\phi}^* = \frac{i}{4}H_0^{(1)}(\sqrt{\alpha}r)$. If there are sources and sinks, Eq (8.45) or (8.46) must be added to (8.50) or (8.51) respectively.

The case of an unsaturated flow in a porous medium which occurs when air fills parts of the pores is very complex and challenging to explain, and will not be discussed here. The reason for the complexity lies in the fact that the surface tension of water reduces the pore pressure on the moisture content; it has rather different values for wetting than for draining.

Note that if compressibility of the granular region is important, then the governing equation becomes the time–dependent diffusion equation

$$s\frac{\partial \phi}{\partial t} = \nabla^2 \phi, \tag{8.52}$$

where s is the 'storativity' of the region. ∎

8.5. Anisotropic Porous Media

Anisotropic porous media require the approach of orthotropic materials. Consider an anisotropic body shown in Fig. 8.8 where X_1, X_2, X_3 denote the directions of orthotropy in a three–dimensional medium.

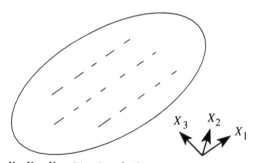

X_1, X_2, X_3 : Direction of orthotropy

Fig. 8.8. A three–dimensional orthotropic medium.

The governing equation in a three–dimensional medium V, referred to the directions X_1, X_2, X_3 of orthotropy, is

$$\nabla^2_\kappa \phi = 0, \tag{8.53}$$

where

$$\nabla^2_\kappa \equiv \kappa_1 \frac{\partial^2}{\partial x_1^2} + \kappa_2 \frac{\partial^2}{\partial x_2^2} + \kappa_3 \frac{\partial^2}{\partial x_3^2}, \tag{8.54}$$

and $\kappa_1, \kappa_2, \kappa_3$ define the material constants in the directions of orthotropy. In order to find the fundamental solution for (8.53), we use the transformation

$$z_j = \frac{x_j}{\sqrt{\kappa_j}}, \quad j = 1, 2, 3. \tag{8.55}$$

Let us assume that the potential is concentrated at a point i. Then the fundamental solution ϕ^* must satisfy

$$\nabla_\kappa^2 \phi^* + \delta(i) = 0, \tag{8.56}$$

where

$$\delta(i) = \delta(x_1 - x_1^i)\delta(x_2 - x_2^i)\delta(x_3 - x_3^i), \tag{8.57}$$

such that the integral of $\delta(i)$ is 1 at the point i. If we substitute (8.55) into (8.56), we get

$$\nabla_\kappa^2 \phi^* + \delta(i) = \frac{\partial \phi^*}{\partial z_1^2} + \frac{\partial \phi^*}{\partial z_2^2} + \frac{\partial^2 \phi^*}{\partial z_3^2} + \delta_\kappa(3) = 0, \tag{8.58}$$

where

$$\delta_\kappa(i) = \delta\left(\sqrt{\kappa_1}(z_1 - z_1^i)\right)\delta\left(\sqrt{\kappa_2}(z_2 - z_2^i)\right)\delta\left(\sqrt{\kappa_3}(z_3 - z_3^i)\right). \tag{8.59}$$

If we integrate Eq (8.58) over a three–dimensional region V, we obtain

$$\iiint_V \phi(\nabla_\kappa^2 \phi^* + \delta_\kappa(i))\, dV = \iiint_V \phi \nabla_\kappa^2 \phi^*\, dV + \iiint_V \phi \delta_\kappa(i)\, dV. \tag{8.60}$$

Now, in view of the translation property (1.55) for the Dirac delta function, we find that

$$\lim_{a \to \infty} \int_{z_j^i - a}^{z_j^i + a} \phi(z_j)\delta\left(\sqrt{\kappa_j}(z_j - z_j^i)\right)\, dz_j$$

$$= \lim_{a \to \infty} \int \frac{1}{\sqrt{\kappa_j}}\phi(x_j/\sqrt{\kappa_j})\delta(x_j - x_j^i)\, dx_j$$

$$= \frac{1}{\sqrt{\kappa_j}}\phi\left(x_j^i/\sqrt{\kappa_j}\right) = \frac{1}{\sqrt{\kappa_j}}\phi(z_j^i) \quad \text{for each } j = 1, 2, 3.$$

Hence the second integral on the right side of (8.60) becomes

$$\iiint_V \phi \delta_\kappa(i)\, dV = \frac{1}{\sqrt{\kappa_1 \kappa_2 \kappa_3}}\phi(i). \tag{8.61}$$

The fundamental solution of Eq (8.58) is of the same as (4.25), viz., $\phi^* = 1/4\pi r$, where in view of (8.61),

$$
r = \sqrt{\kappa_1\kappa_2\kappa_3\left(z_1^2 + z_2^2 + z_3^2\right)} = \sqrt{\kappa_1\kappa_2\kappa_3\left(\frac{x_1^2}{\kappa_1} + \frac{x_2^2}{\kappa_2} + \frac{x_3^2}{\kappa_3}\right)}
$$

$$
\equiv r_0\sqrt{\kappa_1\kappa_2\kappa_3}. \quad (8.62)
$$

Thus, the fundamental solution of D'Arcy equation (8.53) in a three–dimensional region is

$$
\begin{aligned}
\phi^* &= \frac{1}{4\pi}\frac{1}{\sqrt{\kappa_1\kappa_2\kappa_3}r_0} \\
&= \frac{1}{4\pi}\left[\kappa_1\kappa_2\kappa_3\left(\frac{x_1^2}{\kappa_1} + \frac{x_2^2}{\kappa_2} + \frac{x_3^2}{\kappa_3}\right)\right]^{-1/2}.
\end{aligned} \quad (8.63)
$$

Similarly, using (4.26), the fundamental solution for D'Arcy equation in a two–dimensional medium is

$$
\begin{aligned}
\phi^* &= \frac{1}{2\pi\sqrt{\kappa_1\kappa_2}}\ln\left(\frac{1}{r_0}\right) \\
&= -\frac{1}{4\pi\sqrt{\kappa_1\kappa_2}}\ln\left(\frac{x_1^2}{\kappa_1} + \frac{x_2^2}{\kappa_2}\right),
\end{aligned} \qquad r_0 = \sqrt{\frac{x_1^2}{\kappa_1} + \frac{x_2^2}{\kappa_2}}. \quad (8.64)
$$

Again, in a two–dimensional region R, since the Green theorem (1.6) gives

$$
\iint_R \left(\kappa_1\frac{\partial^2\phi}{\partial x_1^2} + \kappa_2\frac{\partial^2\phi}{\partial x_2^2}\right)dx_1\,dx_2 = \int_C \left(\kappa_1\frac{\partial\phi}{\partial n}n_1 + \kappa_2\frac{\partial\phi}{\partial n}n_2\right)ds,
$$
$$(8.65)$$

where $n_1 = \cos(n, x_1)$, $n_2 = \cos(n, x_2)$, we find that the normal derivative (or internal flux) at the boundary C is given by

$$
q = \kappa_1\frac{\partial\phi}{\partial x_1}n_1 + \kappa_2\frac{\partial\phi}{\partial x_2}n_2. \quad (8.66)
$$

In this case we are required to evaluate the boundary integral

$$
I_1 = -\frac{1}{\sqrt{\kappa_1\kappa_2}}\int_C q\ln\left(\frac{x_1}{\kappa_1} + \frac{x_2}{\kappa_2}\right)ds. \quad (8.67)
$$

The integrand in (8.67) becomes unbounded at the singular point i $(0,0)$, but at other points it is integrable since $\ln r$ is integrable on each closed interval $[0, a]$. In the neighborhood of the point i, we can use the logarithm Gauss quadrature (C.6) because the singularity at this point has a nonzero contribution to the value of I_1. To see this, consider a circular cut C_ε with center at i. Then

$$
I_1 = -\frac{1}{\sqrt{\kappa_1\kappa_2}}\lim_{\varepsilon\to 0}\left[\int_{C-C_\varepsilon} q\ln\left(\frac{x_1}{\kappa_1} + \frac{x_2}{\kappa_2}\right)ds + \int_{C_\varepsilon} q\ln\left(\frac{x_1}{\kappa_1} + \frac{x_2}{\kappa_2}\right)ds\right].
$$

In the second integral, take $x_1 + ix_2 = \varepsilon e^{i\theta}$, $\alpha < \theta < \beta$, so that $ds = \varepsilon\, d\theta$; then

$$\lim_{\varepsilon \to 0} \int_{C_\varepsilon} q \ln\left(\frac{x_1}{\kappa_1} + \frac{x_2}{\kappa_2}\right) ds$$

$$= q \lim_{\varepsilon \to 0} \int_\alpha^\beta \ln\left[\varepsilon\left(\frac{\cos^2\theta}{\kappa_1} + \frac{\sin^2\theta}{\kappa_2}\right)\right] \varepsilon\, d\theta = 0.$$

If the medium is isotropic ($\kappa_1 = \kappa_2 = \kappa$), then, e.g., for quadratic elements (Fig. 5.13)

$$
\begin{aligned}
G_{ij} &= -\frac{1}{2\pi\kappa} \int_{-1}^1 \phi_j(\xi) \ln\left(\frac{L_i\xi}{\sqrt{\kappa}}\right) d\xi \\
&= \frac{L_i}{36\pi\kappa}
\begin{bmatrix}
g_{11} & g_{12} & g_{13} \\
& g_{22} & g_{23} \\
\text{sym} & & g_{33}
\end{bmatrix},
\end{aligned}
\tag{8.68}
$$

where ϕ_j, $j = 1, 2, 3$ are the interpolation functions (1.28), L_i the length of the element, and

$$g_{11} = 17 + 3\ln\frac{\kappa}{4}, \quad g_{12} = 4\left(5 + 3\ln\frac{\kappa}{4}\right),$$

$$g_{13} = -1 + 3\ln\frac{\kappa}{4}, \quad g_{22} = 4\left(8 + 3\ln\frac{\kappa}{4}\right),$$

$$g_{23} = 2 + 3\ln\kappa.$$

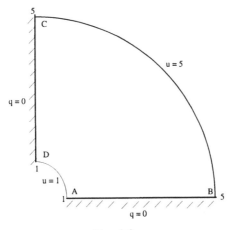

Fig. 8.9.

Example 8.8:
Consider the isotropic medium of Fig. 8.9, and solve the Laplace equation

$\nabla^2 u = 0$, with the boundary conditions shown there. The exact solution is

$$u = 1 + \frac{4\ln r}{\ln 5}, \qquad q = \frac{4}{r\ln 5},$$

where $r = \sqrt{x^2 + y^2}$. The pressure P on the faces AB and CD, and the discharge Q through the entire medium are

$$P_{AB} = 8\left(3 - \frac{2}{\ln 5}\right), \quad P_{CD} = 4\left(3 - \frac{4}{\ln 5}\right), \quad Q = \frac{2\pi\kappa}{\ln 5}.$$

The boundary is discretized in the same manner as in Fig. 5.12 with linear elements with $\kappa = 1.\ \blacksquare$

8.6. Flows with Singularities

In the case of linear elements we use the interpolation functions ϕ_1 and ϕ_2 given by (1.27) between the extreme nodes. If there are singularities present in the medium, these interpolation functions are modified in the neighborhood of a singularity to more closely approximate the exact solution. In solving the Laplace, Poisson, or Helmholtz equation, we sometimes deal with singularities in the manner explained in Example 5.8. While using the BEM, we can either ignore these singularities and expect that the error will be limited to the vicinity of the singularity only and then use the discontinuous elements (§5.5). However, such singularities should not be ignored if an accurate solution is desired.

Example 8.9:
We will consider the flow around a right–angled corner (wedge) as shown in Fig. 8.10.

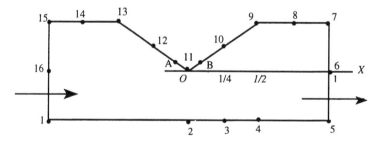

Fig. 8.10. Flow around a right–angled wedge.

The governing equation is the Laplace equation $\nabla^2\phi = 0$. The flow has a singularity at node 11 as the velocity becomes unbounded at that point. At this node which is on a solid boundary we have $q = \partial\phi/\partial n = 0$. We want to compute the potential ϕ specially in the neighborhood of the node 11. We know theoretically that $\phi \sim r^{2/3}$ in the vicinity of this singularity, where r is the distance from the corner. With linear elements we will use the following interpolation functions for the two elements at the corner:

$$\text{For element 10–11}: \quad \phi_1 = \left(1 - \frac{\xi}{L}\right)^{2/3}, \quad \phi_2 = 1 - \left(1 - \frac{\xi}{L}\right)^{2/3},$$

$$\text{For element 11–12}: \quad \phi_1 = 1 - \left(\frac{\xi}{L}\right)^{2/3}, \quad \phi_2 = \left(\frac{\xi}{L}\right)^{2/3},$$

where ξ is the distance from node 11. The program Be2 should be modified accordingly. The other method is to use discontinuous elements and replace node 11 by two adjacent nodes marked A and B. Then the results are: $\phi(0,0) = 0$, $\phi(0.25, 0.25) = 0.505$, $\phi(0.5, 0.5) = 0.72$. ∎

Example 8.10:
Consider the flow through an earth dam with the vertical cut–off and finite depth of tail water, as shown in Fig. 8.11.

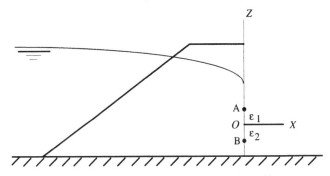

Fig. 8.11. Flow through an earth dam with tail water.

Let the origin be at the point O which is the intersection of the tail water and the downstream face of the dam. This point is a singularity since the velocity tends to infinity there. For the sake of simplicity, we shall assume that the downstream face of the dam is vertical. We expect the solution to have the following properties:

$$\begin{cases} \phi = \phi_0 + z, & \partial\phi/\partial z = 1, \quad \text{for } z > 0 \\ \phi = \phi_0, & \partial\phi/\partial z = 0, \quad \text{for } z < 0. \end{cases} \tag{8.69}$$

This problem is satisfied by

$$\phi = \int_0^a m\tan^{-1}\frac{z - t}{x}\, dt, \tag{8.70}$$

where m is an unknown constant to be determined, and $a \gg 0$ is an arbitrary large number (see Polubarinova–Kochina, 1962). It is easy to see that the solution (8.70) satisfies the Laplace equation since

$$\frac{\partial^2 \phi}{\partial x^2} = \int_0^a \frac{2mx(z-t)}{[x^2 + (z-t)^2]}\, dt = -\frac{\partial^2 \phi}{\partial z^2}.$$

If we set $z-t = \zeta$, use the formula $\int \tan^{-1}\frac{\zeta}{x}\, d\zeta = \zeta \tan^{-1}\frac{\zeta}{x} - \frac{x}{2}\ln\left(x^2 + \zeta^2\right) + c$, where c is an arbitrary constant, and then integrate (8.70), we get

$$\frac{\phi}{m} = z\tan^{-1}\frac{z}{x} - (a-z)\tan^{-1}\frac{a-z}{x} - \frac{x}{2}\ln\frac{1 + (z/x)^2}{1 + [(a-z)/x]^2}, \quad (8.71)$$

and

$$\frac{1}{m}\frac{\partial \phi}{\partial z} = \tan^{-1}\frac{z}{x} + \tan^{-1}\frac{a-z}{x}, \qquad (8.72)$$

which for $a > z$ yields

$$\begin{aligned}
\frac{1}{m}\frac{\partial \phi}{\partial z} &= \frac{\pi}{2} + \frac{\pi}{2} = \pi, \quad \text{for } z > 0, x = 0, \\
\frac{1}{m}\frac{\partial \phi}{\partial z} &= -\frac{\pi}{2} + \frac{\pi}{2} = 0, \quad \text{for } z < 0, x = 0,
\end{aligned} \qquad (8.73)$$

and these satisfy the properties (8.69) if $m = 1/\pi$. This determines m, and from (8.71) we have

$$\frac{1}{m}\frac{\partial \phi}{\partial x} = -\frac{1}{2}\ln\frac{x^2 + z^2}{x^2 + (a-z)^2}, \qquad (8.74)$$

which at $x = 0$ tends to infinity as $\ln(1/|z|)$ near the point O.

Now, for two adjacent nodes A and B on both sides of the point O, the singular boundary integral in the vicinity of the point O is

$$I = \int_A^B q\ln r\, ds. \qquad (8.75)$$

If the distance between the point O and the node A or B is small, we can take $\varepsilon_1 = \varepsilon_2 = \varepsilon$; then r is nearly unchanged over the integration and can be treated like a constant. Since $ds = dz$, Eq (8.75) gives

$$\begin{aligned}
I &= \ln r\left[\int_\varepsilon^0 q_A \ln\frac{e\varepsilon}{z}\, dz + \int_0^{-\varepsilon} q_B \ln\left(-\frac{e\varepsilon}{z}\right)\, dz\right] \\
&= -2\varepsilon \ln r\left[q_A + q_B\right].
\end{aligned} \qquad (8.76)$$

If the point i is at A, then the integral (8.75) can be written as $I = (I_A + I_B) \ln(\varepsilon - z)$, where

$$
\begin{aligned}
I_A &= q_A \int_\varepsilon^0 \ln(\varepsilon - z) \ln\left(\frac{e\varepsilon}{z}\right) dz \\
&= \varepsilon q_A \left[3 - \frac{\pi^2}{6} - 2\ln\varepsilon\right] = \varepsilon q_A \left[1.3550659 - 2\ln\varepsilon\right],
\end{aligned}
\tag{8.77}
$$

by setting $\varepsilon - z = \varepsilon u$, and using the formulas $\int_0^1 \ln u \, du = -1 = \int_0^1 \ln(1 - u) \, du$, and $\int_0^1 \ln u \ln(1 - u) \, du = 2 - \pi^2/6$. We can similarly evaluate

$$
I_B = q_B \int_0^{-\varepsilon} \ln(\varepsilon - z) \ln\left(-\frac{e\varepsilon}{z}\right) dz
$$

by using the formula $\int_0^1 \ln u \ln(1 + u) \, du = 2 - \pi^2/12 - 2\ln 2$. It will be found that $I_A \neq I_B$. ∎

Determination of the free surface in porous media flows is a very sensitive issue because it is always one of the least known problems as to which boundary conditions must be chosen on the free surface. An analysis of convergence and stability of solutions has suggested that we should choose the potential as unknown, and not the velocities, on the free surface. But we will not pursue this topic any further.

8.7. Problems

8.1. Setup and solve for the potential (heat transfer) ϕ on the airfoils in Fig. 8.1.

8.2. Setup and solve the flow problem for the rectangular obstruction shown in Fig. 8.12. Use both stream function and velocity potential formulations.

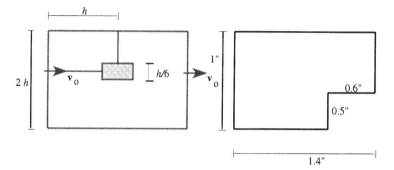

Fig. 8.12. Flow with a solid rectangular obstruction.

8.3. Write a program (call it Be6) for the case of linear elements of the type of Fig. 5.4(b), to solve Poisson equation with up to 5 surfaces.

8.4. Formulate the boundary element method for the case of linear elements of Fig. 5.4(b), to solve the D'Arcy equation in a two–dimensional region.

8.5. Formulate and solve the problem of Example 8.5 by using linear elements.

8.6. Formulate and solve the problem of Example 8.6 by using linear elements.

8.7. Formulate and solve the problem of Example 8.7 by using linear elements.

8.8. Formulate and solve the problem of Example 8.8 by using linear elements.

8.10. Solve the stream function form of the flow problem for a horizontal circular cylinder of radius a which is half–immersed in water of depth $2a$ (see Fig. 8.13).

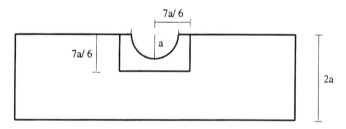

Fig. 8.13.

8.11. Solve the problem of porous flow through the square, shown in Fig. 8.14, with the data $\kappa_x = \kappa_y = 0.1$cm/sec, by dividing the square into three regions R_1, R_2, R_3.

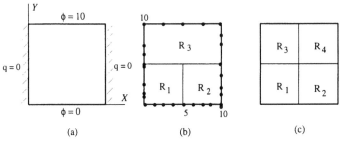

Fig. 8.14.

The exact solution is: $\phi = y$, $\phi_x = 0$, $\phi_y = 0.1$.

8.12. Solve the Problem 8.11 by dividing the square into four symmetric region of Fig. 8.14(c).

8.13. Solve the problem of a homogeneous porous flow in a two–layered medium shown in Fig. 8.15, with $\kappa_A = 1 = \kappa_B$, and the porosity $n_e = 10$ on the interface AB. The exact solution is: $\phi_A = y/2$, $\phi_B = 5 + y/2$, and the velocities $\phi_x = 0$, $\phi_y = 0.5$.

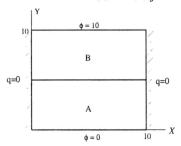

Fig. 8.15.

8.14. By the method that is used to derive the Poisson integral representation from its Fourier series, show that the series solution (8.19) has the integral form

$$\psi(r,\theta) = \frac{(1-r^2)^2}{2\pi} \int_0^{2\pi} \frac{[1 - r\cos(\alpha - \theta)]f(\alpha)}{[1 - 2r\cos(\alpha - \theta) + r^2]^2} -$$
$$- \frac{1}{2}\frac{g(\alpha)}{1 - 2r\cos(\alpha - \theta) + r^2}\bigg] d\alpha.$$

8.15. Let the seepage of groundwater through a homogeneous isotropic aquifer satisfy Eq (8.31). Consider the square region which has a constant head gradient representing a flow in the negative y direction, with three different sets of boundary conditions at the nodes 7, 8 and 9: (a) $\phi_7 = \phi_8 = \phi_9 = 30.0$, (b) $q_7 = q_8 = q_9 = 10.0$, and (c) $Q = 20\kappa$, where ϕ is the piezometric head , κ the aquifer permeability, $q = \partial\phi/\partial n$, and Q the total flow along the nodes 7–9 (Fig. 8.16).

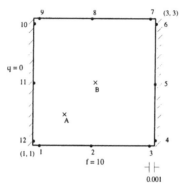

Fig. 8.16.

8.16. (Confined flow about a circular cylinder) The irrotational flow of an ideal fluid about a circular cylinder placed with its axis perpendicular to the plane of the flow between two horizontal walls (see Fig. 5.9) is governed by the Laplace equation $-\nabla^2 u = 0$. Analyze the flow using both stream function and velocity potential formulations of §8.2 and Fig. 8.17 (quarter region shown), where V_0 is the velocity of the fluid parallel to the stream line. The boundary conditions are as follows: (i) For the stream line formulation, $\psi = 0$ on sides AB and BC; $\psi = 2$ on the side DE, $\psi = V_0 y$ on the side AE, and $\partial\psi/\partial n = 0$ on the side CD; (ii) for the velocity potential formulation, $\phi = 0$ on the side CD, $\partial\phi/\partial n = 0$ on the sides AB, BC and DE, and $\partial\psi/\partial n = V_0$ on the side AE.

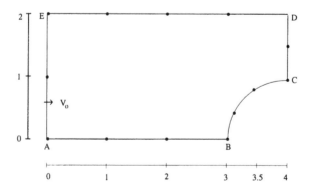

Fig. 8.17. Confined flow about a circular cylinder

8.17. Consider the seepage of groundwater flow through a confined aquifer between two large bodies of water. Fig. 8.18 shows a section where the upper confining layer is penetrated to a depth of 5m by two long trenches 10m wide and 40m apart. There is a steady flow (abstraction) from trench A of $Q = 1.5\kappa$. The piezometric heads in both trenches as well as the distributions of normal velocities are initially not known. Compute the heads at $x = 50$m and 250m for the values of heads as prescribed in Fig. 8.17.

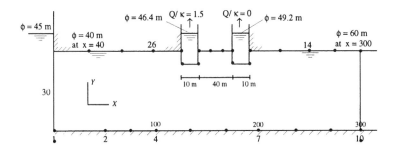

Fig. 8.18.

References and Bibliography for Additional Reading

I. H. Abbott and A. E. van Doenhoff, *Theory of Wing Sections*, Dover, New York, 1959.

H. Antes, *Anwendungen der Methode der Randelemente in der Elastodynamik und der Fluiddynamik*, B. G. Teubner, Stuttgart, 1988.

G. K. Batchelor, *An Introduction to Fluid Dynamics*, Cambridge Univ. Press, Cambridge, 1967.

C. A. Brebbia, J. Telles and L. Wrobel, *Boundary Element Techniques – Theory and Applications in Engineering*, Springer-Verlag, Berlin, 1984.

E. K. Bruch, *The Boundary Element Method for Groundwater Flow*, Lecture Notes in Engineering, Vol. 70, Springer–Verlag, Berlin, 1991.

L. E. Coates, *Including total flow boundary conditions in the BEM with applications to groundwater flow*, in "Boundary Elements in Mechanical and Electrical Engineering (C. A. Brebbia and A. Chaudouet–Miranda, Eds.), Proc. International Boundary Element Symposium, Nice, France, 15–17, May 1990, Computational Mechanics Publications, Southampton and Springer–Verlag, Berlin, 1990, pp. 299–309.

Y. G. Hanna and J. L. Humar, *Boundary Element Analysis of Fluid Domain*, J. Engg. Mech. Div., ASCE **108** (1982).

T. W. Lambe and R. V. Whitman, *Soil Mechanics*, Wiley, New York, 1969.

J. A. Liggett and P. L–F. Liu, *The Boundary Integral Equation Method for Porous Media Flow*, George Allen Unwin, 1983.

P. L–F. Liu and J. A. Liggett, *Boundary solutions to two problems in porous media*, J. Hydraulics Div. ASCE **105** (1979), 171–183.

R. D. Mills, *Computing internal viscous flow problems for the circle by integral methods*, J. Fluid Mech. **79(3)** (1977), 609–624.

P. Ya. Polubarinova–Kochina, *Theory of Groundwater Movement*, Translated from Russian by J. M. R. de Wiest, Princeton Univ. Press, 1962.

H. Schlichting, *Boundary Layer Theory*, McGraw–Hill, New York, 1960.

A. Verruijt, *Theory of Groundwater Flows*, Gordon and Breach Science Publishers, New York, 1970.

A. D. Wheelon, *Tables of Summable Series and Integrals Involving Bessel Functions*, Holden–Day, San Francisco, 1968.

J. C. Wu and J. F. Thompson, *Numerical solutions of time–dependent incompressible Navier–Stokes equations using an integro–differential formulation*, Computers and Fluids **1** (1973), 197–215.

9

Domain Integrals

The domain integrals arise in the BE formulation for the potential, elastostatic, and other problems. Thus, for example, in the Poisson problem (5.65)–(5.66), the function $b(\mathbf{x})$ produces the domain integral (5.71), i.e.,

$$B_i = \iint_R bu^* \, dx \, dy, \qquad (9.1)$$

which appears in the BE Eq (5.70). In elastostatic problems, if the body forces are present, we encounter the domain integral (6.45), i.e.,

$$B_i = \iint_{\tilde{R}} \mathbf{b} \mathbf{u}^* \, dx_1 \, dx_2, \qquad (9.2)$$

which is present in the related BE Eq (6.44). In the case of forced oscillations the governing equation is

$$\left(\nabla^2 + k^2\right) u = b, \qquad (9.3)$$

where $k = \omega/c$ is the wave number, and b is a function of space variables that measures the yield of the sources which may be continuously disturbed or concentrated in a single point. The function b must vanish at infinity. If $b = 0$, Eq (9.3) reduces to the Helmholtz equation (7.37). The BE Eq for (9.3) will also lead to a domain integral of the form (9.1). The domain integral (8.4) which is of the same type as (9.1) also appears in aerodynamic flows around lifting bodies and in flows through porous media defined by (8.31). The interior cell method developed in §5.7 is not a powerful method to numerically compute the domain integrals. This method involves an interior discretization and evaluation of the double integral over each interior cell, which in turn increases the numerical data considerably and thus diminishes the computational advantage that the BEM has over other domain–type methods.

the double integral over each interior cell, which in turn increases the numerical data considerably and thus diminishes the computational advantage that the BEM has over other domain–type methods. In order to avoid the cumbersome situation created by the interior cell method we must have a method that enables a boundary–only solution instead of the domain integrals and is applicable to similar problems whenever they arise. There are three different methods available at present to transform the domain integrals into boundary integrals. These methods have historically evolved as follows: (i) Dual reciprocity method (DRM), (ii) Fourier series expansion method (FSM), and (iii) Multiple reciprocity method (MRM). We shall now present them in that order.

9.1. Dual Reciprocity Method

The direct method of numerically integrating the domain integrals over the interior cells, as developed in §5.7, often becomes difficult or even impossible, to correctly compute, if the fundamental solution u^* is not available in a nice form or the function b is of a complicated form for a problem. The DRM is one of the methods that provides an alternative to the interior cell method.

The DRM for the Poisson problem (5.65)–(5.66) is carried out in three steps:

Step 1: Derive the BE Eq (5.70) and separate the domain integral term.
Step 2: Obtain particular solutions $\hat{u}$, $\hat{q}$ and an approximating function f as follows: The functions $\hat{u}$ and f are chosen such that

$$\nabla^2 \hat{u} = f, \tag{9.4}$$

where f is called the DRM approximating function and is a known nonsingular function dependent on the geometry of the region R. In the case of Eq (5.65) the function $\hat{u}$ is chosen as

$$\hat{u} = \frac{r^2}{4} + \frac{r^3}{9}, \tag{9.5}$$

where $r = \sqrt{x^2 + y^2}$ is the radial distance. Then, since

$$\nabla^2 \hat{u} = \frac{\partial^2 \hat{u}}{\partial r^2} + \frac{1}{r}\frac{\partial \hat{u}}{\partial r}$$

in polar cylinder coordinates, we find that $f = 1 + r$. It should be remarked that (9.5) is not the only choice for $\hat{u}$, although it has been successfully used in computations. The only condition for any choice of $\hat{u}$ is that the function f should be nonsingular, i.e., it has no singularities (see below for more on the choices of f). Once $\hat{u}$ is selected, it is easy to define $\hat{q}$ by

$$
\begin{aligned}
\hat{q} = \frac{\partial \hat{u}}{\partial n} &= \frac{\partial \hat{u}}{\partial x}\frac{\partial x}{\partial n} \frac{\partial \hat{u}}{\partial y}\frac{\partial y}{\partial n} \\
&= \frac{\partial \hat{u}}{\partial r}\frac{\partial r}{\partial x}\frac{\partial x}{\partial n} + \frac{\partial \hat{u}}{\partial r}\frac{\partial r}{\partial y}\frac{\partial y}{\partial n} \\
&= \left(\frac{\partial r}{\partial x}\frac{\partial x}{\partial n} + \frac{\partial r}{\partial y}\frac{\partial y}{\partial n} \right) \frac{\partial \hat{u}}{\partial r}.
\end{aligned}
\tag{9.6}
$$

Thus, in this case

$$
\hat{q} = \left(\frac{\partial r}{\partial x}\frac{\partial x}{\partial n} + \frac{\partial r}{\partial y}\frac{\partial y}{\partial n} \right) \left(\frac{r}{2} + \frac{r^2}{3} \right).
\tag{9.7}
$$

Step 3: Note that the BI Eq (5.69) without the domain term $\iint_R bu^* \, dx \, dy$ is the same as (5.14). Since this domain term leads to the vector B_i defined by (9.1), we shall transform the domain integral to the boundary integral over $\tilde{C}$ and obtain a DRM expression which is equivalent to the vector B_i. This DRM expression is then added to the right side of (5.17), or (5.32), or (5.47), or (5.61), depending on the type of boundary elements. This will result in a matrix equation of the form $B + HU = GQ$ which is solved as in Chapter 5.

The procedure to determine this DRM BE Eq starts with defining an approximation of the function b by

$$
b = \sum_{j=1}^{N+M} \beta_j f_j,
\tag{9.8}
$$

where N is the total number of boundary nodes, M the total number of interior nodes, $\{\beta_j\}$ a set of undetermined coefficients, and f_j the DRM approximating functions which are related to the particular solutions $\hat{u}_j$ by

$$
\nabla^2 \hat{u}_j = f_j.
\tag{9.9}
$$

The interior nodes, unlike the nodes of the interior cells of §5.7, are important in the DRM because often times we may require to determine solution at one or all of them, or the function b may vary rapidly there, or the boundary conditions

are homogeneous. However, only a few interior nodes are sufficient in most
of the problems to obtain the desired accuracy in the DRM.

If we substitute (9.9) into (9.8) we get

$$b \approx \sum_{j=1}^{N+M} \beta_j \nabla^2 \hat{u}_j = \sum_{j=1}^{N+M} \beta_j f_j, \tag{9.10}$$

which, upon substitution into (5.65), gives

$$\nabla^2 u \approx \sum_{j=1}^{N+M} \beta_j \nabla^2 \hat{u}_j. \tag{9.11}$$

Notice that both left and right sides in (9.11) contain the Laplacian. Hence
we follows the basic procedure, as in §5.1, i.e., multiply both sides of (9.11)
by the fundamental solution u^* and integrate over R:

$$\iint_R \left(\nabla^2 u \right) u^* \, dx \, dy \approx \sum_{j=1}^{N+M} \beta_j \iint_R \left(\nabla^2 \hat{u}_j \right) u^* \, dx \, dy. \tag{9.12}$$

Use of (1.5c) leads to the following equation for each i:

$$c(i)u(i) + \int_C u q_i^* \, ds - \int_C u_i^* q \, ds =$$

$$= \sum_{j=1}^{N+M} \beta_i \left\{ c(i)\hat{u}_{ij} + \int_C \hat{u}_j q_i^* \, ds - \int_C u_i^* \hat{q}_j \, ds \right\}. \tag{9.13}$$

Eq (9.13) can also be obtained directly by applying Green's second identity
or reciprocity theorem (1.8) to (9.11) — hence the name DRM. (See Problem
9.1). Notice that there is no domain integral in (9.13). Now we can discretize
Eq (9.13) for each boundary and interior node, thus obtaining

$$c(i)u(i) + \sum_{k=1}^{N} \int_{\tilde{C}_k} u q^* \, ds - \sum_{k=1}^{N} \int_{\tilde{C}_k} u^* q \, ds =$$

$$= \sum_{j=1}^{N+M} \beta_j \left\{ c(i)\hat{u}_{ij} + \sum_{k=1}^{N} \int_{\tilde{C}_k} \hat{u}_j q^* \, ds - \sum_{k=1}^{N} \int_{\tilde{C}_k} u^* \hat{q}_j \, ds \right\}. \tag{9.14}$$

Using the functions H_{ij} and G_{ij} defined in (5.17)–(5.19), the above equation
can be written as

$$c(i)u(i) + \sum_{k=1}^{N} H_{ik} u_k - \sum_{k=1}^{N} G_{ik} q_j =$$

$$= \sum_{j=1}^{N+M} \left\{ \beta_j \left[c(i)\hat{u}_{ij} + \sum_{k=1}^{N} H_{ik} \hat{u}_{kj} - \sum_{k=1}^{N} G_{ik} \hat{q}_k \right] \right\}. \tag{9.15}$$

Note that the value of $c(i)$ is unity at the M interior points. Recall that both $\hat{u}$ and $\hat{q}$ are known once f is defined. After applying the boundary conditions, Eq (9.15) goes through the collocation process just like the one used in §5.3, and we obtain the DRM BE Eq

$$Hu - Gq = \sum_{j=1}^{N+M} \beta_j \left(H\hat{u}_j - G\hat{q}_j \right), \qquad (9.16)$$

which in the matrix form is

$$H\mathbf{u} - G\mathbf{q} = \left(H\hat{U} - G\hat{Q} \right) \beta, \qquad (9.17)$$

where the matrices $H, G, \hat{U}, \hat{Q}$ are computed from $u^*, q^*, \hat{u}$ and $\hat{q}$ respectively. Eq (9.16) in its form is also valid for three–dimensional Poisson problems.

Eq (9.17) is the basis for the application of the DRM, where, by taking the value of b at $(N + M)$ different (collocation) points, the vector β is expressed in the matrix form as

$$\mathbf{b} = \mathbf{F}\beta. \qquad (9.18)$$

In (9.18) each column of $\mathbf{F}$ contains the values of f_j which are the values of f at the $(N + M)$ DRM collocation points. On inversion Eq (9.18) yields the value of β as

$$\beta = \mathbf{F}^{-1}\mathbf{b}, \qquad (9.19)$$

which defines the values of β in (9.17) since b is a known function. We can now rewrite the DRM BE Eq (9.17) in the form

$$H\mathbf{u} - G\mathbf{q} = \left(H\hat{U} - G\hat{Q} \right) \mathbf{F}^{-1}\mathbf{b}, \qquad (9.20\text{a})$$

or

$$H\mathbf{u} - G\mathbf{q} = \mathbf{W}\mathbf{b}, \qquad (9.20\text{b})$$

where

$$\mathbf{W} = \left(H\hat{U} - G\hat{Q} \right) \mathbf{F}^{-1}, \qquad (9.21)$$

and the matrix $\mathbf{F}$ depends on the geometric data and is independent of the governing equation or the boundary conditions. The matrix $\mathbf{W}$ is obtained by matrix multiplication of the known matrices. From computational point, the matrix $\mathbf{F}$ needs be inverted only once and is stored for all subsequent use with the same type of discretization. The vector b will be different, yet known, for different problems. A schematic representation of Eq (9.17) is given in Appendix D to help write the computer code for the DRM.

Different choices of f: The choice of the function f in the DRM formulation requires that the resulting matrix $\mathbf{F}$ in (9.18) be nonsingular (otherwise $\mathbf{F}$ cannot be inverted). The different choices of f are:

(i) Elements of the Pascal triangle, viz.,

$$
\begin{array}{ccccccc}
& & & 1 & & & \\
& & x & & y & & \\
& x^2 & & xy & & y^2 & \\
x^3 & & x^2 y & & xy^2 & & y^3 \\
& & & \vdots & & &
\end{array}
$$

(ii) Trigonometric series, either of the type used in Example 2.5, viz.,

$$\sum_j \sum_k \alpha_{jk} \sin \frac{j\pi x}{a_1} \sin \frac{k\pi y}{a_2},$$

where a_1, a_2 are constants, or those used in the Fourier series.

(iii) Test functions developed in §2.4.

(iv) The distance function r used in the definition of the fundamental solution, which gives an infinitely many choices for f as

$$1, \quad r, \quad 1+r, \quad 1+r+r^2, \quad \ldots, \quad 1+r+\cdots+r^n+\ldots.$$

This is the widely used choice as we have shown above in (9.5). In fact, $f = r$, or $f = 1 + r$ is just one component of the general choice

$$f = 1 + r + \cdots + r^m. \tag{9.22}$$

Then

$$\hat{u} = \frac{r^2}{4} + \frac{r^3}{9} + \cdots + \frac{r^{m+2}}{(m+2)^2}, \tag{9.23}$$

and

$$q = \left(\frac{\partial r}{\partial x} \frac{\partial x}{\partial n} + \frac{\partial r}{\partial y} \frac{\partial y}{\partial n} \right) \left(\frac{r}{2} + \frac{r^2}{3} + \cdots + \frac{r^{m+1}}{m+2} \right). \tag{9.24}$$

(v) Since the Laplacian is satisfied by the Bessel function J_0 in the polar cylindrical coordinate system with the origin at node j for axisymmetric problems, we can choose f_j for the Helmholtz equation as

$$f_j = J_0(k|x - x_j|).$$

In spherical coordinate system with the origin at j, the equation $\left(\nabla^2 + k^2\right) f_j = 0$ in axisymmetric problems is satisfied by

$$f_j = \frac{\sin(k|x - x_j|)}{k|x - x_j|}.$$

Notice that these approximating functions are of a totally different type than those represented by (9.22) and are rarely used.

Some particular cases:

1. A simple case where the function $b(x, y)$ is constant, linear, or harmonic in R has been treated in §5.9. We shall now consider another simple case when the function b satisfies the Poisson equation

$$\nabla^2 b = \gamma b, \tag{9.25}$$

where γ is a nonzero constant. Then the domain integral (9.1) is given by

$$B_i = \frac{1}{\gamma} \iint_R u^* \nabla^2 b \, dx \, dy, \tag{9.26}$$

which, by using the Green's second identity (1.8), becomes

$$B_i = \frac{1}{\gamma} \left[\int_C u^* \frac{\partial b}{\partial n} \, ds - \int_C q^* b \, ds + \iint_R b \nabla^2 u^* \, dx \, dy \right]. \tag{9.27}$$

Since u^* is the fundamental solution, the last integral over R in (9.27) reduces to $\iint_R b \nabla^2 u^* \, dx \, dy = -b(i)$. This leads to

$$B_i = \frac{1}{\gamma} \left[\int_C u^* \frac{\partial b}{\partial n} \, ds - \int_C b q^* \, ds - b(i) \right]. \tag{9.28}$$

If the point i is on the boundary C, the second integral on the right side is evaluated as a Cauchy p–v integral. Thus,

$$B_i = \frac{1}{\gamma} \left[\int_C u^* \frac{\partial b}{\partial n} \, ds - PV \int_C b q^* \, ds - c(i) b(i) \right], \tag{9.29}$$

where $c(i)$ is defined by (5.15). The domain integral B_i in (9.29) is now transformed into a boundary integral.

2. If $b = \lambda u$, then Eq (9.20a) becomes

$$H\mathbf{u} - G\mathbf{q} = \lambda \left(H\hat{U} - G\hat{Q} \right) \mathbf{F}^{-1} \mathbf{u}. \tag{9.30a}$$

This can be written as

$$Hu - Gq = \lambda Wu, \tag{9.30b}$$

or

$$(H - \lambda W)\,u = Gq, \tag{9.30c}$$

which can be solved like an eigenvalue problem.

3. For the Helmholtz equation (7.37), note that $b = -k^2 u$, and thus (9.19) becomes

$$\beta = -k^2 F^{-1} u, \tag{9.31}$$

and the associated DRM BE Eq is

$$Hu - Gq = -k^2 Wu. \tag{9.32}$$

Example 9.1:
Consider the case of a vibrating cantilever beam (fixed–free) shown in Fig. 9.1. In this case $k^2 = \rho \omega^2 / E$, where ρ is the mass density, ω the radian frequency, and E the Young's modulus.

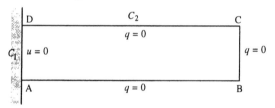

Fig. 9.1. Cantilever Beam

Since the boundary part AD is clamped, $u = 0$ there, and $q = 0$ on the remaining free part $ABCD$. Before we use (9.32), we can partition the matrices in (9.32) and eliminate the terms in q by writing (9.32) as

$$\begin{bmatrix} H_{11} & H_{12} \\ H_{21} & H_{22} \end{bmatrix} \begin{Bmatrix} u_1 \\ u_2 \end{Bmatrix} - \begin{bmatrix} G_{11} & G_{12} \\ G_{21} & G_{22} \end{bmatrix} \begin{Bmatrix} q_1 \\ q_2 \end{Bmatrix} =$$
$$= -k^2 \begin{bmatrix} W_{11} & W_{12} \\ W_{21} & W_{22} \end{bmatrix} \begin{Bmatrix} u_1 \\ u_2 \end{Bmatrix},$$

which for this problem is written as

$$\begin{bmatrix} H_{11} & H_{12} \\ H_{21} & H_{22} \end{bmatrix} \begin{Bmatrix} 0 \\ u \end{Bmatrix} - \begin{bmatrix} G_{11} & G_{12} \\ G_{21} & G_{22} \end{bmatrix} \begin{Bmatrix} q \\ 0 \end{Bmatrix} =$$
$$= -k^2 \begin{bmatrix} W_{11} & W_{12} \\ W_{21} & W_{22} \end{bmatrix} \begin{Bmatrix} 0 \\ u \end{Bmatrix},$$

or

$$H_{12}u - G_{11}q = -k^2 W_{12}u$$
$$H_{22}u - G_{21}q = -k^2 W_{22}u,$$

which, after eliminating q, becomes

$$\left(H_{22} - G_{21}G_{11}^{-1}H_{12}\right)u = -k^2 \left(W_{22} - G_{21}G_{11}^{-1}W_{12}\right)u,$$

or

$$Su = k^2 Tu, \tag{9.33}$$

where $S = H_{22} - G_{21}G_{11}^{-1}H_{12}$ and $T = G_{21}G_{11}^{-1}W_{12} - W_{22}$ are the stiffness and mass matrices. Inverting the matrix T, we obtain from (9.33) the system

$$Au = k^2 u, \tag{9.34}$$

where $A = T^{-1}S$. The eigenvalue problem (9.34) can now be solved easily. Note that since the matrix A is nonsymmetric, some of the eigenvalues will be complex. But these complex eigenvalues are found to be of higher order and may not be of much interest. The BE discretization of the beam is taken to be the same as in Fig. 9.3 with $N = 24$ and $M = 9$. For the vibrating beam the exact solutions for the eigenvectors (displacement) and eigenfrequencies are given by

$$u_m = C \sin \frac{(2m-1)\pi x}{2l},$$
$$k_m = \frac{(2m-1)\pi}{2l}, \quad m = 1, 2, \ldots,$$

where C is a constant (usually taken as unity) and l is the length of the beam. Thus, $k_1 = 0.196$, $k_2 = 0.589$, $k_3 = 0.981$, $k_4 = 1.374$. The corresponding eigenfrequencies are $\omega_m = k_m \sqrt{E/\rho}$, $m = 1, 2, \ldots$. The eigenvectors (displacements) for k_1, k_2, k_3 and k_4 are shown in Fig. 9.2. ∎

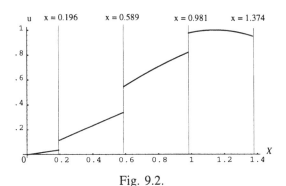

Fig. 9.2.

Example 9.2:

Consider the problem of the vibration of a rectangular membrane of length a and width b and fixed all along its edges. The eigenfrequencies ω_{nm} and the normal modes θ_{nm} are given by

$$\omega_{nm} = ck_{nm} = c\pi\sqrt{\frac{n^2}{a^2} + \frac{m^2}{b^2}},$$

$$\theta_{nm} = \sin\frac{n\pi x}{a}\sin\frac{m\pi y}{b}, \quad n, m = 1, 2, \ldots,$$

where c denotes the sound velocity. The BE discretization is shown in Fig. 9.3, with $N = 24$ and $M = 9$. ∎

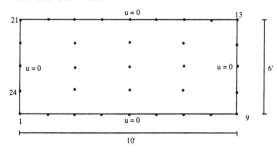

Fig. 9.3.

4. In elastostatics, the Cauchy–Navier equations are

$$\mu\frac{\partial^2 u_i}{\partial x_j \partial x_j} + (\lambda + \mu)\frac{\partial^2 u_j}{\partial x_i \partial x_j} + b_i = 0, \tag{9.35}$$

where u_i are displacements and b_i the body forces (see §4.4). We shall consider the deformation of the body in the plane $x_3 = $ const, since the deformation in the state of plane stress is independent of the x_3 coordinate. The general solution of (9.35) can be written as

$$u_i = u_i^c + u_i^p, \tag{9.36}$$

where the superscripts 'c' and 'p' refer to the complementary function and the particular integral. Let the gravitational acceleration be directed along the x_2-axis. Then the components of the body forces are $b_1 = 0$, $b_2 = -\rho g$, where ρ is the mass density and g the acceleration due to gravity. The particular integrals in this case are given by (see Sokolnikoff, p. 258)

$$u_1^p = -\frac{\lambda}{4\mu(\lambda + \mu)}\rho g x_1 x_2,$$

$$u_2^p = \frac{\lambda}{8\mu(\lambda + \mu)}\rho g x_1^2 + \frac{\lambda + 2\mu}{4\mu(\lambda + \mu)}\rho g x_2^2, \tag{9.37}$$

which yield the stress components

$$\sigma_{11}^{\mathrm{P}} = 0 = \sigma_{12}^{\mathrm{P}}, \quad \sigma_{22}^{\mathrm{P}} = \rho g x_2, \tag{9.38}$$

and the related tractions $p_i^{\mathrm{P}} = \sigma_{ij}^{\mathrm{P}} n_j$ are given by

$$p_1^{\mathrm{P}} = 0, \quad p_2^{\mathrm{P}} = \rho g x_2 n_2. \tag{9.39}$$

In view of (9.36), the general solution for stresses and tractions will also be of the same form, viz., $\sigma_{ij} = \sigma_{ij}^{\mathrm{c}} + \sigma_{ij}^{\mathrm{P}}$, and $p_i = p_i^{\mathrm{c}} + p_i^{\mathrm{P}}$. In the DRM formulation we shall use the above particular integrals since they are independent of the geometry and the boundary conditions for any problem. The fundamental solutions are defined by (6.28)–(6.33). For a three–dimensional body the BI Eq (6.40) reduces to

$$c_l(i)u_l(i) = \iint_S \left[\left(p_k^{\mathrm{c}} + p_k^{\mathrm{P}} \right) u_{lk}^* - \left(u_k^{\mathrm{c}} + u_k^{\mathrm{P}} \right) p_{lk}^* \right] dS. \tag{9.40}$$

Since

$$G p^{\mathrm{c}} - H u^{\mathrm{c}} = 0, \tag{9.41}$$

where G and H are defined in (6.47) and (6.49), we obtain the DRM BE Eq

$$G \left(p - p^{\mathrm{P}} \right) - H \left(u - u^{\mathrm{P}} \right) = 0, \tag{9.42}$$

which, after rearrangement, can be written as

$$AX = F + G p^{\mathrm{P}} - H u^{\mathrm{P}}, \tag{9.43}$$

where F is the vector obtained from the prescribed displacements and tractions, A is the coefficient matrix, and X the vector that contains unknown displacements and tractions. Example 6.3 can be solved using this DRM formulation.

Example 9.3:
Consider the Poisson equation $-\nabla^2 u = 2$ in a region R which is the right isosceles triangle of hypotenuse a (see Fig. 9.4). This problem describes the back and forth movement (seiche) of water waves in a right isosceles triangular basin of finite depth. The eigenvalues are $k_m = m\pi, m\pi a$, where $m = 1, 2, \ldots$ ∎

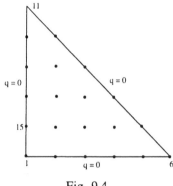

Fig. 9.4.

The DRM also goes by a lesser known name of the particular integral method. It appears that the DRM is a technique to approximate the particular integrals by restoring the method of undetermined coefficients through a truncated series of finitely many terms which are defined by collocation at an arbitrary number of N boundary and M interior nodes (M may be zero in many problems). Although the computation reduces considerably, there are yet many questions that remain unanswered, e.g., those related to the criterion for the choice of collocation points $N + M$, or the shape functions to be chosen. This method also involves the inversion of a full nonsymmetric matrix which may sometimes cause computational problems.

9.2. Fourier Series Method

An arbitrary function b does not, in general, satisfy Eq (9.25). However, we can choose a set of functions $\{b_m(x)\}$, $m = 1, 2, \ldots$, such that each b_m satisfies (9.25), i.e., $\nabla^2 b_m = \gamma_m b_m$. Then the function b can be expanded as a series of this set, i.e.,

$$b = \sum_{m=1}^{\infty} k_m b_m(\mathbf{x}). \qquad (9.44)$$

If the set $\{b_m(\mathbf{x})\}$ is a complete set, then we can take only a finite part of the series (9.44), instead of the whole infinite series, to represent the function b (this will ensure sufficient accuracy). Then substitution of such a finite series into the domain integral (9.1) gives

$$B_i \approx \sum_{m=1}^{N} \iint_R u^* b_m(\mathbf{x}) \, dx \, dy. \tag{9.45}$$

Analogous to (9.29), the domain integral in this case is transformed into a boundary integral:

$$B_i = \sum_{m=1}^{N} \frac{k_1}{\gamma_m} \left[\int_C u^* \frac{\partial b_m}{\partial n} \, ds - PV \int_C b_m q^* \, ds - c(i) b_m(i) \right]. \tag{9.46}$$

The Fourier series method (FSM) is based on using the complete set of trigono-metric functions and represent $b(\mathbf{x})$ approximately by a finite part of the Fourier series as

$$b(x, y) \approx a_0 f_0 + \sum_{k=1}^{4} \sum_{n=1}^{N} a_n^k f_n^k + \sum_{k=1}^{4} \sum_{n=1}^{N} \sum_{m=1}^{M} a_{nm}^k f_{nm}^k, \tag{9.47}$$

where
$$f_0 = 1,$$

$$
\begin{aligned}
f_n^1 &= \cos(n\pi x/\alpha), & f_{nm}^1 &= f_n^1 f_m^3, \\
f_n^2 &= \sin(n\pi x/\alpha), & f_{nm}^2 &= f_n^1 f_m^4, \\
f_n^3 &= \cos(n\pi y/\beta), & f_{nm}^3 &= f_n^2 f_m^3, \\
f_n^4 &= \sin(n\pi y/\beta), & f_{nm}^4 &= f_n^2 f_m^4.
\end{aligned} \tag{9.48}
$$

$$a_0 = \frac{1}{4\alpha\beta} \iint_{R_f} b(x, y) \, dx \, dy,$$

$$a_k^n = \frac{1}{2\alpha\beta} \iint_{R_f} b(x, y) f_n^k(x, y) \, dx \, dy, \tag{9.49}$$

$$a_{nm}^k = \frac{1}{2\alpha\beta} \iint_{R_f} b(x, y) f_{nm}^k(x, y) \, dx \, dy,$$

where $2\alpha, 2\beta$ are the dimensions of the rectangle R_f containing the region R (Fig. 9.5). Substituting the Fourier series (9.47) into the domain integral (9.1) we get

$$B_i = \iint_R u^* b \, dx \, dy$$

$$= a_0 \iint_R u^* f_0 \, dx \, dy + \sum_{k=1}^{4} \sum_{n=1}^{N} a_n^k \iint_R u^* f_n^k \, dx \, dy + \tag{9.50}$$

$$+ \sum_{k=1}^{4} \sum_{n=1}^{N} \sum_{m=1}^{M} a_{nm}^k \iint_R u^* f_{nm}^k \, dx \, dy.$$

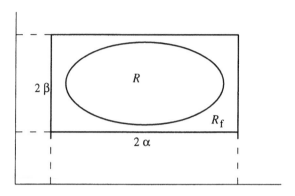

Fig. 9.5.

Let us now assume that for each of the functions f_0, f_n^k or f_{nm}^k there exists a corresponding function u_0, u_n^k or u_{nm}^k such that

$$\nabla^2 u_0 = f_0 \, (= 1), \quad \nabla^2 u_n^k = f_n, \quad \nabla^2 u_{nm}^k = f_{nm}, \qquad (9.51)$$

for $k = 1, 2, 3, 4$, and $n, m = 1, 2, \ldots$. Solving the equations in (9.51), we get

$$u_0 = \frac{1}{4} \left[(x - x_0)^2 + (y - y_0)^2 \right],$$

$$u_n^k = \begin{cases} -\left(\dfrac{\alpha}{n\pi} \right)^2 f_n^k & \text{for } k = 1, 2, \\[2ex] -\left(\dfrac{\beta}{n\pi} \right)^2 f_n^k & \text{for } k = 3, 4, \end{cases} \qquad (9.52)$$

$$u_{nm}^k = -\frac{1}{\pi^2 \left[\left(\dfrac{n}{\alpha} \right)^2 + \left(\dfrac{m}{\beta} \right)^2 \right]} f_{nm}^k,$$

where (x_0, y_0) is an arbitrary point in R (which can be taken at the origin of coordinates or at the center of R such that the two points (x, y) and (x_0, y_0) are not very close to each other). We shall now take Eqs (9.51) and apply Green's second identity to them. This yields

$$\iint_R u^* f_0 \, dx \, dy = \iint_R u^* \nabla^2 u_0 \, dx \, dy$$

$$= \int_C u^* \frac{\partial u_0}{\partial n} \, ds - \int_C q^* u_0 \, ds - c(i) u_0(i), \qquad (9.53)$$

$$\iint_R u^* f_n^k \, dx \, dy = \iint_R u^* \nabla^2 u_n^k \, dx \, dy$$

$$= \int_C u^* \frac{\partial u_n^k}{\partial n} \, ds - \int_C q^* u_n^k \, ds - c(i) u_n^k(i), \qquad (9.54)$$

$$\iint_R u^* f_{nm}^k \, dx \, dy = \iint_R u^* \nabla^2 u_{nm}^k \, dx \, dy$$

$$= \int_C u^* \frac{\partial u_{nm}^k}{\partial n} \, ds - \int_C q^* u_{nm}^k \, ds - c(i) u_{nm}^k(i). \qquad (9.55)$$

We shall use the notation:

$$g_0 = \frac{\partial u_0}{\partial n}, \quad g_n^k = \frac{\partial u_n^k}{\partial n}, \quad g_{nm}^k = \frac{\partial u_{nm}^k}{\partial n}. \qquad (9.56)$$

Then Eq (9.50) becomes

$$B_i = \iint_R u^* b \, dx \, dy$$

$$= \gamma_0 \left[\int_C u^* g_0 \, ds - \int_C q^* u_0 \, ds - c(i) u_0(i) \right] +$$

$$+ \sum_{k=1}^4 \sum_{n=1}^N \gamma_n^k \left[\int_C u^* g_n^k \, ds \int_C q^* f_n^k \, ds - c(i) f_n^k(i) \right] +$$

$$+ \sum_{k=1}^4 \sum_{n=1}^N \sum_{m=1}^M \gamma_{nm}^k \left[\int_C u^* g_{nm}^k \, ds - \int_C q^* f_{nm}^k \, ds - c(i) f_{nm}^k(i) \right], \qquad (9.57)$$

where

$$\gamma_0 = a_0,$$

$$\gamma_n^k = \begin{cases} -\dfrac{a_n^k}{\left(\frac{n\pi}{\alpha}\right)^2}, & k = 1, 2, \\[4mm] -\dfrac{a_n^k}{\left(\frac{n\pi}{\beta}\right)^2}, & k = 3, 4, \end{cases} \qquad (9.58)$$

$$\gamma_{nm}^k = -\dfrac{a_{nm}^k}{\left(\frac{n\pi}{\alpha}\right)^2 + \left(\frac{n\pi}{\beta}\right)^2}.$$

Another method to transform the domain integral B_i into boundary integrals is to choose a function v^* such that $\nabla^2 v^* = u^*$, where v^* is given by (5.78) (see §5.9). Then, as in (5.75),

$$\iint_R u^* \, dx \, dy = \iint_R \nabla^2 v^* \, dx \, dy = \int_C \frac{\partial v^*}{\partial n} \, ds,$$

and, as in (9.57),

$$
\begin{aligned}
B_i &= \iint_R bu^* \, dx \, dy \\
&= \gamma_0 \int_C \frac{\partial v^*}{\partial n} \, ds + \\
&+ \sum_{k=1}^{4} \sum_{n=1}^{N} \gamma_n^k \left[\int_C u^* g_n^k \, ds - \int_C q^* f_n^k \, ds - c(i) f_n^k(i) \right] + \\
&+ \sum_{k=1}^{4} \sum_{n=1}^{N} \sum_{m=1}^{M} \gamma_{nm}^k \left[\int_C u^* g_{nm}^k \, ds - \int_C q^* f_{nm}^k \, ds - c(i) f_{nm}^k(i) \right].
\end{aligned}
$$

$$(9.59)$$

We have, therefore, two formulas (9.57) and (9.59), either of which will transform the domain integral into boundary integrals.

Since the function $b(x, y)$ in the Poisson equation is known, its Fourier coefficients (9.49), and hence its Fourier series (9.47), are easily determined. This makes known the domain integral B_i, given by (9.57) or (9.59), in which all line integrals are numerically evaluated by Gauss quadrature. This is done as follows:

Case 1: The source point x' (or i) is inside R. Let $\phi(x, y)$ denote the integrand in any line integral in (9.57) or (9.59). Let P denote the total number of Gauss points, and Q the total number of boundary elements on the boundary C. Then

$$
\begin{aligned}
\int_C \phi(x, y) \, ds &= \sum_{j=1}^{Q} \int_{\tilde{C}_j} \phi(x, y) \, ds \\
&= \sum_{j=1}^{Q} \left\{ \sum_{p=1}^{P} \phi(\zeta_p) W_p \, |J| \right\},
\end{aligned}
$$

$$(9.60)$$

where ζ_p and W_p are the Gauss points and weights, defined in Appendix C, and $|J| = L_j/2$ is the value of the Jacobian, with L_j being the length of the element $\tilde{C}_j$.

Case 2: If the source point x' (or i) is on the element $\tilde{C}_j$, then the Gauss quadrature (9.60) does not work in this case because of the singularities of the function u^*, q^*, v^*, and $\partial v^*/\partial n$, where

$$
\begin{aligned}
u^* &= \frac{1}{2\pi} \ln\left(\frac{1}{r}\right), & v^* &= \frac{r^2}{8\pi}\left[1 + \ln\left(\frac{1}{r}\right)\right], \\
q^* &= -\frac{1}{2\pi r} \frac{\partial r}{\partial n}, & \frac{\partial v^*}{\partial n} &= \frac{r}{8\pi}\left[1 + 2\ln\left(\frac{1}{r}\right)\right] \frac{\partial r}{\partial n}.
\end{aligned}
$$

$$(9.61)$$

Consider an integral

$$I = \int_{-1}^{1} f(x)\, dx, \qquad (9.62)$$

where $f(x)$ has a singularity at a point $x_0 \in [-1, 1]$. We shall choose a cubic Galerkin–type approximation

$$x(z) = a + bz + cz^2 + dz^3, \qquad (9.63)$$

such that

$$x_0 = x(z_0),\ x(1) = 1,\ x(-1) = -1,\ \left.\frac{dx}{dz}\right|_{z_0} = 0 = \left.\frac{d^2 x}{dz^2}\right|_{z_0}. \qquad (9.64)$$

The last two conditions in (9.64) imply respectively that the Jacobian of the transformation is zero at the singularity z_0 and the Jacobian is an extremum which can be chosen to have an arbitrarily small value. The transformation (9.63) permits us to smoothen the singularity by the Jacobian, thus transforming integral (9.62) into

$$I = \int_{-1}^{1} f(x, z) \frac{dx}{dz}\, dz. \qquad (9.65)$$

Also, the conditions (9.64) yield $d = 1/(1 + 3z_0^2)$, $a = 3z_0 d = -c$, and $b = 3z_0^2 d$. Since $x_0 = x(z_0) = a + bz_0 + cz_0^2 + dz_0^3$, solving this cubic equation in z_0 by Cardan's method we get

$$z_0 = (x^* x_0 + |x^*|)^{1/3} + (x^* x_0 - |x^*|)^{1/3} + x_0,$$

where $x^* = x_0^2 - 1$. Hence, the integral (9.62) becomes

$$I = \int_{-1}^{1} f\left(\frac{(z - z_0)^3 + z_0(3 + z_0^2)}{1 + 3z_0^2}\right) \frac{3(z - z_0)^2}{1 + 3z_0^2}\, dz. \qquad (9.66)$$

With this expression for I, the Gauss quadrature can now be applied to each integral in (9.57) or (9.59) in this case.

Example 9.4:
Let the regions R and R_f be rectangles defined by

$$R = \{[2, 7] \times [5, 15]\}, \qquad R_f = \{[0, \alpha] \times [0, \beta]\},$$

where $\alpha = 10$, $\beta = 20$ (see Fig. 9.6).

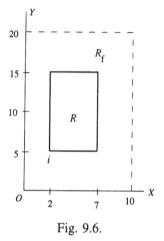

Fig. 9.6.

Let

$$b(x, y) = \left| \sin \frac{\pi x}{\alpha} \sin \frac{\pi y}{\beta} \right| .$$

It has the Fourier series

$$b(x, y) \approx \frac{4}{\pi^2} + \frac{8}{\pi^2} \sum_{n=1}^{N} \frac{1}{1 - 4n^2} \left[\cos \frac{2n\pi x}{\alpha} + \cos \frac{2n\pi y}{\beta} \right] +$$

$$+ \frac{16}{\pi^2} \sum_{n,m=1}^{N} \frac{1}{(1 - 4n^2)(1 - 4m^2)} \cos \frac{2n\pi x}{\alpha} \cos \frac{2n\pi y}{\beta} .$$

Then the domain integral B_i, given by (9.57), is computed at the source point $i\,(2, 5)$, and the results are as follows:

Source point	N	B_i
$(2, 5)$	2	-67.851
	5	-67.134
	10	-67.108

Exact $B_i = -67.10896$ ∎

Example 9.5:
Let the regions R and R_f be as in Fig 9.6, and let $b(x, y) = 1$. This is an even function and its Fourier series is

$$b(x, y) \approx \frac{4}{\pi} \sum_{n=1}^{N} \frac{1}{2n - 1} \sin \frac{(2n - 1)\pi x}{\alpha} ,$$

which is used to compute the domain integral B_i given by (9.59). The results at the source point $i\,(2,5)$ are:

Source point	N	B_i
$(2,5)$	2	-83.683
	5	-82.716
	10	-82.707
	15	-82.728

Exact $B_i = -82.7296$

If we consider the periodic extension of $b(x,y) = 1$ as an odd function in both x and y, then its Fourier series is

$$b(x,y) = 1$$

$$\approx \frac{16}{\pi^2} \sum_{n=1}^{M} \sum_{m=1}^{M} \frac{1}{(2n-1)(2m-1)} \sin \frac{(2n-1)\pi x}{\alpha} \sin \frac{(2m-1)\pi y}{\beta}.$$

Using this expansion we can compute the values of B_i given by (9.59) and get the following results:

Source point	N	B_i
$(2,5)$	2	-84.742
	5	-83.515
	10	-82.723
	15	-82.725

Exact $B_i = -82.7296$

Note that in this case the results are a little less accurate. ∎

Example 9.6:
Let $b(x,y) = a^2 - x^2$, $x \in R$. The exact solution of the Poisson equation (5.65) in this case is

$$u = \frac{1}{2}a^2 x^2 - \frac{1}{12}x^4.$$

Let the region R be a quarter circle of radius $a = 10$ (Fig. 9.7).

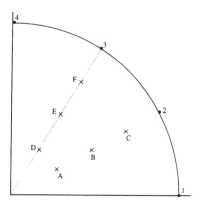

Fig. 9.7.

Assume that $u = 0$ on the boundary C (Dirichlet type boundary condition). The Fourier series for this function is

$$b(x, y) = a^2 - x^2 \approx \frac{2}{3}a^2 - \frac{4a^2}{\pi^2} \sum_{n=1}^{N} \frac{(-1)^n}{n^2} \cos \frac{n\pi x}{a}.$$

Using (9.57) with $N = 5$, the results for u are as follows:

Interior Point	Potentail u	Exact u
$A \, (2.5 \cos \pi/6, 2.5 \sin \pi/6)$	224.93	232.5439
$B \, (5 \cos \pi/6, 5 \sin \pi/6)$	918.42	908.2031
$C \, (7.5 \cos \pi/6, 7.5 \sin \pi/6)$	1974.17	1861.0595
$D \, (2.5 \cos \pi/3, 2.5 \sin \pi/3)$	103.80	77.9155
$E \, (5 \cos \pi/3, 5 \sin \pi/3)$	351.31	309.2448
$F \, (7.5 \cos \pi/3, 7.5 \sin \pi/3)$	746.11	686.6455 ∎

9.3. Multiple Reciprocity Method

Before we start with the two–dimensional BI Eq (5.69), we shall change the notation slightly and denote the function b in (5.65) by b_0, the fundamental solution u^* by u_0^*, and the domain integral B_i by B_0. With this notation, the domain integrals (9.1) becomes

$$B_0 = \iint_R b_0 u_0^* \, dx \, dy. \tag{9.67}$$

Let us introduce a function u_1^* related with $u^* \equiv u_0^*$ such that

$$\nabla^2 u_1^* = u_0^*. \tag{9.68}$$

Then the domain integral (9.67) becomes

$$
\begin{aligned}
B_0 &= \iint_R b_0 \nabla^2 u_1^* \, dx \, dy \\
&= \iint_R u_1^* \nabla^2 b_0 \, dx \, dy - \int_C \left(u_1^* \frac{\partial b_0}{\partial n} - b_0 \frac{\partial u_1^*}{\partial n} \right) ds \quad \text{by (1.8)} \quad (9.69) \\
&= B_1 + \int_C \left(b_0 \frac{\partial u_1^*}{\partial n} - u_1^* \frac{\partial b_0}{\partial n} \right) ds.
\end{aligned}
$$

Now define b_1 and u_2^* such that

$$b_1 = \nabla^2 b_0, \quad \nabla^2 u_2^* = u_1^*. \tag{9.70}$$

Then the integral B_1 becomes

$$
\begin{aligned}
B_1 &= \iint_S b_1 u_1^* \, dx \, dy \\
&= \iint_R u_2^* \nabla^2 b_1 \, dx \, dy - \int_C \left(u_2^* \frac{\partial b_1}{\partial n} - b_1 \frac{\partial u_2^*}{\partial n} \right) ds \quad \text{by (1.8)} \quad (9.71) \\
&= B_2 + \int_C \left(b_1 \frac{\partial u_2^*}{\partial n} - u_2^* \frac{\partial b_1}{\partial n} \right) ds.
\end{aligned}
$$

Continuing this process successively, we get

$$b_{k+1} = \nabla^2 b_k, \tag{9.72}$$

$$\nabla^2 u_{k+1}^* = u_k^*, \quad k = 0, 1, 2, \ldots . \tag{9.73}$$

Thus, by iteration,

$$B_0 = \sum_{k=0}^{\infty} \int_C \left(b_k \frac{\partial u_{k+1}^*}{\partial n} - u_{k+1}^* \frac{\partial b_k}{\partial n} \right) ds. \tag{9.74}$$

Substituting (9.74) into (9.67), the MRM BI Eq is given by

$$c(i)u(i) + \int_C uq^* \, ds = \int_C qu^* \, ds -$$

$$- \sum_{k=0}^{\infty} \int_C \left(b_k \frac{\partial u_{k+1}^*}{\partial n} - u_{k+1}^* \frac{\partial b_k}{\partial n} \right) ds, \tag{9.75}$$

which leads to the MRM BE Eq

$$c(i)u(i) + \sum_{j=1}^{N} u_j \int_{\tilde{C}_j} q^* \, ds = \sum_{j=1}^{N} q_j \int_{\tilde{C}_j} u^* \, ds -$$

$$- \sum_{j=1}^{N+M} \sum_{k=0}^{\infty} \left\{ b_k \int_{\tilde{C}_k} \frac{\partial u_{k+1}^*}{\partial n} \, ds - \frac{\partial b_k}{\partial n} \int_{\tilde{C}_k} u_{k+1}^* \, ds \right\}, \quad (9.76a)$$

or, in matrix form,

$$HU - GQ = \sum_{j=1}^{\infty} (H_j R_j - G_j S_j), \quad (9.76b)$$

where H and G are the same matrices as defined in (5.73), and R_j and S_j are the vectors which contain the functions b_k and their normal derivatives respectively. The upper limit of summation in (9.76b) is taken as a small positive integer instead of infinity because the terms in the series decrease to zero very rapidly. The convergence of this series can be controlled since all the terms are known (see Problem 9.10).

The recurrence formula (9.73) expressing $\nabla^2 u_{k+1}^*$ in terms of u_k^*, $k = 0, 1, 2, \ldots$, demands that the higher order fundamental solutions u_k^* must be specified. In the case of the Laplacian in two–dimensional problems the function u_k^* is, in general, defined by

$$u_k^* = \frac{1}{2\pi} r^{2k} \left(C_k \ln r - D_k \right), \quad r = |\mathbf{x} - \mathbf{x}'|, \quad (9.77)$$

where C_k and D_k are determined by the recurrence relations

$$C_0 = 1, \quad C_{k+1} = \frac{C_k}{4(k+1)^2},$$

$$D_0 = 0, \quad D_{k+1} = \frac{1}{4(k+1)^2} \left(\frac{C_k}{k+1} + D_k \right) \quad (9.78)$$

$$\text{for } k = 0, 1, 2, \ldots.$$

Note that the case $k = 0$ corresponds to the original fundamental solution u^*. Moreover, the presence of factorials in the denominator in the formula (9.78) for each C_k and D_k ensures fast convergence. The functions u_k^* and q_k^* for $k = 1, 2, \ldots$ do not have any singularities. This means that the integrations in H_{j+1} and G_{j+1} in (9.76b) are straightforward for $j = 1, 2, \ldots$. Thus, the computer implementation of (9.76b) is easy and follows the standard BE routines. Note that the virtual flux q_j^* is given by

$$q_j^* = \frac{\partial u_j^*}{\partial n} = \frac{1}{2\pi} r^{2j-1} [(1 + 2j \ln r) C_j - 2j D_j] \frac{\partial r}{\partial n}. \qquad (9.79)$$

Example 9.7:
For the Helmholtz equation (7.37), we have ϕ instead of u; thus, $u_0 = \phi_0$, and $b_0 = -k^2\phi$. Then Eq (9.74) becomes

$$B_0 = \sum_{j=0}^{\infty} (-1)^j k^{2(j+1)} \int_C (\phi q_{j+1}^* - u_{j+1}^* q) \, ds, \qquad (9.80)$$

where $q = \partial\phi/\partial n$. Substituting this into (9.67) gives

$$-k^2 c(i) u(i) + \sum_{j=0}^{\infty} (-1)^j k^{2j} \int_C \phi q_j^* \, ds = \sum_{j+0}^{\infty} (-1)^j k^{2j} \int_C q u_j^* \, ds,$$

which yields the BE Eq

$$\left\{ \sum_{j=0}^{\infty} (-1)^j k^{2j} H_j \right\} U = \left\{ \sum_{j=0}^{\infty} (-1)^j k^{2j} G_j \right\} Q, \qquad (9.81)$$

or, in matrix form

$$HU = GQ. \qquad (9.82)$$

Since this problem is similar to solving the eigenvalue problem, we rewrite (9.82) with the Dirichlet boundary conditions in the first partition of the vectors U and Q and the Neumann boundary conditions in the second partition of these vectors. This leads to

$$[H_u \quad H_q] \left\{ \begin{array}{c} 0 \\ u_q \end{array} \right\} = [G_u \quad G_q] \left\{ \begin{array}{c} Q_u \\ 0 \end{array} \right\}, \qquad (9.83)$$

which after the collocation process of §5.3 reduces to the matrix form

$$AX = F. \qquad (9.84)$$

If $\det A = 0$, then we obtain nontrivial solutions of the system (9.84). ∎

Example 9.8:

For the heat conduction problems, we have $-\frac{1}{k}b_0$ instead of b in the Poisson equation, where k is the thermal conductivity. Then the BE Eq (9.76b) becomes

$$HU - GQ = \frac{1}{k}\sum_{j=0}^{\infty}(G_{j+1}S_j - H_{j+1}R_j),\qquad(9.85)$$

which can be transformed into the form $AX = F.$ ∎

Example 9.9:

Consider the rectangular plate of dimensions $1' \times 0.5'$ with the BE discretization shown in Fig. 9.8 ($N = 20, M = 0$).

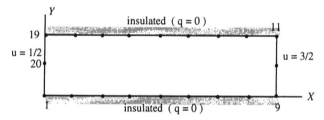

Fig. 9.8.

Take $k = 1$, $b_0 = -\frac{\pi^2}{2}\cos(\pi x)$, which is a harmonic heat source, and the boundary conditions are: $u(0,y) = 1/2$, $u(1,y) = 3/2$, $\frac{\partial u}{\partial y}(x,0) = 0 = \frac{\partial u}{\partial y}(x,1)$. The exact solution is $u = 1 - \frac{1}{2}\cos(\pi x).$ ∎

9.4. Problems

9.1. Derive Eq (9.13) directly from (5.65)–(5.66) by applying the Green's second identity (1.8).

9.2. Modify the program Be1 or Be5 for the DRM by supplementing the matrix (9.20) or (9.21) to it.

9.3. Solve the problem of Example 9.1 for (a) a fixed–fixed beam, and (b) a free–free beam.

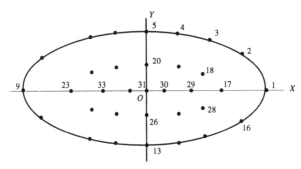

Fig. 9.9.

9.4. Solve the problem of the harbor resonance, governed by the Helmholtz equation $(\nabla^2 + k^2)\,\phi = 0$, with $k^2 = \omega^2/hg$, where h is the mean water depth from a reference datum, ω the natural frequency, and $\phi = h + \eta$ the sum of mean water depth h and the free surface elevation η. Use the method of Example 9.1 in a rectangular channel with one closed end.

9.5. Use the DRM to solve the problem of wave propagation in a rectangular cavity of dimensions a along the x–axis and b along the y–axis under the Neumann conditions, i.e., $q = 0$ along the boundary.

9.6. Use the DRM to solve the problem of free vibrations of a circular membrane of radius $r = 1$, which is fixed along its circumference with the Dirichlet boundary conditions $u = 0$ all along the circumference.

9.7. Use the DRM or the MRM to solve the torsion problem (6.61) with $G\theta = 1$ (i.e., $\nabla^2 u = -2$) for the elliptic region of semi–major axis $a_1 = 2$ and semi–minor axis $a_2 = 1$, shown in Fig. 9.9.

9.8. Solve the torsion problem $\nabla^2\phi = -b$ for the elliptic region of Fig. 9.9, where (i) $b = -x$, (ii) $b = -x^2$, and (iii) $b = 4 - x^2$.

9.9. Solve the torsion problem $\nabla^2 u = -2$ for the mesh shown in Fig. 9.10, where only a quarter circle of Fig. 9.7 is considered because of symmetry. The shear stresses are given by $\tau_{zy} = \dfrac{\partial u}{\partial x}$, $\tau_{zx} = \dfrac{\partial u}{\partial y}$. Use FSM by extending $b(x, y)$ as an even function and take $a_0 = 2$, $a_n^1 = 0 = a_n^2$ for the Fourier coefficients.

9.10. Establish the convergence of the infinite series in Eq (9.66b).

9.11. Use MRM to solve the heat conduction problem in a long rectangular slab of dimensions $10' \times 1'$, insulated at one end, with a constant heat source of $b_0 = 1$, and $k = 0.5$. Compute the potential at the middle of the slab if the boundary conditions are $u(x,0) = 0$, $u(x,1) = 1$, $q(0,y) = 0$, and $q(10,y) = 10$.

9.12. Use MRM to solve Problem 5.11.

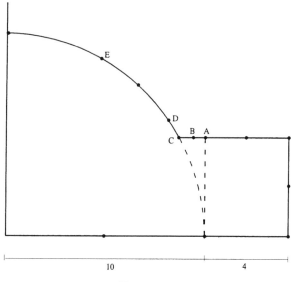

Fig. 9.10.

References and Bibliography for Additional Reading

C. Alessandri and A. Trelli, *An alternative technique for reducing domain integrals to the boundary*, Boundary Element Methods in Mechanical and Electrical Engineering, C. A. Brebbia and A. Chaudouet–Miranda, Eds., Computational Mechanics Publications, Southampton, Springer–Verlag, Berlin, 1990, pp. 517–529.

M. M. Aral and Y. Tang, *A boundary only procedure for time dependent diffusion equations*, Applied Math. Modeling **12** (1988), 610–618.

J. P. S. Azevedo and C. A. Brebbia, *An efficient technique for reducing domain integrals to the boundary*, Proc. 10th BEM Conf., Southampton (1988), Springer–Verlag, Berlin, 347–363.

C. A. Brebbia and L. C. Wrobel, *The solution of parabolic problems using the dual reciprocity boundary elements*, Proc. of the IUTAM Symp. (T. A. Cruise, Ed.) (1987), Springer–Verlag, Heidelberg, 55–73.

————, J. Jefferson Rego Silva and P. W. Partridge, *Computational Formulation*, Ch. 2 in Boundary Element Methods in Acoustics (R. D. Ciskowski and C. A. Brebbia, Eds.), Computational Mechanics Publications, Southampton, and Elsevier Applied Science, London, 1993.

L. E. Coates, *Integral equation failure with Poisson equation*, Boundary Element

Methods in Mechanical and Electrical Engineering, C. A. Brebbia and A. Chaudouet–Miranda, Eds., Computational Mechanics Publications, Southampton, Springer–Verlag, Berlin, 1990, pp. 517–529.

M. Itagaki and C. A. Brebbia, *Boundary element method applied to neutron diffusion problems*, Proc. 10th BEM Conf., Southampton (1988), Springer–Verlag, Berlin, 45–53.

D. P. Kontoni, P. W. Partridge and C. A. Brebbia, *The dual reciprocity boundary element method for the eigenvalue analysis of Helmholtz problems*, Adv. Eng. Software **13** (1991), 1–16.

C. A. Loeffer and W. J. Mansur, *Analysis of time integration schemes for boundary element applications to transient wave propagation problems*, Boundary Element Techniques: Applications in Stress Analysis and Heat Transfer, (C. A. Brebbia and W. S. Venturini, Eds.), Computational Mechanics Publications, Southampton, 1987, pp. 105–122.

D. Nardini and C. A. Brebbia, *A new approach to free vibration analysis using boundary elements*, Boundary Element Methods in Engineering, C. A. Brebbia, Ed., Springer–Verlag, Berlin, 1982.

A. C. Neves and C. A. Brebbia, *The multiple reciprocity boundary elements method for transforming domain integrals to the boundary*, Int, J. for Numer. Methods in Engg. **31** (1991), 709–727.

S. M. Niku and C. A. Brebbia, *Dual reciprocity boundary element formulation for potential problems with arbitrarily distributed sources*, Engg. Anal. **5** (1988), 46–48.

A. J. Nowak, *Solving linear heat conduction problems by the multiple reciprocity method*, Ch. 3 in Boundary Element Methods in Heat Transfer (L. C. Wrobel and C. A. Brebbia, Eds.), Computational Mechanics Publications, Southampton, and Elsevier Applied Science, London, 1992.

A. J. Nowak and C. A. Brebbia, *The multiple–reciprocity method. A new approach for transforming BEM domain integrals to boundary*, Engg. Anal. with Boundary Elements **6** (1989), 164–167.

———, *The Multiple Reciprocity Method*, Ch. 3 in Advanced Formulations in Boundary Element Methods (M. H. Aliabadi and C. A. Brebbia, Eds.), Computational Mechanics Publications, Southampton, and Elsevier Applied Science, London, 1993.

———, *Boundary integral formulation of mass matrices for dynamic analysis*, Topics in Boundary Element Research, Vol. 2, Springer–Verlag, Berlin, 185.

D. A. Pape and P. K. Banerjee, *Treatment of boundary forces in 2D elastostatic BEM using particular integrals*, J. Appl. Mech., Trans. ASME **54** (1987), 866–871.

P.W. Partridge and C. A. Brebbia, *Computer implementation of the BEM dual reciprocity method for the solution of the Poisson type equations*, Software for Engg. Workstations **5** (1989), 199–206.

———, *The dual reciprocity boundary element method for the Helmholtz equation*, Boundary Element Methods in Mechanical and Electrical Engineering, C. A. Brebbia and A. Chaudouet–Miranda, Eds., Computational Mechanics Publications, Southampton, Springer–Verlag, Berlin, 1990, pp. 543–555.

———, *The Dual Reciprocity Method*, Ch. 2 in Advanced Formulations in Boundary Element Methods (M. H. Aliabadi and C. A. Brebbia, Eds.), Computational Mechanics Publications, Southampton, and Elsevier Applied Science, London, 1993.

———and L. C. Wrobel, *The Dual Reciprocity Boundary Element Method*, Computational Mechanics Publications, Southampton, 1990.

I. S. Skolnikoff, *Mathematical Theory of Elasticity*, 2nd ed., McGraw–Hill, New York, 1956.

———— and L. C. Wrobel, *The Dual Reciprocity Boundary Element Method*, Computational Mechanics Publications, Southampton, 1990.

W. Tang, *Transforming Domain into Boundary Integrals in Boundary Element Method*, Lecture Notes in Engineering, vol. 35, Springer–Verlag, Berlin, 1988.

L. C. Wrobel and C. A. Brebbia, *Boundary elements for non–linear heat conduction problems*, Communications in Applied Numerical Methods **4** (1988), 617–622.

10

Transient Problems

Consider the partial differential equation governing the transient heat transfer (and similar) problems

$$a\frac{\partial u}{\partial t} - \frac{\partial}{\partial x}\left(k_1\frac{\partial u}{\partial x}\right) - \frac{\partial}{\partial y}\left(k_2\frac{\partial u}{\partial y}\right) + b = 0, \quad \text{in } R \times T, \qquad (10.1)$$

with the boundary conditions

$$u = \hat{u} \quad \text{on } C_1, \quad t \geq 0, \qquad (10.2a)$$

$$k_1 u_x n_x + k_2 u_y n_y + \beta(u - u_\infty) = \hat{q} \quad \text{on } C_2, \quad t \geq 0, \qquad (10.2b)$$

and the initial condition

$$u = u_0 \quad \text{in } R \text{ for } t = 0, \qquad (10.3)$$

where R is a two–dimensional region with boundary $C = C_1 \cup C_2$, T denotes the time interval $(0, t_0]$; a, k_1, k_2 and b are given functions of position and/or time, and β, u_∞ and u_0 are prescribed quantities. Here k_1 and k_2 are thermal conductivities along the x and y directions, and b denotes the internal heat generation per unit volume. For $a = 1$, $k_1 = k_2 = k$, Eq (10.1) reduces to the unsteady Poisson equation

$$k\nabla^2 u = \frac{\partial u}{\partial t} + b. \qquad (10.4)$$

If $b = 0$, we obtain the transient Fourier equation

$$\nabla^2 u = \frac{1}{k}\frac{\partial u}{\partial t}. \qquad (10.5)$$

We shall derive the BE formulation for Eq (10.5), and develop the forward time marching process for the transient Fourier equation (10.5). We shall then develop the Laplace transform BEM for the unsteady Poisson equation (10.4). These two methods, combined with the DRM, FSM, or MRM of Chapter 9, will enable us to get the BEM for the unsteady Poisson equation (10.4), although we will provide no details since it will be repetitious. The last section is devoted to the transient Fourier–Kirchhoff and the Helmholtz equations which are numerically solved with both the transient DRM and MRM, by using the θ–scheme.

10.1. Transient Fourier Equation

For the two–dimensional Fourier equation (10.5), the fundamental solution is given by (see Example 4.5)

$$u^*(\mathbf{x}, t; \mathbf{x}', t') = \frac{1}{4\pi k(t - t')} e^{-r^2/4k(t - t')}, \quad \text{for } \mathbf{x} \in R, \, t > t', \quad (10.6)$$

where $\mathbf{x} = (x, y)$, $\mathbf{x}' = (x', y')$, and $r = |\mathbf{x} - \mathbf{x}'|$. Note that the function u^* represents the temperature field produced by an instantaneous source (of heat) at the point $\mathbf{x}'$ and time t'. An application of the weak variational formulation (§2.1) on Eq (10.5) gives

$$0 = \int_{t_0}^{t} \iint_{R} u \left(k\nabla^2 u^* - \frac{\partial u^*}{\partial t} \right) dA(\mathbf{x}') \, dt -$$

$$- k \int_{t_0}^{t} \int_{C} (u^* q - u q^*) \, ds(\mathbf{x}') \, dt' - \iint_{R} [u u^*]_{t'=t_0}^{t} \, dt' \, dA(\mathbf{x}')$$

$$= \int_{t_0}^{t} \iint_{R} u \left(k\nabla^2 u^* - \frac{\partial u^*}{\partial t} \right) dA(\mathbf{x}') \, dt -$$

$$- k \int_{t_0}^{t} \int_{C} (u^* q - u q^*) \, ds(\mathbf{x}') \, dt' - \iint_{R} \int_{t_0}^{t} \frac{\partial}{\partial t'} [u u^*] \, dt' \, dA(\mathbf{x}')$$

$$= \int_{t_0}^{t} \iint_{R} u \left(k\nabla^2 u^* - \frac{\partial u^*}{\partial t} \right) dA(\mathbf{x}') \, dt -$$

$$- k \int_{t_0}^{t} \int_{C} (u^* q - u q^*) \, ds(\mathbf{x}') \, dt' - \iint_{R} u_0 u_0^* \, dA(\mathbf{x}').$$

Thus,

$$0 = \int_{t_0}^t \iint_R u\delta(\mathbf{x}, \mathbf{x}')\, dA(\mathbf{x}')\, dt - k \int_{t_0}^t \int_C (u^*q - uq^*)\, ds(\mathbf{x}')\, dt' -$$
$$- \iint_R u_0 u_0^*\, dA(\mathbf{x}'), \quad (10.7)$$

which, in view of the translation property (1.55), yields the BI Eq

$$cu = \iint_R u_0 u_0^*\, dA(\mathbf{x}') + k \int_C \int_{t_0}^t (u^*q - uq^*)\, dt'\, ds(\mathbf{x}'). \qquad (10.8)$$

Now we shall present a simple case of the transient DRM to transform the domain integral in (10.8) into boundary integrals. Let us choose a function v such that

$$\nabla^2 v = u_0^*. \qquad (10.9)$$

Then, since u_0 is a harmonic function in R, i.e., $\nabla^2 u_0 = 0$, we find that

$$\iint_R u_0 u_0^*\, dA(\mathbf{x}') = \iint_R u_0 \nabla^2 v\, dA(\mathbf{x}')$$
$$= \iint_R \left(u_0 \nabla^2 v - v\nabla^2 u_0 \right) dA(\mathbf{x}') \qquad (10.10)$$
$$= \int_C \left(u_0 \frac{\partial v}{\partial n} - v\frac{\partial u_0}{\partial n} \right) ds(\mathbf{x}'),$$

where we have subtracted $\nabla^2 u_0$ (which is $= 0$) from the integral and then applied Green's second identity (1.8). The relation (10.10) is equivalent to the DRM for (10.8). Recall that the function v has been obtained in §5.9 for the steady–state case. For the present transient case the derivation of the function v is given below in Example 10.1.

We shall now derive the BE Eq for the heat conduction problem defined by (10.5) with the initial condition (10.3) and the essential boundary condition (10.2a) and a simplified natural boundary condition $q = q_0$ on C_2. Then the BI Eq (10.8) for this problem becomes

$$cu = \iint_R u_0 u_0^*\, dA + \int_{C_1} \int_{t_0}^t (u^*q_1 - \hat{u}q^*)\, dt'\, ds + \int_{C_2} \int_{t_0}^t (u^*q_0 - u_2 q^*)\, dt'\, ds,$$
$$(10.11)$$

where q_1 denotes the (unknown) value of q on C_1 and u_2 the (unknown) value of u on C_2. Note that if q_1 and u_2 are known, then we could compute the value of u directly at an interior point of R (with $c = 1$) by evaluating the right side of (10.11). But, since u_2 and q_1 are not known, we shall, in order to compute the right side of (10.11), introduce the following notation:

$$\hat{H}u_2 = \int_{C_2} \int_{t_0}^{t} u_2 q^* \, dt' \, ds,$$

$$Gq_1 = \int_{C_1} \int_{t_0}^{t} u^* q_1 \, dt' \, ds, \tag{10.12}$$

$$\hat{F} = \iint_{R} u_0 u_0^* \, dA - \int_{C_1} \int_{t_0}^{t} \hat{u} q^* \, dt' \, ds + \int_{C_2} \int_{t_0}^{t} u^* q_0 \, dt' \, ds.$$

Then the BI Eq (10.11) can be written as

$$cu = -\hat{H}u_2 + Gq + \hat{F}, \tag{10.13}$$

Note that the quantities u_2 and q_1 in (10.13) are unknown; also, Eq (10.13) is valid at points on the boundary $C = C_1 \cup C_2$. Thus, if we take $\mathbf{x} \in C_1$, then (10.13) gives

$$\hat{H}u_2 - Gq_1 = \hat{F} - c\hat{u}. \tag{10.14}$$

But if we take $\mathbf{x} \in C_2$, then (10.13) gives

$$cu_2 + \hat{H}u_2 - Gq_1 = \hat{F}. \tag{10.15}$$

Define H and F by

$$Hu_2 = c_2 u_2 + \hat{H}u_2, \quad F = \hat{F} - c_1 u, \tag{10.16}$$

where $c_1 = c$, $c_2 = 0$ if $\mathbf{x} \in C_1$, and $c_1 = 0$, $c_2 = c$ if $\mathbf{x} \in C_2$. Then (10.14)–(10.15) yield the BE Eq

$$Hu_2 = Gq_1 + F, \tag{10.17}$$

which holds on the boundary $C \times T$ of the region $R \times T$ with u_2 and q_1 as unknown quantities.

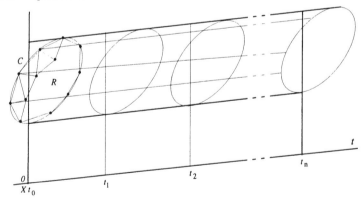

Fig. 10.1. Geometry and discretization of the region $R \times T$.

A numerical solution of the BE Eq (10.17) is carried out as follows: We will denote the boundary element approximations of u_2 and q_1 over $C \times T$ by $\hat{u}_2$ and $\hat{q}_1$ respectively. Note that the function F on the right side of (10.17) contains the domain integral $\iint_R u_0 u_0^* \, dA$ which, in general, cannot be evaluated analytically. This integral can, however, be evaluated numerically by the interior cell method of §5.9. The region R is therefore discretized into a finite number of interior cells (elements). As shown in Fig. 10.1, this discretization of an arbitrary region R and its boundary C into cells produces an approximate domain $\tilde{R}$ and an approximate boundary $\tilde{C}$. Let us denote by $\tilde{R}_j$ a typical cell j, by $M(\tilde{R})$ the total number of such cells $R_j, j = 1, \cdots, M$, and by $N(\tilde{R})$ the total number of nodes in $\tilde{R}$. Similarly for the boundary $\tilde{C}$, $M(\tilde{C})$ will denote the total number of boundary elements and $N(\tilde{C})$ the total number of boundary nodes. Let $\phi_i(\mathbf{x})$, $i = 1, \cdots, N(\tilde{C})$, denote the shape functions over $\tilde{C}$ and $\psi_\alpha(t)$, $\alpha = 1, 2, \cdots$, the shape functions over T. Since these shape functions are interpolation functions, they satisfy the conditions

$$\phi_i(\mathbf{x}_j) = \delta_{ij}, \quad \psi_\alpha(t_\beta) = \delta_{\alpha\beta}. \tag{10.18}$$

Hence, we have the following approximations for $\hat{u}_2$ and $\hat{q}_1$:

$$\hat{u}_2 = \sum_{i,\alpha}{}^2 (\hat{u}_2)_{i\alpha} \phi_i(\mathbf{x})\psi_\alpha(t), \tag{10.19}$$

$$\hat{q}_1 = \sum_{i,\alpha}{}^1 (\hat{q}_1)_{i\alpha} \phi_i(\mathbf{x})\psi_\alpha(t), \tag{10.20}$$

where the superscripts 1 and 2 over the summation sign indicate sum over the nodes of $\tilde{C}_1$ and $\tilde{C}_2$ respectively , and the parameters $(\hat{u}_2)_{i\alpha}$ and $(\hat{q}_1)_{i\alpha}$ represent the nodal values of $\hat{u}_2$ and $\hat{q}_1$ at node i on C and node α on T. In view of (10.19) and (10.20), Eq (10.17) becomes

$$\sum_i{}^2 \sum_\alpha \left(H\phi_i\psi_\alpha\right)(\hat{u}_2)_{i\alpha} = \sum_i{}^1 \sum_\alpha \left(G\phi_i\psi_\alpha\right)(\hat{q}_1)_{i\alpha} + F. \tag{10.21}$$

In order to evaluate $(\hat{u}_2)_{i\alpha}$ and $(\hat{q}_1)_{i\alpha}$, we shall collocate (10.21) at points $\mathbf{x}_j$ on $\tilde{C}$ and at points t_β on T. Define

$$\left(H\phi_i\psi_\alpha\right)(\mathbf{x}_j, t_\beta) = H_{j\beta i\alpha},$$
$$\left(G\phi_i\psi_\alpha\right)(\mathbf{x}_j, t_\beta) = G_{j\beta i\alpha}, \tag{10.22}$$
$$F(\mathbf{x}_j, t_\beta) = F_{j\beta}.$$

Then (10.21) becomes

$$\sum_i {}^2 \sum_\alpha H_{j\beta i\alpha}(\hat{u}_2)_{i\alpha} = \sum_i {}^1 \sum_\alpha G_{j\beta i\alpha}(\hat{q}_1)_{i\alpha} + F_{j\beta}, \qquad (10.23a)$$

or in matrix form

$$H\hat{u}_2 = G\hat{q}_1 + F, \qquad (10.23b)$$

which, when solved, yields the nodal values of $(\hat{u}_2)_{i\alpha}$ and $(\hat{q}_1)_{i\alpha}$. However, in order to successfully solve (10.23a), we must evaluate the coefficients $H_{j\beta i\alpha}$, $G_{j\beta i\alpha}$, and $F_{j\beta}$. To do this, we take constant elements in time T, and divide the time interval $[t_0, t]$ into subintervals $(t_0, t_1), (t_1, t_2), \cdots, (t_{N-1}, t_N)$ such that $t_\alpha = t_0 + \alpha\Delta t$, $\alpha = 1, 2, \cdots, N$, where Δt is a constant time step. The shape function ψ_α is defined by

$$\psi_\alpha = \begin{cases} 1 & \text{for } t_{\alpha-1} < t \leq t_\alpha, \\ 0 & \text{otherwise.} \end{cases}$$

Choose the collocation points to coincide with the nodes $(\mathbf{x}_j, t_1)$ on $\tilde{C} \times (t_0, t_1)$. This reduces (10.23b) to the following system of linear equations:

$$H^{11}\hat{u}_2^{11} - G^{11}\tilde{q}_1^{11} = F^1, \qquad (10.24)$$

where H^{11} is an $N(\tilde{C}) \times N(\tilde{C}_2)$ matrix, G^{11} an $N(\tilde{C}) \times N(\tilde{C}_1)$ matrix, and F^1 an $N(\tilde{C})$ vector. Since $N(\tilde{C}_1) + N(\tilde{C}_2) = N(\tilde{C})$, the system (10.24) is compatible. Note that at each node on C either the temperature $\hat{u}_2$ or the flux $\hat{q}_1$ (but not both) is unknown. The evaluation of integrals in H^{11}, G^{11}, and F^1 is carried out as follows:

$$H_{ji}^{11} \equiv H_{j1i1} = \left(H\phi_i\psi_\alpha\right)(\mathbf{x}_j, t_1)$$

$$= \int_{\tilde{C}_2} \int_{t_0}^{t_1} q^*(\mathbf{x}_j, t; \mathbf{x}', t')\phi_i(\mathbf{x}')\, dt'\, ds(\mathbf{x}')$$

$$= \int_{\tilde{C}_2} \int_{t_0}^{t_1} \frac{1}{2(t_1 - t')}(\mathbf{r} \cdot \hat{n})u^*\phi_i\, dt'\, ds(\mathbf{x}')$$

$$= -\frac{1}{8k\pi} \int_{\tilde{C}_2} (\mathbf{r} \cdot \hat{n})\phi_i \int_{t_0}^{t_1} \frac{1}{(t_1 - t')^2}e^{-r^2/4k(t_1 - t')}\, dt'\, ds(\mathbf{x}')$$

using (10.6).

$$(10.25)$$

where $q^* = -\dfrac{1}{2(t - t')}(\mathbf{r}\cdot\hat{n})\, u^*$. If we set $\sigma = r^2/4k(t_1 - t')$, so that $\sigma = r^2/4k(t_1 - t_0) = r^2/4k\Delta t \equiv \sigma_0$ at $t' = t_0$, we find that

$$\int_{t_0}^{t_1} \frac{1}{(t_1 - t')^2}e^{-r^2/4k(t_1 - t')}\, dt' = \frac{4k}{r^2} \int_{\sigma_0}^{\infty} e^{-\sigma}\, d\sigma$$

$$= \frac{4ke^{-\sigma_0}}{r^2} = \frac{4k}{r^2}e^{-r^2/4k\Delta t}. \qquad (10.26)$$

Substituting the value of the integral (10.26) into (10.25) we get

$$H_{ji}^{11} = -\frac{1}{2\pi} \int_{\tilde{C}_2} \frac{(\mathbf{r} \cdot \hat{\mathbf{n}})}{r^2} e^{-r^2/4k\Delta t} \phi_i \, ds(\mathbf{x}'). \qquad (10.27)$$

Since the integrand in (10.27) is regular as $r \to 0$, this integral is evaluated by Gauss quadrature, with the following empirical guidelines for accuracy and speed: For constant cells in R, use 1 or 2 points Gauss rule; for linear cells in R, use 2 or 3 points Gauss rule; and for quadratic cells in R, use 3 or 4 points Gauss rule in (10.27). Thus,

$$
\begin{aligned}
G_{ji}^{11} = G_{j1i1} &= \left(G\phi_i \psi_\alpha \right)(\mathbf{x}_j, t_1) \\
&= \int_{\tilde{C}_1} \int_{t_0}^{t_1} u^*(\mathbf{x}_j, t; \mathbf{x}', t') \phi_i(\mathbf{x}') \, dt' \, ds(\mathbf{x}') \\
&= \frac{1}{4k\pi} \int_{\tilde{C}_1} \phi_i \int_{t_0}^{t_1} \frac{1}{t_1 - t'} e^{-r^2/4k(t_1-t')} \, dt' \, ds(\mathbf{x}') \quad \text{using (10.6)} \\
&= \frac{1}{4k\pi} \int_{\tilde{C}_1} \phi_i \int_{\sigma_0}^{\infty} e^{-\sigma} \, d\sigma \, ds(\mathbf{x}') \\
&= \frac{1}{4k\pi} \int_{\tilde{C}_1} \phi_i E_1(\sigma_0) \, ds(\mathbf{x}') \\
&= \frac{1}{4k\pi} \int_{\tilde{C}_1} \phi_i [-\gamma - \ln \sigma_0 + e_1(\sigma_0)] \, ds(\mathbf{x}'),
\end{aligned}
$$
$$(10.28)$$

where we have used the same substitution for σ as in (10.26). Since the line integral (10.28) consists of a singular part (involving $\ln \sigma_0$) and the remaining regular part, the singular part must be evaluated for linear cells as in (5.41), while the regular part can be computed by Gauss quadrature. Finally,

$$
\begin{aligned}
F^1 = F_{ji}^1 &= \iint_R u_0 u_0^* \, dA(\mathbf{x}') \\
&= \iint_R u_0(\mathbf{x}') u_0^*(\mathbf{x}_j, t_1; \mathbf{x}', t_0) \, dA(\mathbf{x}') \\
&= \iint_R u_0(\mathbf{x}') \frac{1}{4k\pi(t_1 t_0)} e^{-r^2] 4k(t_1-t_0)} \, dA(\mathbf{x}') \quad \text{using (10.6)} \\
&= \frac{1}{4k\pi\Delta t} \iint_R u_0(\mathbf{x}') e^{-r^2/4k\Delta t} \, dA(\mathbf{x}').
\end{aligned}
$$
$$(10.29)$$

If $u_0(\mathbf{x}')$ is harmonic, then the value of this integral is given by

$$F^1 = \frac{1}{4k\pi\Delta t} \int_C \left(u_0 \frac{\partial v}{\partial n} - v \frac{\partial u_0}{\partial n} \right) ds, \qquad (10.30)$$

where v is defined by (10.9). Otherwise, Gauss quadrature is used to evaluate (10.29) over each cell in $\tilde{R}$.

The formulas (10.27)–(10.30) help us compute the unknowns $\hat{u}_2$ and $\hat{q}_1$ only in the first time interval $[t_0, t_1]$. In order to compute these unknowns over the entire interval $T = [t_0, t_n]$, we shall use the forward marching process, as explained in the next section.

Example 10.1:
To find the function v defined in (10.9) we shall first transform Eq (10.9) into polar cylindrical coordinates as

$$\frac{1}{r}\frac{d}{dr}\left(r\frac{dv}{dr}\right) = \frac{1}{4\pi k(t - t_0)}e^{-r^2/4k(t-t_0)}. \tag{10.31}$$

If we set $z = r^2/4k(t - t_0)$, so that $dz/dr = r/2k(t - t_0)$, we find on integrating (10.31) that

$$r\frac{dv}{dr} = 2z\frac{dv}{dz} = -\frac{1}{2\pi}e^{-z} + A_1,$$

which after integrating again with respect to z gives

$$v = -\frac{1}{4\pi}E_1(z) + A_1 \ln z + A_2, \tag{10.32}$$

where $E_1(z) = \displaystyle\int_z^\infty \frac{e^{-t}}{t}\,dt = \int_1^\infty \frac{e^{-zt}}{t}\,dt$, $\Re z > 0$, $|\arg z| < \pi$, is the exponential integral function, with the properties:

(i) $\quad E_1(z) = -\gamma - \ln z + e_1(z), \quad e_1(z) = \displaystyle\sum_{n=1}^\infty \frac{(-1)^{n+1}z^n}{nn!}.$

(ii) $\quad \dfrac{d}{dz}E_n(z) = -E_{n-1}(z), \ n = 1, 2, \cdots; \quad E_0(z) = e^{-z}/z.$

(iii) $\quad \dfrac{d}{dz}\left[e^z E_1(z)\right] = e^z E_1(z) - \dfrac{1}{z}.$

Since $E_1(z)$ has a logarithmic singularity as $z \to 0$ (i.e., $r \to 0$), we use the property (i) to determine the integration constants A_1, A_2 such that this singularity vanishes. Thus, since from (10.32)

$$\begin{aligned}
v &= -\frac{1}{4\pi}E_1(z) + A_1 \ln z + A_2 \\
&= -\frac{1}{4\pi}\left[-\gamma - \ln z + e_1(z)\right] + A_1 \ln z + A_2 \\
&= \frac{\gamma}{4\pi} + \left(A_1 + \frac{1}{4\pi}\right)\ln z - \frac{1}{4\pi}e_1(z) + A_2,
\end{aligned} \tag{10.33}$$

we must have $A_1 = -1/4\pi$ and $A_2 = -\gamma/4\pi$. Hence from (10.33) we obtain

$$v = -\frac{1}{4\pi} e_1(z) = -\frac{1}{4\pi} e_1\left(\frac{r^2}{4k(t-t_0)}\right). \tag{10.34}$$

We can check this function v by substituting it in (10.9). Then

$$u^* = \nabla^2 v = -\frac{1}{4\pi} \frac{1}{r} \frac{d}{dr}\left(r\frac{dv}{dr}\right) e_1\left(\frac{r*2}{4k(t-t_0)}\right)$$

$$= -\frac{1}{4\pi k(t-t_0)} \frac{d}{dz}\left[z\frac{de_1}{dz}\right]$$

$$= \frac{1}{4\pi k(t-t_0)} \sum_{n=1}^{\infty} \frac{(-1)^{n+1} z^{n-1}}{(n-1)!},$$

which is the same as (10.6). ∎

10.2. Forward Time Marching Process

The procedure to numerically compute the unknown nodal vectors $\hat{u}_2$ and $\hat{q}_1$ is to start all time integrations at the initial time $t = t_0$ (which may be 0) and march forward with the time step Δt. Suppose that we have already computed $\hat{u}_2$ and $\hat{q}_1$ at $t_\alpha, \alpha = 1, 2, \cdots, n-1$. Then formula (10.23a) allows us to advance the solution to t_n. Since the collocation time t_β in (10.23a) is taken the same as t_α, we get from (10.23a)

$$\sum_{\alpha=1}^{n} \left[H^{n\alpha}\hat{u}_2^\alpha - G^{n\alpha}\hat{q}_1^\alpha\right] = F^n, \tag{10.35}$$

which, after transferring the unknowns $\hat{u}_2$ and $\hat{q}_1$ to the left side, leads to the system of linear equations

$$\sum_{\alpha=1}^{n-1} \left[H^{n\alpha}\hat{u}_2^\alpha - G^{n\alpha}\hat{q}_1^\alpha\right] + H^{nn}\hat{u}_2^n - G^{nn}\hat{q}_1^n = F^n,$$

or

$$H^{nn}\hat{u}_2^n - G^{nn}\hat{q}_1^n = F^n - \sum_{\alpha=1}^{n-1} \left[H^{n\alpha}\hat{u}_2^\alpha - G^{n\alpha}\hat{q}_1^\alpha\right]. \tag{10.36}$$

This equation, when solved, yields $\hat{u}_2^n$ and $\hat{q}_1^n$ (recall that the quantities $\hat{u}_2^\alpha$ and $\hat{q}_1^\alpha$ for $\alpha = 1, 2, \cdots, n-1$, are already determined). Note that t_β is the collocation time and t_α are the nodes in the time interval $T = [t_0, t_N]$. The diagonal elements $H^{\alpha\alpha}$ and $G^{\alpha\alpha}$, $\alpha = 1, \ldots, n$, have the same, yet their respective value, in the matrices H and G. Thus, at the n–th step we are required to compute only the matrices H^{n1} and G^{n1} and the right side vector F^n. This leads to the linear algebraic system (10.23b) which can be easily solved by Gauss elimination or other effective methods. Note that if we use the subroutine SGECO of Linpack to estimate the condition number of the resulting matrices for several test cases, it will be found that in all of them the condition number is typically less than 10^2, indicating good stability for the system (10.23b). This process consists in marching forward in time from $t = 0$, step–by–step with time increments of Δt to any specified time $t = N\Delta t$. The important question that often arises is to ask what the optimal step size Δt should be. Empirically, it is known that $\Delta t \leq 0.005$ for $\Delta x = l/10$. Some research has successfully used $\Delta t = 0.005$ to 0.04 for a rectangular region, and $\Delta t = 0.05$ for circular regions. Others have also used Δt from 0.002 to 0.05 in plane regions.

10.3. Laplace Transform BEM

For the three–dimensional unsteady Poisson equation (10.4), a well -posed mixed boundary value problem must satisfy the following initial and boundary conditions:

(i) $u(\mathbf{x}, 0) = f_1(\mathbf{x})$ at $t = 0$ (also referred to as the initial source);

(ii) $u(\mathbf{x}_0, t) = g(\mathbf{x}_0, t)$ on S_1 ($\mathbf{x}_0 \in S = S_1 \cup S_2$, where S is the boundary surface of a three–dimensional region V);

(iii) $q(\mathbf{x}_0, t) = h(\mathbf{x}_0, t)$ on S_2 ;

(iv) or a mixed boundary condition, like $Au(\mathbf{x}_0, t) + Bq(\mathbf{x}_0, t) = k(\mathbf{x}_0, t)$.

The fundamental solution u^* is, in general, given by

$$u^*(\mathbf{x}, t; \mathbf{x}', t') = \frac{e^{-r^2/4(t-t')}}{[4\pi(t - t')]^m}, \qquad (10.37)$$

where $r = |\mathbf{x} - \mathbf{x}'|$, and $2m$ denotes the spatial dimension of the problem. We shall use the weak variational formulation (§2.1) on Eq (10.4). Since, in view of (1.5c),

$$\iiint_V u^* \nabla^2 u \, dV = \iiint_V u \nabla^2 u^* \, dV + \iint_S [u^* q - u q^*] \, dS, \quad (10.38)$$

the weak variational form of (10.4) is

$$0 = \iiint_V \left(k\nabla^2 u - \frac{\partial u}{\partial t} - b \right) u^* \, dV$$

$$= \iiint_V ku\nabla^2 u^* \, dV + k \iint_S [u^* q - u q^*] \, dS - \iiint_V u^* \left(\frac{\partial u}{\partial t} + b \right) dV,$$

which, in view of the fact that $k\nabla^2 u^* - \partial u^*/\partial t = -\delta(\mathbf{x},t;\mathbf{x}',t')$, reduces to

$$\iiint_V u^* \left(\frac{\partial u}{\partial t} + b \right) dV = k \iint_S [u^* q - u q^*] \, dS +$$

$$+ \iiint_V u \left(\frac{\partial u^*}{\partial t} - \delta \right) dV. \quad (10.39)$$

Now, integration with respect to t' gives

$$\iiint_V \int_0^t u^* (\frac{\partial u}{\partial t} + b) \, dt' \, dV = k \iint_S \int_0^t [u^* q - u q^*] \, dt' \, dS +$$

$$+ \iiint_V \int_0^t u (\frac{\partial u^*}{\partial t} - \delta) \, dt' \, dV,$$

or

$$\iiint_V \left([uu^*]_0^t - \int_0^t u \frac{\partial u^*}{\partial t} \, dt' + u^* \star b \right) dV = k \iint_S [u^* \star q - u \star q^*] \, dS +$$

$$+ \iiint_V \int_0^t u \frac{\partial u^*}{\partial t} \, dt' \, dV - cu(\mathbf{x}',t'), \quad (10.40)$$

where $g_1 \star g_2$ denotes the convolution defined by

$$g_1 \star g_2 = \int_0^t g_1(\mathbf{x}, t - t') g_2(\mathbf{x}, t') \, dt', \quad (10.41)$$

and c is defined by (5.15). By taking limit in (10.36) as $t \to t'$ from left, we find that

$$u^*(\mathbf{x}, t; \mathbf{x}', t'-) = 0,$$

which, in view of the boundary condition (i), implies that $u(\mathbf{x}, 0) = f_1(\mathbf{x})$, $\dfrac{\partial u^*}{\partial t} = -\dfrac{\partial u^*}{\partial t'}$, and then (10.40) leads to the BI Eq

$$- cu(\mathbf{x}', t') = \iiint_V \left([uu^*]_0^t + \int_0^t u \frac{\partial u^*}{\partial t'} \, dt' + (u^* \star b) \right) dV -$$

$$- k \iint_S [u^* \star q - u \star q^*] \, dS - \int_0^t \iiint_V u \frac{\partial u^*}{\partial t'} \, dV \, dt', \quad (10.42)$$

in which the integral terms over $(0, t)$ cancel and we get

$$cu(\mathbf{x}', t') = k \iint_S [u^* \star q - u \star q^*] \, dS - \iiint_V [u^*(\mathbf{x}, t; \mathbf{x}', t')u(\mathbf{x}, t) -$$

$$- u^*(\mathbf{x}, t; \mathbf{x}', 0)u(\mathbf{x}, 0) + (u^* \star b)] \, dV$$

$$= k \iint_S [u^* \star q - u \star q^*] \, dS + \iiint_V [u^* f_1 - u^* \star b] \, dV,$$

$$(10.43)$$

where the term $u^*(\mathbf{x}, t; \mathbf{x}', t') = 0$. The BI Eq (10.43) computes the potential $u(\mathbf{x}', t')$ at any time t', due to the initial source $f_1(\mathbf{x})$, the time–dependent source density $b(\mathbf{x}, t)$ throughout V, and all (both known and unknown) boundary potentials and fluxes over S. The following points should be noted:

(i) The kernel functions involving u^* and q^* are singular as $\mathbf{x}' \to \mathbf{x}$ and $t' \to t$, but the integrals involving them (i.e., the integrals over V and S) exist in the ordinary and the Cauchy p-v sense.

(ii) The kernel functions are well behaved on all infinite boundaries. The initially unknown boundary data can be computed from (10.43) by taking $\mathbf{x}'$ to a boundary point $\mathbf{x}_0$. This yields

$$cu(\mathbf{x}', t) = \frac{\omega}{4\pi} u(\mathbf{x}_0, t) + \iint_S [u^* \star q - q^* \star u] \, dS + \iiint_V [u^* f_1 - u^* \star b] \, dV,$$

$$(10.44)$$

where ω is the solid angle enclosed by the boundary at the point $\mathbf{x}'_0$. Note that $\omega = 2\pi$ in the case when the boundary is smooth.

Let us denote the Laplace transform of $u(\mathbf{x}, t)$ by

$$\mathcal{L}\{u(\mathbf{x}, t)\} \equiv \bar{u}(\mathbf{x}, s) \overset{\text{def}}{=} \int_0^\infty u(\mathbf{x}, t)e^{-st} \, dt, \qquad (10.45)$$

where s is the transform variable. If we apply the Laplace transform to Eq (10.5) and the mixed boundary condition (iv), we get

$$k \frac{d^2 \bar{u}(\mathbf{x}, s)}{dx^2} - s\bar{u}(\mathbf{x}, s) = \bar{b}(\mathbf{x}, s), \qquad (10.46)$$

$$A\bar{u} + B\bar{q} = \bar{k}, \tag{10.47}$$

where $\bar{q}^* = \mathcal{L}\{q^*\}$. The fundamental solution $\bar{u}^*$ for Eq (10.46), where $\bar{u}^* = \mathcal{L}\{u^*\}$ and u^* is defined by (10.37), is

$$\bar{u}^*(\mathbf{x}, s) = \frac{\pi}{2} K_0(\sqrt{s}r), \tag{10.48}$$

where K_0 is the modified Bessel function of second kind and zero order. Then the BE solution of (10.46)–(10.47) is given by

$$k\bar{u}(\mathbf{x}', s) = \iint_S \left(\bar{u}^* \bar{q} - \bar{u}\bar{q}^* \right) dS. \tag{10.49}$$

The major difficulty for this method lies in the inversion of (10.49) to get $u(\mathbf{x}', t')$. However, computer software for inversion of the Laplace transforms with complex s is available, but how good such software is may still be doubtful.

The forward time marching can be used with the Laplace transform technique. We will illustrate the time–marching process by considering a one–dimensional example.

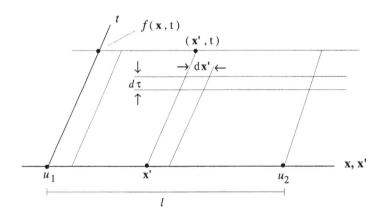

Fig. 10.2. One–dimensional Example.

Example 10.2:
Consider the (x, t)–plane of Fig. 10.2 and a uniform one–dimensional field extending from $x = 0$ to $x = l$ with the initial prescribed potential $f_1(x, 0)$ along it. In this figure, we have lines (u_1) and (u_2) drawn parallel to the time axis. We will consider the potentials u_1, u_2 to be constant (if they are linear, a different computation is required) and prescribed along the lines parallel to the

time axis. Then, in this case, the boundary fluxes q_1 and q_2 are both unknown functions of time.

In one–dimensional case, the BI Eq (10.43) with $k = 1$ becomes

$$u(x,t) = \left[u^* \star q - q^* \star u\right]_0^l + \int_0^l (f_1 u^* - u^* \star b)\, dx. \qquad (10.50)$$

From (10.37) we get (with $2m = 1$)

$$u^*(x,t;x',t') = \frac{e^{r^2/4(t-t')}}{2\sqrt{\pi(t-t')}}, \qquad r = |x - x'|,$$

$$q^*(x,t;x',t') = \frac{u^*}{2(t-t')}\, r \operatorname{sgn} r.$$

Since u_1 and u_2 are constant, we evaluate

$$q^* \star u_1 = \int_0^t q^*(x,t-t')u_1(x,t')\, dt'$$

$$= \int_0^t \frac{e^{r^2/4(t-t')}}{2\sqrt{\pi(t-t')}}\frac{1}{2(t-t')}\, r \operatorname{sgn} r\, u_1\, dt'$$

$$= u_1 r \operatorname{sgn} r \frac{1}{4\sqrt{\pi}} \int_0^t \frac{e^{r^2/4(t-t')}}{(t-t')^{3/2}}\, dt'$$

$$= \frac{1}{2}u_1 r \operatorname{sgn} r \operatorname{erfc}\left(\frac{r}{2\sqrt{t}}\right),$$

where $\operatorname{erfc} z = \dfrac{2}{\sqrt{\pi}}\displaystyle\int_z^\infty e^{-u^2}\, du$. If q_1 is constant (it may be specified in lieu of u_1), then

$$u^* \star q_1 = \int_0^t u^*(x,t-t')q_1(x,t')\, dt'$$

$$= \int_0^t \frac{e^{r^2/4(t-t')}}{2\sqrt{\pi(t-t')}}q_1\, dt'$$

$$= q_1\left[\sqrt{\frac{t}{\pi}}\, e^{-r^2/4t} - \frac{|r|}{2}\operatorname{erfc}\left(\frac{r}{2\sqrt{t}}\right)\right],$$

$$u_1 = \lim_{\varepsilon \to 0} u(0 + \varepsilon, t) = \left[\,q^*(l, 0 + \varepsilon)\quad q^*(0, 0 + \varepsilon)\,\right] \star \begin{Bmatrix} u_1 \\ u_2 \end{Bmatrix} +$$

$$+ \left[\,-u^*(l, 0 + \varepsilon)\quad -u^*(0, 0 + \varepsilon)\,\right] \star \begin{Bmatrix} q_1 \\ q_2 \end{Bmatrix} +$$

$$+ \int_0^l (f_1 u^* - u^* \star b)(x, 0 + \varepsilon)\, dx,$$

$$u_2 = \lim_{\varepsilon \to 0} u(l - \varepsilon, t) = [\, q^*(l, l - \varepsilon) \quad q^*(0, l - \varepsilon)\,] \star \begin{Bmatrix} u_1 \\ u_2 \end{Bmatrix} +$$

$$+ [-u^*(l, l - \varepsilon) \quad -u^*(0, l - \varepsilon)\,] \star \begin{Bmatrix} q_1 \\ q_2 \end{Bmatrix} +$$

$$+ \int_0^l (f_1 u^* - u^* \star b)(x, l - \varepsilon) \, dx.$$

If, for example, we take $f_1 = f_0 \, (= \text{const})$, then the integral in (10.50) becomes

$$\int_0^l (f_1 u^*) \, dx = f_0 \int_0^l u^* \, dx = 2 f_0 \left[\text{erf} \left(\frac{l - x'}{2\sqrt{t}} \right) + \text{erf} \left(\frac{x'}{2\sqrt{t}} \right) \right].$$

As $\varepsilon \to 0$, we get

$$\begin{Bmatrix} u_1 \\ u_2 \end{Bmatrix} = \begin{bmatrix} q_1^* & q_2^* \\ q_2^* & q_1^* \end{bmatrix} \begin{Bmatrix} u_1 \\ u_2 \end{Bmatrix} + \begin{bmatrix} -u_1^* & u_2^* \\ -u_2^* & u_1^* \end{bmatrix} \begin{Bmatrix} q_1 \\ q_2 \end{Bmatrix} +$$

$$+ 2 f_0 \, \text{erf} \left(\frac{l}{2\sqrt{t}} \right) \begin{Bmatrix} 1 \\ 1 \end{Bmatrix}, \quad (10.51)$$

where

$$q_1^* = q^*(l, t; 0 + \varepsilon, t') = q^*(0, t; l - \varepsilon, t') = \frac{1}{2},$$

$$q_2^* = q^*(0, t; 0 + \varepsilon, t') = q^*(l, t; l - \varepsilon, t') = \frac{1}{2} \, \text{erf} \left(\frac{l}{2\sqrt{t}} \right),$$

$$u_1^* = u^*(l, t; 0 + \varepsilon, t') = u^*(0, t; l - \varepsilon, t'),$$

$$u_2^* = u^*(0, t; 0 + \varepsilon, t') = u^*(l, t; l - \varepsilon, t') = \frac{1}{2},$$

Now, in order to discretize the $u^* \star q$ terms along the time axis, we have

$$\int_0^{N\Delta t} u^*(x, t; x', t') q(x, t') \, dt' = \int_0^{(N-1)\Delta t} u^*(x, t; x', t') q(x, t') \, dt' +$$

$$+ \int_{(N-1)\Delta t}^{N\Delta t} u^*(x, t; x', t') q(x, t') \, dt',$$

where the first integral on the right side involves the known data obtained from the solution from the $(N - 1)$st time steps. Thus, if we assume that both q_1 and q_2 are constant over the time step Δt, we can write (10.51) as

$$\left(\int_{(N-1)\Delta t}^{N\Delta t} \begin{bmatrix} u_1^* & -u_2^* \\ u_2^* & -u_1^* \end{bmatrix} dt' \right) \begin{Bmatrix} q_1 \\ q_2 \end{Bmatrix}$$

$$= \int_0^{(N-1)\Delta t} \left(\begin{bmatrix} u_1^* & -u_2^* \\ u_2^* & -u_1^* \end{bmatrix} \begin{Bmatrix} q_1 \\ q_2 \end{Bmatrix} \right) dt' +$$

$$+ \begin{bmatrix} -1/2 & \frac{1}{2} \, \text{erf} \left(\frac{l}{2\sqrt{t}} \right) \\ \text{sym} & -1/2 \end{bmatrix} \begin{Bmatrix} u_1 \\ u_2 \end{Bmatrix} + 2 f_0 \, \text{erf} \left(\frac{l}{2\sqrt{t}} \right) \begin{Bmatrix} 1 \\ 1 \end{Bmatrix}, \quad N = 2, 3, \ldots.$$

This equation can be solved for q_1, q_2 at time t since the right side has only the known quantities in succession $(N = 2, 3, \cdots)$. ∎

10.4. Transient DRM

We shall assume that the thermal properties are constant and the heat conduction is governed by the Fourier–Kirchhoff equation

$$\nabla^2 u = \frac{1}{k}\frac{\partial u}{\partial t} = \frac{1}{k}\dot{u}, \tag{10.52}$$

where a dot over a quantity denotes its derivative with respect to time t. Then the DRM BE Eq (9.20a) becomes

$$H\mathbf{u} - G\mathbf{q} = \frac{1}{k}\left(H\hat{U} - G\hat{Q}\right)\mathbf{F}^{-1}\dot{u}. \tag{10.53}$$

We will use the θ–scheme which approximates the mean value of $\mathbf{u}$ and $\mathbf{q}$ at two consecutive time steps t_n and t_{n+1} by using a weighted average of $\dot{u}$ at these time steps. This scheme is given by

$$\dot{u} = \frac{\{u\}_{n+1} - \{u\}_n}{\Delta t_{n+1}}, \tag{10.54}$$

where $\Delta t_{n+1} = t_{n+1} - t_n$, and the suffix n denotes the value of the quantity at time t_n for $n = 0, 1, 2, \ldots$. Eq (10.53) is then written as

$$H\mathbf{u} - G\mathbf{q} = \frac{1}{k\Delta t_{n+1}}\mathbf{W}\left(\{u\}_{n+1} - \{u\}_n\right), \tag{10.55}$$

where the matrix $\mathbf{W}$ is defined by (9.21). In the θ–scheme the variations of $\mathbf{u}$ and $\mathbf{q}$ with each time step are defined by

$$\begin{aligned}
\mathbf{u} &= \theta_u \{\dot{u}\}_{n+1} + (1 - \theta_u)\{\dot{u}\}_n, \\
\mathbf{q} &= \theta_q \{\dot{q}\}_{n+1} + (1 - \theta_q)\{\dot{q}\}_n,
\end{aligned} \tag{10.56}$$

where $0 \leq \theta_u, \theta_q \leq 1$. The weights θ_u and θ_q refer to some well known schemes which are:

$$\theta_u \text{ or } \theta_q = \begin{cases} 0, & \text{Forward Difference} \\ 1, & \text{Backward Difference} \\ 1/2, & \text{Crank–Nicolson} \\ 2/3, & \text{Galerkin.} \end{cases} \tag{10.57}$$

Of these, the forward and backward difference schemes are conditionally stable, whereas the other two schemes are unconditionally stable. Thus, the transient DRM BE Eq (10.55) becomes

$$\hat{A}\{u\}_{n+1} + \hat{B}\{u\}_n = \hat{C}\{q\}_{n+1} + \hat{D}\{q\}_n, \qquad (10.58)$$

where

$$\hat{A} = \theta_u H - \frac{1}{k\Delta t_{n+1}} \mathbf{W},$$

$$\hat{B} = (1 - \theta_u) H + \frac{1}{k\Delta t_{n+1}} \mathbf{W}, \qquad (10.59)$$

$$\hat{C} = \theta_q G,$$

$$\hat{D} = (1 - \theta_u) G.$$

After substituting the boundary conditions on **u** and **q** at time t and using the rearrangement technique of §5.3, Eq (10.58) reduces to

$$AX = F. \qquad (10.60)$$

The system of equations (10.60) determines the unknown solutions (of **u** and **q**) at time $t = t_{n+1}$ in terms of known solutions at time $t = t_n, n = 0, 1, 2, \ldots$. Since the solutions are known at $t = 0$, we can take $t_{n+1} = n\Delta t$, where Δt is the equi–spaced time step, start at $t_n = 0$, and compute the solutions for $t_{n+1} = n\Delta t$. This process is then continued successively forward in time, moving with the time step Δt each time, until the desired time is reached.

For the unsteady Helmholtz equation, with constant variation of displacement with time, the governing equation is

$$\left(\nabla^2 + k^2\right) u = \dot{u}. \qquad (10.61)$$

The DRM BE Eq (9.32), with $b = -k^2 u + \dot{u}$ in this case, becomes

$$H\mathbf{u} - G\mathbf{q} = -k^2 \mathbf{Wu} + \dot{u}. \qquad (10.62)$$

If we use the θ–scheme (10.56), the transient DRM BE Eq (10.62) reduces to

$$\hat{E}\{u\}_{n+1} + \hat{F}\{u\}_n = \hat{C}\{q\}_{n+1} + \hat{D}\{q\}_n, \qquad (10.63)$$

where

$$\hat{E} = \theta_u \left(H + k^2 \mathbf{W}\right) - \frac{1}{\Delta t_{n+1}},$$

$$\hat{F} = (1 - \theta_u) \left(H + k^2 \mathbf{W}\right) + \frac{1}{\Delta t_{n+1}}, \qquad (10.64)$$

and $\hat{C}$ and $\hat{D}$ are defined in (10.59). After substituting the boundary conditions for **u** and **q** at time t and rearranging the terms, the system (10.63) reduces to the form (10.60) which can be solved for unknown **u** and **q** by the above successive technique with time step Δt until the desired time t_0 is reached. Since the FSM is not widely used, we shall move onto the transient MRM in the next section.

Example 10.3:
Consider the transient heat conduction problem, governed by Eq (10.5), with $k = 1$, subject to the boundary conditions:

$$\frac{\partial u}{\partial x}(0, y, t) = 0 = \frac{\partial u}{\partial y}(x, 0, t),$$
$$u(1, y, t) = 0 = u(x, 1, t), \quad \text{for } t > 0,$$

and the initial condition $u(x, y, 0) = 0$ for all (x, y) in the square plate of side 1 shown in Fig. 10.3. We choose the Crank–Nicolson scheme with $\theta = 1/2$, and since this scheme is unconditionally stable, we can choose any value of Δt. With $\Delta t = 0.01$, the results are as follows:

Location	DRM	Exact
Node 1	0.2947	0.2923
Node 2, 16	0.2789	0.2759
Node 3, 15	0.2293	0.2276
Node 4, 14	0.0.1397	0.1432
Nodes 5–13	0.0	0.0
A	0.2642	0.2613
B	0.2178	0.2214
C	0.1333	0.1405
D	0.1811	0.1765
E	0.1127	0.1119
F	0.0.0728	0.0707

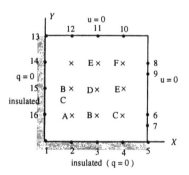

Fig. 10.3.

The symmetry about the line $y = x$ is used in the problem. It should be noted that for such problems the programs Bel–Bell are modified by introducing a parameter Item, such that Item $= 0$ refers to the steady–state case, while Item $= 1$ carries out the transient analysis. The algorithm is as follows:

If Item $= 1$, read $\Delta t, \theta, t_0$;
write $\Delta t, \theta, t_0$;
Initiatize $t = 0.0$;
Define $\theta \Delta t$ and $(1.0 - \theta)\Delta t$;
$t = t + \Delta t$,
If $t > t_0$, stop;
Initialize the matrices H, G, and assemble Eq (10.23b). ■

Example 10.4:
Consider the initial value problem for the Fourier equation (10.52) with $k = 1$;
the initial condition is $u(x, y, 0) = 100 \sin \pi x \sin \pi y$, $0 < x, y < 1$, and the
boundary conditions are: $u(0, y, t) = u(1, y, t) = 0, u(x, 0, t) = u(x, 1, t) =$
$0, 0 < x, y < 1, t > 0$. The exact solution is $u = 100 \sin \pi x \sin \pi y \, e^{-2\pi^2 t}$.
Use $\Delta t = 0.02$, and $\theta = 1/2$. Because of symmetry, the results at $t = 0.5$
are given below for $0 \le y \le x, 0 \le x \le 1/2$ (Fig. 10.4).

Node	Exact	DRM
1	1.32648	1.346
2	2.52312	2.559
3	3.47277	3.523
4	4.08249	4.143
5	4.29259	4.356
6	8.16499	8.287
7	11.23815	11.312
8	13.21123	13.515
9	13.89112	14.127
10	12.56463	12.769
11	9.09185	9.235
12	4.79926	4.872

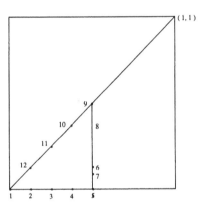

Fig. 10.4.

10.5. Transient MRM

We shall again assume that the transient heat conduction is governed by the
Fourier–Kirchhoff equation (10.52), where we shall denote $b_0 = \dfrac{1}{k}\dot{u}$. Then

$$b_1 = \nabla^2 b_0 = \frac{1}{k} \nabla^2 \dot{u}$$

$$= \frac{1}{k} \frac{d}{dt} \nabla^2 u = \frac{1}{k} \frac{d}{dt} \left(\frac{1}{k} \nabla^2 \dot{u} \right) \tag{10.65}$$

$$= \frac{1}{k^2} \ddot{u},$$

and, in general,

$$b_j = \nabla^2 b_{j-1} = \frac{1}{k^{j+1}} \frac{d^{j+1} u}{dt^{j+1}}, \tag{10.66}$$

$$\frac{\partial b_j}{\partial n} = \frac{1}{k^{j+1}} \frac{d^{j+1} q}{dt^{j+1}}. \tag{10.67}$$

The domain integral B_0, defined by (9.74), becomes

$$B_0 = \sum_{j=0}^{\infty} \int_C \left(b_j q_{j+1}^* - u_{j+1}^* \frac{db_j}{dn} \right) ds$$

$$= \sum_{j=0}^{\infty} \int_C \frac{1}{k^{j+1}} \left[q_{j+1}^* \frac{d^{j+1} u}{dt^{j+1}} - u_{j+1}^* \frac{d^{j+1} q}{dt^{j+1}} \right] ds \tag{10.68}$$

$$= \sum_{j=1}^{\infty} \int_C \frac{1}{k^j} \left[q_j^* \frac{d^j u}{dt^j} - u_j^* \frac{d^j q}{dt^j} \right] ds.$$

Substituting (10.68) into the MRM BI Eq (9.75), we obtain

$$kc(i)u(i) + \int_C uq^* \, ds = \int_C u^* q \, ds -$$

$$- \sum_{j=1}^{\infty} \frac{1}{k^j} \int_C \left[q_j^* \frac{d^j u}{dt^j} - u_j^* \frac{d^j q}{dt^j} \right] ds,$$

or

$$kc(i)u(i) + \sum_{j=0}^{\infty} \frac{1}{k^j} \int_C u_j^* \frac{d^j u}{dt^j} \, ds = \sum_{j=0}^{\infty} \frac{1}{k^j} \int_C u_j^* \frac{d^j q}{dt^j} \, ds, \tag{10.69}$$

where the derivatives of zero order of the functions u and q are the functions themselves. Then the MRM BE Eq (9.76b) becomes

$$HU - GQ = \sum_{j=1}^{\infty} \left(H_j \frac{d^j u}{dt^j} - G_j \frac{d^j q}{dt^j} \right),$$

or

$$\sum_{j=0}^{\infty}\left(H_j\frac{d^j u}{dt^j} - G_j\frac{d^j q}{dt^j}\right) = 0, \tag{10.70}$$

where

$$H_0 = H, \quad G_0 = G,$$

$$H_j = \frac{1}{k^j}\int_{\tilde{C}_j} q_j^* \, ds, \quad G_j = \frac{1}{k^j}\int_{\tilde{C}_j} u_j^* \, ds. \tag{10.71}$$

Eq (10.70), when expanded, is written as

$$H_0 U + \frac{1}{k}H_1\dot{U} + \frac{1}{k^2}H_2\ddot{U} + \cdots$$

$$= G_0 Q + \frac{1}{k}G_1\dot{Q} + \frac{1}{k^2}G_2\ddot{Q} + \cdots, \tag{10.72}$$

where j dots over a quantity denote the j-th derivative with respect to t. We shall use the θ-scheme (10.56) again. Thus,

$$U = \theta_U\{U\}_{n+1} + (1 - \theta_U)\{U\}_n,$$

$$Q = \theta_Q\{Q\}_{n+1} + (1 - \theta_Q)\{Q\}_n,$$

$$\dot{U} = \frac{\{U\}_{n+1} - \{U\}_n}{\Delta t_{n+1}}, \quad \dot{Q} = \frac{\{Q\}_{n+1} - \{Q\}_n}{\Delta t_{n+1}}, \tag{10.73}$$

$$\ddot{U} = \frac{\{\dot{U}\}_{n+1} - \{\dot{U}\}_n}{\Delta t_{n+1}}, \quad \ddot{Q} = \frac{\{\dot{Q}\}_{n+1} - \{\dot{Q}\}_n}{\Delta t_{n+1}},$$

and so on, where $0 \le \theta_U, \theta_Q \le 1$. The first order approximation of (10.72) is given by

$$\hat{H}_1\{U\}_{n+1} - \hat{G}_1\{U\}_n = \hat{H}_2\{Q\}_{n+1} - \hat{G}_2\{Q\}_n, \tag{10.74}$$

where

$$\hat{H}_1 = \theta_U H_0 + \frac{1}{\Delta t_{n+1}}H_1,$$

$$\hat{H}_2 = -(1 - \theta_U) H_0 + \frac{1}{\Delta t_{n+1}}H_1,$$

$$\hat{G}_1 = \theta_Q G_0 + \frac{1}{\Delta t_{n+1}}H_1, \tag{10.75}$$

$$\hat{H}_2 = (1 - \theta_Q) G_0 - \frac{1}{\Delta t_{n+1}}H_1.$$

Higher order approximations of (10.72) can be similarly obtained.

The transient MRM for the Helmholtz equation (10.61) can be similarly developed by taking $b_0 = -k^2 u + \dot{u}$ (see Problem 10.2).

Example 10.5:
We shall consider the same problem as in Example 10.3, to compare the results. Thus, for example, $u = 0.2987$ at Node 1; $u = 0.2756$ at node 16; $u = 0.2567$ at A; $u = 0.1729$ at D; and $u = 0.0694$ at F. ∎

Example 10.6:
The unsteady heat conduction in a relatively thin plate or slab of material which is heated (or cooled) on one or both sides can be approximated, away from the edges, by the one–dimensional Fourier equation

$$\frac{\partial u}{\partial t} = \frac{1}{k}\frac{\partial^2 u}{\partial x^2}. \tag{10.76}$$

This case often occurs in engineering applications. Consider Eq (10.76) with $k = 1$; the initial condition is $u(x,0) = 100\sin \pi x$, $0 < x < 1$, and the Dirichlet boundary conditions are: $u(0,t) = u(1,t) = 0$, $t > 0$. The exact solution is $u = 100\sin \pi x\, e^{-\pi^2 t}$. The solutions at $t = 0.5$ are presented in Fig. 10.5, with $\Delta t = 0.005$ and $\theta = 0.5$. The solutions show symmetry about the line $x = 1/2$. ∎

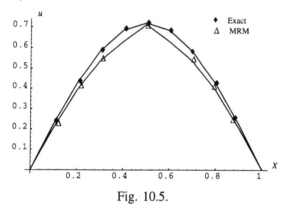

Fig. 10.5.

10.6. Problems

10.1. Since $u = u(\mathbf{x}, t')$ implies $\dfrac{\partial u}{\partial t'} = k\nabla^2 u$, and $u^* = u^*(\mathbf{x}, \mathbf{x}', t, t')$ implies $\dfrac{\partial u^*}{\partial t'} = -k\nabla^2 u^*$, show that

(a) the functions u and u^* are solutions of these two adjoint partial differential equations, and

$$\frac{\partial}{\partial t'}[uu^*] = u^*\frac{\partial u}{\partial t'} + u\frac{\partial u^*}{\partial t'} = k(u^*\nabla^2 u - u\nabla^2 u^*); \qquad (10.77)$$

(b) integrate both sides of Eq (10.77) over the cylinder $C \times (0, t_0]$ and derive the BI Eq (10.8).

10.2. Derive the transient MRM with the θ–scheme for the Helmholtz equation (10.61).

10.3. Solve the one–dimensional form of Eq (10.4) for a metal slab extending from $x = -l$ to $x = l$, which has a prescribed initial temperature T_0. The slab is brought into contact with an agitated liquid both having a uniform temperature T_a which may be higher than the melting point of the bar T_m (see Fig. 10.6).

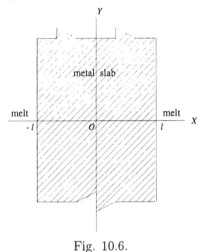

Fig. 10.6.

Thus, the initial and boundary conditions are: $T(x, 0) = T_0, T(\pm l, t) = T_m, \kappa\frac{\partial T}{\partial x}(\pm l, t) = c_s(T_a - T_m) + \rho\lambda\frac{\partial}{\partial t}(\pm l, t)$, where κ is the thermal conductivity, ρ the density, λ the latent heat in melting, and c_s the coefficient of surface heat transfer.

10.4. Solve the unsteady heat conduction problem in a square plate of side $l = 6$ feet with the Dirichlet boundary condition $u = 0$ all along the boundary, and the initial condition $u = 30$ at $t = 0$ for all x, y. Divide the square plate into four quarters, with origin at the center. Use the top right quarter plate (because of symmetry) with 8 linear elements, and take time step $\Delta t = 0.05$. The exact solution is given by

$$u = \sum_{n=1}^{\infty} \sum_{m=1}^{\infty} A_{nm} \sin \frac{n\pi x}{l} \sin \frac{m\pi y}{l} e^{-k(n^2+m^2)\pi^2 t/l},$$

where

$$A_{nm} = \frac{120}{nm\pi^2} \left[(-1)^n - 1 \right] \left[(-1)^m - 1 \right].$$

10.5. Consider the heat conduction in a rectangular plate having three circular holes as shown in Fig. 10.7(a) with the boundary conditions noted there. Because of symmetry, take the BE mesh in Fig. 10.7(b) with $N = 20$, $M = 1$, and initial condition $U_0 = 10$.

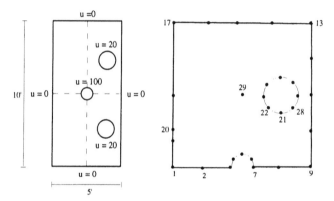

Fig. 10.7.

10.6. Consider a circular opening in an infinite region with the initial condition $u_0 = 10$ on the circular boundary. Investigate the surface temperature $u(\mathbf{x}, t)$ of the cooling hole in the region shown in Fig. 10.8, where only a quarter sector is considered with $N = 25$ and $M = 4$.

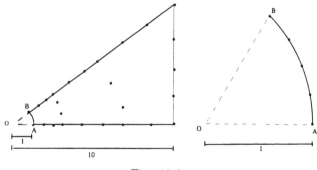

Fig. 10.8.

10.7. Consider the heat conduction problem in a rectangle, with $k = 1$ W/m K; the boundary conditions are: $u(x,0) = u(0,y) = 0$ (insulated) and the Robin–type boundary condition is satisfied on the sides $x = 0.5$ and $y = 1$. The above set of boundary conditions cause a thermal shock at the sides $x = 0.5$ and $y = 1$. For computations, use $\Delta t = 0.02$ and $\theta = 1/2$.

10.8. Solve the Fourier–Kirchhoff equation (10.52) with $k = 1$, for the initial value problem: $u(x,0) = t$, $0 < x < 1$, $t > 0$, and $u_x(0,t) = u_x(1,t) = 0$, $t > 0$.

10.9. Find the voltage distribution of a lossless transmission line of length l, if the end $x = 0$ is connected at time $t = 0$ to a source of variable e.m.f. $Ee^{-\alpha t}$ and the end $x = l$ is kept open. It is assumed that the current and voltage in the line are initially zero.

References and Bibliography for Additional Reading

R. A. Adey and A. Elzein, *Accuracy of Boundary Element Computations*, in Boundary Elements in Mechanical and Electrical Engineering (C. A. Brebbia and A. Chaudouet–Miranda, Eds.), Proc. of the International Boundary Element Symposium, Nice, France, 15–17 May, 1990, Computational Mechanics Publications, Southampton and Springer–Verlag, Berlin, 1990.

M. H. Aliabadi and C. A. Brebbia (Eds.), *Advanced Formulations in Boundary Element Methods*, Computational Mechanics Publications, Southampton and Elsevier Applied Science, London, 1993.

C. A. Brebbia, *Progress in Boundary Element Methods*, Vols. 1 and 2, Pentech Press, London, 1981, 1983.

———— (Ed.), *Topics in Boundary Element Research*, Vol. 1, Springer–Verlag, Berlin, 1984.

———— and A. Chaudouet–Miranda (Eds.), *Boundary Elements in Mechanical and Electrical Engineering*, Proc. of the International Boundary Element Symposium, Nice, France, 15–17 May, 1990, Computational Mechanics Publications, Southampton and Springer–Verlag, Berlin, 1990.

————, J. C. F. Telles and L. C. Wrobel, *Boundary Element Techniques*, Springer–Verlag, Berlin, 1984.

H. S. Carslaw and J. C. Jaeger, *Conduction of Heat in Solids*, 2nd Ed., Clarendon Press, Oxford, 1959.

J. J. Dongarra, C. B. Moler, J. R. Bunch and G. W. Stewart, *LINPACK Users' Guide*, SIAM, Philadelphia, 1979.

B. Ould Sidi Fall and J. P. Barrand, *Solution of a Transient Heat Conduction Problem by the BEM Code involving only Boundary integrals*, in Boundary Elements in Mechanical and Electrical Engineering (C. A. Brebbia and A. Chaudouet–Miranda, Eds.), Proc. of the International Boundary Element Symposium, Nice, France, 15–17 May, 1990, Computational Mechanics Publications, Southampton and Springer–Verlag, Berlin, 1990.

A. J. Nowak, *Solving Linear Heat Conduction Problems by the Multiple Reciprocity*

Method, Ch. 3 in Boundary Element Methods in Heat Transfer (L. C. Wrobel and C. A. Brebbia, Eds.), Computational Mechanics Publications, Southampton and Elsevier Applied Science, London, 1992.

—————and C. A. Brebbia, *The Multiple Reciprocity Method*, Ch. 3 in Advanced Formulations in Boundary Element Methods (M. H. Aliabadi and C. A. Brebbia, Eds.), Computational Mechanics Publications, Southampton and Elsevier Applied Science, London, 1993.

P. D. Lax and R. D. Richtmyer, *Survey of the stability of finite difference equations*, Comm. in Pure Appl. Math. **9** (1956), 267–293.

J. W. Leech, *Stability of finite-difference equations for the transient response of a flat plate*, J. AIAA **3** (1965), 1772–1773.

N. N. Lebedev, I. P. Skalskaya and Y. S. Uflyand, *Worked Problems in Applied Mathematics*, Trans. from Russian by R. S. Silverman, Dover, New York, 1965.

P. W. Partridge and C. A. Brebbia, The Double Reciprocity Method (1993), Computational Mechanics Publications, Southampton and Elsevier Applied Science, London.

H. L. G. Pina and J. L. M. Fernandez, *Applications in Transient Heat Conduction*, Ch. 2 in Topics in Boundary Element Research, Vol. 1 (C. A. Brebbia, Ed.), Springer–Verlag, Berlin, 1984.

J. N. Reddy, *An Introduction to the Finite Element Method*, McGraw–Hill, New York, 1984.

F. J. Rizzo and D. J. Shippy, *A method of solution for certain problems of transient heat conduction*, AIAA J. **8** (1970), 2004–2009.

A

Variational Calculus

A.1. Line Integrals

In many types of boundary value problems, the variational methods are used to provide precise formulations which can be applied in any prescribed system of coordinates. We shall derive the Euler equation which is the necessary condition for the solution of the following problem: Find an unknown function $y(x)$ for which the integral of a function F of x, $u(x)$, and one or more derivatives of $u(x)$, is a minimum. Let two fixed points $A(a, c)$ and $B(b, d)$ in the xy–plane be joined by a curve $C : u = u(x)$ (see Fig. A.1 (a)). Then $c = u(a)$, $d = u(b)$, and $u' = du/dx = u'(x)$ at each point Q of C. Thus, the curve C determines a value of

$$I = \int_a^b F(x, u, u') \, dx. \qquad (A.1)$$

The value of the integral I will change if we replace C by a new curve joining A and B. In order to investigate this variation of I with C, i.e., to determine as to for what curve C the integral I has a minimum (or a maximum) value, we shall confine to a set of curves C_α which are defined as follows: First we select any curve $u = \phi(x)$ such that

$$\phi(a) = 0 = \phi(b) \qquad (A.2)$$

(see Fig. A.1(b)). Then for any value of the parameter α, the curve C_α is, in view of (A.2), defined by

$$U = U(x) = u(x) + \alpha\phi(x), \quad U(a) = c, \quad U(b) = d. \qquad (A.3)$$

Hence it passes through A and B. From (A.3) we find that on the curve C_α

$$U' = \frac{dU}{dx} = u'(x) + \alpha\phi'(x). \qquad (A.4)$$

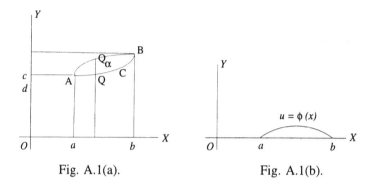

Fig. A.1(a).   Fig. A.1(b).

For any value of α, we thus have a curve C_α and may form the value of I along C_α by substituting the values from (A.3) and (A.4) as

$$I = \int_a^b F(x, U, U')\, dx. \qquad (A.5)$$

Using the differentiation formula

$$\frac{\partial F}{\partial \alpha} = \frac{\partial F}{\partial U}\frac{\partial U}{\partial \alpha} + \frac{\partial F}{\partial U'}\frac{\partial U'}{\partial \alpha},$$

and noticing from (A.3) and (A.4) that $\dfrac{\partial U}{\partial \alpha} = \phi(x),\ \dfrac{\partial U'}{\partial \alpha} = \phi'(x)$, we find that

$$\frac{\partial F}{\partial \alpha} = \phi(x)\frac{\partial F}{\partial U} + \phi'(x)\frac{\partial F}{\partial U'}. \qquad (A.6)$$

Hence,

$$\frac{\partial I}{\partial \alpha} = \int_a^b \left[\phi(x)\frac{\partial F}{\partial U} + \phi'(x)\frac{\partial F}{\partial U'}\right] dx. \qquad (A.7)$$

If we integrate by parts the second term in the integrand in (A.7), and use (A.2), we get

$$\frac{\partial I}{\partial \alpha} = \int_a^b \phi(x)\left[\phi(x)\frac{\partial F}{\partial U} - \frac{d}{dx}\left(\frac{\partial F}{\partial U'}\right)\right] dx. \qquad (A.8)$$

Let us assume that there is a twice–differentiable curve, say C, for which the value of I is a minimum. Then, the value of I on C will be less than the value of I on any other curve C_α. Thus, $I(\alpha)$ will assume a minimum value for $\alpha = 0$, since $\partial I/\partial \alpha$ is continuous. But, then $U = u$ and $U' = u'$ when $\alpha = 0$. Hence, taking $\alpha = 0$ in (A.8), and $I'(0) = 0$, we find that

$$\int_a^b \phi(x)\left[\frac{\partial F}{\partial u} - \frac{d}{dx}\left(\frac{\partial F}{\partial u'}\right)\right] dx = 0. \tag{A.9}$$

Since the solution $\phi(x)$ is arbitrary, except for the conditions (A.2), the factor in the square brackets in (A.9) is continuous, which implies that

$$\frac{\partial F}{\partial u} - \frac{d}{dx}\left(\frac{\partial F}{\partial u'}\right) = 0 \tag{A.10}$$

for all $x \in [a, b]$. (If the expression within the square brackets in (A.9) were nonzero at any point, say x_0, there would be some small interval including this point, say $x_1 < x_0 < x_2$, in which it remains nonzero; then, by taking the function $\phi(x)$ in Fig. A.2, the integrand in (A.10) would be zero except for the interval (x_1, x_2) and nonzero for this interval.) Thus we have proved: If the integral I, defined by (A.1), has a minimum (or a maximum) along any sufficiently smooth curve C joining A and B, then $u = u(x)$ will be a solution of the differential Eq (A.10). This is known as the Euler equation, and its solution $u(x)$ as the extremal. Note that the derivative d/dx in (A.10) is computed by recalling that $u = u(x)$ and $u' = u'(x) = du/dx$ are functions of x. After this differentiation is carried out, Eq (A.10) becomes

$$\frac{\partial F}{\partial u} - \frac{\partial^2 F}{\partial x \partial u'} - \frac{\partial^2 F}{\partial u \partial u'}\frac{du}{dx} - \frac{\partial^2 F}{\partial u'^2}\frac{d^2 u}{dx^2} = 0. \tag{A.11}$$

This differential equation is of order 2, and its solution contains two arbitrary constants which must be determined from the conditions that the curve passes through A and B.

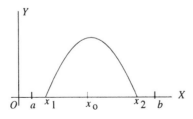

Fig. A.2.

Example A.1:
We shall find the arc C that minimizes the area S of the surface of revolution

generated by revolving $C : u = y(x)$ about x–axis. Since

$$S = 2\pi \int_a^b y\, ds = 2\pi \int_a^b y\sqrt{1 + y'^2}\, dx, \qquad (A.12)$$

we shall therefore minimize $S/2\pi$. Replacing u by y in (A.1) we get

$$I = \int_a^b y\sqrt{1 + y'^2}\, dx. \qquad (A.13)$$

Here $F(x, u, u') = F(x, y, y') = y\sqrt{1 + y'^2}$. Thus, $\partial F/\partial u = \partial F/\partial y = \sqrt{1 + y'^2}$, $\partial F/\partial u' = \partial F/\partial y' = yy'/\sqrt{1 + y'^2}$, and the Euler equation (A.10) or (A.11) becomes

$$\sqrt{1 + y'^2} - \frac{d}{dx}\frac{yy'}{\sqrt{1 + y'^2}} = 0, \quad \text{or } 1 + y'^2 - yy'' = 0. \qquad (A.14)$$

Taking $y' = p$, we find that $y'' = p\,dp/dy$, and the solution of (A.14) is

$$y = c_1 \cosh \frac{x - c_2}{c_1}, \qquad (A.15)$$

which represents a family of catenaries with x–axis as base (see Fig. A.3.). The constants c_1 and c_2 can be determined so as to make (A.15) pass through A and B. It will then be the minimizing arc. Since the integral (A.12) is with respect to arc–length s, the extremal (A.15) is called the geodesic for this problem. ■

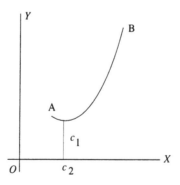

Fig. A.3.

By following the above procedure and using integration by parts twice in the third term in the integrand in

$$\frac{dI}{d\alpha} = \int_a^b \left[\phi(x)\frac{\partial F}{\partial U} + \phi'(x)\frac{\partial F}{\partial U'} + \phi''(x)\frac{\partial^2 F}{\partial U''}\right] dx. \qquad (A.16)$$

and taking the additional requirement that $\phi'(a) = 0 = \phi'(b)$, we find that the Euler equation which provides a necessary condition for the functional

$$I = \int_a^b F(x, u, u', u'') \, dx \qquad (A.17)$$

to be a minimum is

$$\frac{\partial F}{\partial u} - \frac{d}{dx}\left(\frac{\partial F}{\partial u'}\right) + \frac{d^2}{dx^2}\left(\frac{\partial F}{\partial u''}\right) = 0 \qquad (A.18)$$

(see Prob. A.1.).

If the integral I depends on two functions u and v, i.e.,

$$I = \int_a^b F(x, u, v, u', v') \, dx, \qquad (A.19)$$

then there are two Euler equations:

$$\frac{\partial F}{\partial u} - \frac{d}{dx}\left(\frac{\partial F}{\partial u'}\right) = 0, \quad \frac{\partial F}{\partial u} - \frac{d}{dx}\left(\frac{\partial F}{\partial u'}\right) = 0. \qquad (A.20)$$

To derive (A.20), introduce two functions $\phi(x)$ and $\psi(x)$ and two parameters α and β such that $U = u + \alpha\phi(x)$, $V = v + \beta\psi(x)$. Then $\partial I/\partial\alpha = 0$ and $\partial I/\partial\beta = 0$.

A.2. Variational Notation

Eq (A.3) implies that the difference $U - u = \alpha\phi(x)$, or QQ_α in Fig. A.1, is the change in U, starting from u and the value of $\alpha = 0$. Hence $\alpha\phi(x) = U - u$ is called the (first) variation in u and is denoted by δu. It may also be regarded as the differential dU, for at $\alpha = 0$, $d\alpha = \alpha - 0 = \alpha$; $dU/d\alpha = \phi(x)$ so that

$$dU = \left(\frac{dU}{d\alpha}\right) d\alpha = \phi(x)\alpha = \delta u. \qquad (A.21)$$

Also, Eq (A.4) may be rewritten as $U' = u' + \alpha\phi'(x)$, or

$$U' - u' = \alpha\phi'(x). \qquad (A.22)$$

The term $\alpha\phi'(x) = U' - u'$ is the variation of u' and is denoted by $\delta u'$. It is also the differential

$$dU' = \left(\frac{dU'}{d\alpha}\right) d\alpha = \phi'(x)\alpha = \delta u'. \tag{A.23}$$

From (A.21) and (A.23) it follows that

$$\frac{d(\delta u)}{dx} = \delta\left(\frac{du}{dx}\right),$$

or

$$(\delta u') = \delta(u'), \tag{A.24}$$

i.e., the operator δ and d/dx are commutative. The variation of u'' and higher derivatives of u are defined analogously.

If u is changed to $u + \delta u = u + \alpha w$, then the corresponding change in a function $F = F(x, u, u')$ at u is defined by

$$\delta F = \frac{\partial F}{\partial u} \delta u + \frac{\partial F}{\partial u'} \delta u'. \tag{A.25}$$

Since δ behaves like a differential operator, it has the following properties:

(i) $\delta(f_1 \pm f_2) = \delta f_1 \pm \delta f_2$.
(ii) $\delta(f_1 f_2) = f_1 \delta f_2 + f_2 \delta f_1$.
(iii) $\delta(f_1/f_2) = (f_2 \delta f_1 - f_1 \delta f_2)/f_2^2$.
(iv) $\delta(f)^n = n(f)^{n-1} \delta f$.
(v) $D(\delta u) = \delta(Du)$,
(vi) $\delta \int_a^b u(x)\, dx = \int_a^b \delta u(x)\, dx$

where $' \equiv D \equiv d/dx$, and f_1, f_2 are functions of x, u, u'. Note that (v) and (vi) are commutative properties under differential and integral operators. The above six properties (i)–(vi) can be easily proved by using (A.25).

Example A.2:
Show that

$$\nabla(\delta u) \cdot \nabla u = \frac{1}{2}\delta|\nabla u|^2. \tag{A.26}$$

Note that (in matrix notation)

$$\nabla(\delta u) = \left[\frac{\partial}{\partial x}(\delta u) \quad \frac{\partial}{\partial y}(\delta u)\right] = [\delta(u_x) \quad \delta(u_y)] = \delta(\nabla u).$$

Thus,

$$
\nabla(\delta u) \cdot \nabla u = \delta(\nabla u) \cdot \nabla u
$$

$$
= \left[\frac{\partial}{\partial x}(u_x)\delta x + \frac{\partial}{\partial y}(u_x)\delta y \quad \frac{\partial}{\partial x}(u_y)\delta x + \frac{\partial}{\partial y}(u_y)\delta y \right] [u_x \quad u_y]^T
$$

$$
= \left\{ \frac{\partial}{\partial x}(u_x)u_x + \frac{\partial}{\partial x}(u_y)u_y \right\}\delta x + \left\{ \frac{\partial}{\partial y}(u_x)u_x + \frac{\partial}{\partial y}(u_y)u_y \right\}\delta y
$$

$$
= \frac{1}{2}\frac{\partial}{\partial x}|\nabla u|^2 \delta x + \frac{1}{2}\frac{\partial}{\partial y}|\nabla u|^2 \delta y
$$

$$
= \frac{1}{2}\delta|\nabla u|^2. \blacksquare
$$

From (A.8) we find that

$$
\delta I = \left(\frac{dI}{d\alpha} \right)_{\alpha=0}
$$

$$
= \int_a^b \left[\frac{\partial F}{\partial u} - \frac{d}{dx}\left(\frac{\partial F}{\partial u'} \right) \right] \delta u \, dx, \tag{A.27}
$$

where the variations at the end points are zero. Similarly, for I defined by (A.17) we have

$$
\delta I = \int_a^b \left[\frac{\partial F}{\partial u} - \frac{d}{dx}\left(\frac{\partial F}{\partial u'} \right) + \frac{d^2}{dx^2}\left(\frac{\partial F}{\partial u''} \right) \right] \delta u \, dx, \tag{A.28}
$$

where the variations are restricted by the conditions: $\phi'(a) = 0 = \phi'(b)$.

For the integral I in (A.19) whose integrand $F(x, u, u', v, v')$ contains two independent variables u and v, we define the variations as total differentials so that

$$
\delta u = \alpha\phi(x), \quad \delta u' = \alpha\phi'(x), \quad \delta v = \alpha\psi(x), \quad \delta v' = \alpha\psi'(x).
$$

Then the definition in this case becomes

$$
\delta F = \frac{\partial F}{\partial u}\delta u + \frac{\partial F}{\partial u'}\delta u' + \frac{\partial F}{\partial v}\delta v + \frac{\partial F}{\partial v'}\delta v'. \tag{A.29}
$$

For the integral I in (A.19) we have

$$
\delta I = \int_a^b \left[\frac{\partial F}{\partial u} - \frac{d}{dx}\left(\frac{\partial F}{\partial u'} \right) \right] \delta u \, dx + \int_a^b \left[\frac{\partial F}{\partial v} - \frac{d}{dx}\left(\frac{\partial F}{\partial v'} \right) \right] \delta v \, dx. \tag{A.30}
$$

In the parametric case, we can introduce a new parameter t as the independent variable and regard x and u (or x, u and v) as the dependent variables. For example, let us consider the integrand $F(x, u, u')$. Using the hat to denote differentiation with respect to the parameter t, we have: $u' = \hat{u}/\hat{x}$, and $F(x, u, u')\, dx = F(x, u, \hat{u}/\hat{x})\hat{x}\, dt$. Let $x = a$ when $t = t_1$, and $x = b$ when $t = t_2$, and denote $F(x, u, \hat{u}/\hat{x})\hat{x}$ by $G(x, u, \hat{x}, \hat{u})$. Then, the integral (A.1) can be written as

$$I = \int_a^b F(x, u, u')\, dx = \int_{t_1}^{t_2} G(x, u, \hat{x}, \hat{u})\, dt,$$

and (A.27) will become (using (A.30) with x, u, v replaced by t, x, u)

$$\delta I = \int_{t_1}^{t_2} \left[\frac{\partial G}{\partial x} - \frac{d}{dt}\left(\frac{\partial G}{\partial \hat{x}}\right) \right] \delta x\, dt + \int_{t_1}^{t_2} \left[\frac{\partial G}{\partial u} - \frac{d}{dt}\left(\frac{\partial G}{\partial u}\right) \right] \delta u\, dt, \quad (A.31)$$

where δx and δu are zero at t_1 and t_2.

A.3. Multiple Integrals

Let $F(a, y, u, p, q)$ be a twice differentiable function of five variables, where the dependent variable u is a function of x and y and $p = u_x = \partial u/\partial x$, $q = u_y = \partial u/\partial y$. We shall study the variation of the integral

$$I = \iint_S F(x, y, u, p, q)\, dx\, dy \qquad (A.32)$$

over a surface S which passes through a fixed boundary curve B, and determine for which surface S the integral I is a minimum. Following the method of §A.1, we first choose any surface S defined by $u = \phi(x, y)$ such that $\phi(x, y) = 0$ on the curve C, where C is the projection of the boundary curve B in the xy-plane. Since $u = \phi(x, y)$ passes through C, we define any surface S_α for any α by

$$U(x, y) = u(x, y) + \alpha\phi(x, y). \qquad (A.33)$$

This surface S_α also passes through C. On this surface

$$P \equiv U_x = p + \alpha\phi_x, \quad Q \equiv U_y = q + \alpha\phi_y, \qquad (A.34)$$

so that the integral I is written as

$$I = \iint_{S_\alpha} F(x, y, U, P, Q)\, dx\, dy. \qquad (A.35)$$

Hence

$$\frac{dI}{d\alpha} = \int\int \left[\frac{\partial F}{\partial U}\phi + \frac{\partial F}{\partial P}\phi_x + \frac{\partial F}{\partial Q}\phi_y\right] dx\, dy. \qquad (A.36)$$

Using (1.5c) we find from (A.36) that

$$\left.\frac{dI}{d\alpha}\right|_{\alpha=0} = \int\int \left[\frac{\partial F}{\partial U} - \frac{\partial}{\partial x}\left(\frac{\partial F}{\partial P}\right) - \frac{\partial}{\partial y}\frac{\partial F}{\partial Q}\right]\phi\, dx\, dy, \qquad (A.37)$$

since $U = u$, $P = p$, $Q = q$ when $\alpha = 0$. If we multiply the value of $dI/d\alpha|_{\alpha=0}$ by $d\alpha = \alpha - 0 = \alpha$ and write $\delta\alpha$ for $\alpha\phi$, we find, as in (A.25), that

$$\delta\alpha = \int\int \left[\frac{\partial F}{\partial u} - \frac{\partial}{\partial x}\left(\frac{\partial F}{\partial p}\right) - \frac{\partial}{\partial y}\left(\frac{\partial F}{\partial q}\right)\right]\delta u\, dx\, dy. \qquad (A.38)$$

If $u(x,y)$ is the surface for which I is a minimum, then $\delta I = 0$ for arbitrary δu. Since the expression in the square brackets in (A.38) is continuous, it must be zero. This gives a necessary condition for I to be a minimum as

$$\frac{\partial F}{\partial u} - \frac{\partial}{\partial x}\left(\frac{\partial F}{\partial p}\right) - \frac{\partial}{\partial y}\left(\frac{\partial F}{\partial q}\right) = 0. \qquad (A.39)$$

This is the Euler equation for (A.32), and any surface corresponding to a solution of this equation is an extremal.

If $F = F(x, y, u, p, q, r, s, t)$, where p and q are defined above and $r = u_{xx}$, $s = u_{xy}$, and $t = u_{yy}$, then the Euler equation to minimize the integral $I = \int\int F(x, y, u, p, q, r, s, t)\, dx\, dy$ is given by

$$\frac{\partial F}{\partial u} - \frac{\partial}{\partial x}\left(\frac{\partial F}{\partial p}\right) - \frac{\partial}{\partial y}\left(\frac{\partial F}{\partial q}\right) + \frac{\partial^2}{\partial x^2}\left(\frac{\partial F}{\partial r}\right) + \frac{\partial^2}{\partial x\partial y}\left(\frac{\partial F}{\partial s}\right) + \frac{\partial^2}{\partial y^2}\left(\frac{\partial F}{\partial t}\right) = 0. \qquad (A.40)$$

Example A.3:
The condition that the Dirichlet integral

$$I = \int\int \left[\left(\frac{\partial u}{\partial x}\right)^2 + \left(\frac{\partial u}{\partial y}\right)^2\right] dx\, dy \qquad (A.41)$$

be a minimum is

$$\nabla^2 u = \frac{\partial^2 u}{\partial x^2}\frac{\partial^2 u}{\partial y^2} = 0, \qquad (A.42)$$

which is the Laplace equation. ∎

A.4. Problems

A.1. Fill in the details in the derivation of the Euler equation (A.18).

A.2. Fill in the details in the derivation of the Euler equation (A.20).

A.3. Find the geodesics for the following problems:

(a) On the xy–plane, take $I = \int ds = \int \sqrt{1 + y'^2}\, dx$.

(b) On the xy–plane, take $I = \int ds = \int \sqrt{1 + r^2(d\theta/dr)^2}\, dr$.

(c) On the cylinder $x^2 + y^2 = a^2$, $-\infty < z < \infty$, take $x = a\cos t$, $y = a\sin t$, and $I = \int ds = \int \sqrt{a^2 + (dz/dt)^2}\, dt$.

A.4. A ray of light moves between two fixed points in the xy–plane with variable velocity $v(x, y)$. By Fermat's law, its travel time is $\int \dfrac{ds}{v} = \int \dfrac{\sqrt{1 + y'^2}}{v}\, dx$. Show that the paths for a minimum travel time are given by

$$\frac{vy''}{\sqrt{1 + y'^2}} - \frac{\partial v}{\partial x} y' + \frac{\partial v}{\partial y} = 0.$$

A.5. Find the extremal when the following integral is minimized:

(a) $\int (y'^2 + y^2)\, dx$.

(b) $\int (y''^2 + y^2)\, dx$.

A.6. If the end points are not fixed, derive

$$\int_a^b F(x, u, u')\, dx = \delta u \frac{\partial F}{\partial u'}\Big|_a^b + \int_a^b \Big[\frac{\partial F}{\partial u} - \frac{d}{dx}\Big(\frac{\partial F}{\partial u'}\Big)\Big]\delta u\, dx.$$

A.7. Take $\delta u = \alpha\phi(x) + \beta\psi(x)$ in the integral (A.1), and let δI denote the total differential of I at $\alpha = 0 = \beta$, where $\delta\alpha = \alpha$, $\delta\beta = \beta$. Set $H \equiv \dfrac{\partial F}{\partial u} - \dfrac{d}{dx}\Big(\dfrac{\partial F}{\partial u'}\Big)$, and show that the variation δI in this case can be written as

$$\delta I = d\alpha \int_a^b H\phi\, dx + d\beta \int_a^b H\psi\, dx.$$

A.8. Find the extremal for the problem of determining a curve C of prescribed length l joining AB and maximizing the area $A = \int y\,dx$, bounded by C, x–axis and two fixed ordinates.

References and Bibliography for Additional Reading

M. Becker, *The principles and Applications of Variational Methods*, MIT Press, Cambridge, MA, 1964.

M. J. Forray, *Variational Calculus in Science and Engineering*, Academic Press, New York, 1972.

P. Franklin, *Methods of Advanced Calculus*, McGraw–Hill, New York, 1944.

W. Fulks, *Advanced Calculus*, 3rd ed., Wiley, New York, 1978.

F. B. Hildebrand, *Methods of Applied Mathematics*, 2nd ed., Prentice–Hall, New York, 1965.

S. G. Mikhlin, *Variational Methods in Mathematical Physics*, Pergamon Press, New York, 1964.

R. E. Williamson, R. H. Crowell and H. F. Trotter, *Calculus of Vector Functions*, 2nd ed., Prentice–Hall, Englewood Cliffs, NJ, 1968.

B

Potential Theory

B.1. Electrostatic Potential

In the three–dimensional space, let $\mathbf{x} = (x_1, x_2, x_3)$ denote a point in a domain V with its boundary (surface) S. Let V_{int} and V_{ext} denote the regions interior and exterior to S respectively. We will find the fundamental solution (or free–space solution) $u^*(\mathbf{x}, \mathbf{x}')$ of the problem

$$-\nabla^2 u^* = \delta(\mathbf{x}, \mathbf{x}'), \qquad u^*(\infty) = 0, \tag{B.1}$$

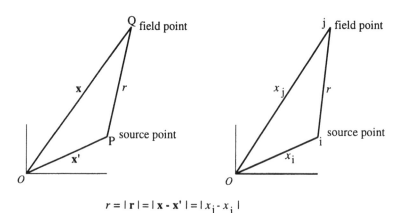

$$r = |\mathbf{r}| = |\mathbf{x} - \mathbf{x}'| = |x_j - x_i|$$

Fig. B.1.

where $\mathbf{x}' = (x_1', x_2', x_3')$ is a fixed point in V. Note that u^* can be interpreted as the electrostatic potential at an arbitrary point $\mathbf{x}$ (called the field point) due to a unit charge at the point $\mathbf{x}'$ (called the source point), both points lying on the surface S (see Fig. B.1). Such a potential is given by

$$u^*(\mathbf{x}; \mathbf{x}') = \frac{1}{4\pi r} = \frac{1}{4\pi|\mathbf{x} - \mathbf{x}'|}, \qquad (B.2)$$

and in the two–dimensional case by

$$u^*(\mathbf{x}; \mathbf{x}') = \frac{1}{2\pi} \ln \frac{1}{r} = \frac{1}{2\pi} \ln \frac{1}{|\mathbf{x} - \mathbf{x}'|}, \qquad (B.3)$$

where $\mathbf{x} = (x_1, x_2)$ and $\mathbf{x}' = (x_1', x_2')$. See Chapter 4 for details of derivation of fundamental solutions (B.2) and (B.3). Note that u^* has a special place in the development of the BEM. It can also be used to obtain fundamental solutions for other boundary value problems. This is explained in the following example.

Example B.1:
The fundamental solution (B.2) can be used to obtain the fundamental solution for the Poisson equation

$$-\nabla^2 u = 4\pi b, \qquad (B.4)$$

where b is a given function of position $\mathbf{x}$. To do so: multiply Eq (B.1) by $u(\mathbf{x})$, Eq (B.4) by $u^*(\mathbf{x}; \mathbf{x}')$, subtract, integrate over the region V, and use Green's second identity (1.8) and the translation property (1.55) of the Dirac delta function. This leads to

$$\iiint_V \left(u^* \nabla^2 u - u \nabla^2 u^* \right) dV = \iiint_V \left(u\delta(\mathbf{x}, \mathbf{x}') - 4\pi b u^* \right) dV$$

or

$$\iint_S \left(u^* \frac{\partial u}{\partial n} - u \frac{\partial u^*}{\partial n} \right) dS = u(\mathbf{x}') - 4\pi \iiint_V b u^* \, dV,$$

or

$$\frac{1}{4\pi} \iint_S \frac{1}{r} \frac{\partial u}{\partial n} \, dS - \frac{1}{4\pi} \iint_S u \frac{\partial}{\partial n} \left(\frac{1}{r} \right) dS = u(\mathbf{x}') - \iiint_V \frac{b}{r} \, dV.$$

Thus,

$$u(\mathbf{x}') = \iiint_V \frac{b}{r} \, dV - \frac{1}{4\pi} \iint_S u \frac{\partial}{\partial n} \left(\frac{1}{r} \right) dS + \frac{1}{4\pi} \iint_S \frac{1}{r} \frac{\partial u}{\partial n} \, dS. \quad (B.5)$$

Let $u|_S = \tau$, $\dfrac{\partial u}{\partial n}\Big|_S = \sigma$ be prescribed. Then (B.5) is rewritten as

$$u(P) = \iiint_V \frac{b}{r} \, dV + \frac{1}{4\pi} \iint_S \frac{\sigma(Q)}{r} \, dS - \frac{1}{4\pi} \iint_S \tau(Q) \frac{\partial}{\partial n} \left(\frac{1}{r} \right) dS, \qquad (B.6)$$

where P is the source point $\mathbf{x}'$ and Q the field point $\mathbf{x}$ on S (see Fig. B.1). ∎

Note that (B.5) and (B.6) lead to many interesting properties of harmonic functions. In the language of electrostatics, the first term on the right side of (B.6) is called the volume or Newtonian potential with boundary density b; the second term is called the single–layer potential with charge (or source) density σ; the third term is called the double–layer potential with dipole density τ. These three potentials have the following properties:

Newtonian Potential: (i) $\nabla^2 u = 0$ for points P in V_{ext}; (ii) For points P within V_{int}, the integral is improper but it converges and admits two differentials under the integral sign if f is sufficiently smooth. The result is $\nabla^2 u = -4\pi f(P)$.

Single–Layer Potential: (i) $\nabla^2 u = 0$ outside S; (ii) The integral is improper at the surface S but converges uniformly if S is regular. Also, the integral remains continuous as we pass through S; (iii) Derivatives of u in the outward direction normal to S are:

$$\frac{\partial u}{\partial n}\bigg|_{P_+} = -2\pi\sigma(Q) + \iint_S \sigma(Q)\frac{\cos(\mathbf{x}'-\mathbf{x},n)}{|\mathbf{x}-\mathbf{x}'|^2}\,dS, \qquad (B.7)$$

$$\frac{\partial u}{\partial n}\bigg|_{P_-} = 2\pi\sigma(Q) + \iint_S \sigma(Q)\frac{\cos(\mathbf{x}'-\mathbf{x},n)}{|\mathbf{x}-\mathbf{x}'|^2}\,dS, \qquad (B.8)$$

where P_+ and P_- mean that we approach S from V_{int} and V_{ext} respectively, and $\mathbf{x}$ and $\mathbf{x}'$ are both on S. From (B.7) and (B.8) we find that the jump discontinuity of the normal derivative of u across S is

$$\sigma = \frac{1}{4\pi}\left(\frac{\partial u}{\partial n}\bigg|_{P_-} - \frac{\partial u}{\partial n}\bigg|_{P_+}\right). \qquad (B.9)$$

Double–Layer Potential: (i) $\nabla^2 u = 0$ outside S; (ii) The integral is improper at the surface but converges if S is regular; (iii) The integral undergoes a discontinuity when passing through S such that

$$u\big|_{P_+} = 2\pi\tau(P) - \iint_S \tau(Q)\frac{\cos(\mathbf{x}-\mathbf{x}',n)}{|\mathbf{x}-\mathbf{x}'|^2}\,dS, \qquad (B.10)$$

$$u\big|_{P_-} = -2\pi\tau(P) - \iint_S \tau(Q)\frac{\cos(\mathbf{x}-\mathbf{x}',n)}{|\mathbf{x}-\mathbf{x}'|^2}\,dS, \qquad (B.11)$$

and jump of u across S is, from (B.10) and (B.11), given by

$$\tau = \frac{1}{4\pi}\left(u\big|_{P_+} - u\big|_{P_-}\right); \qquad (B.12)$$

(iv) The normal derivatives remain continuous as S is covered.
Note that Eqs (B.7), (B.8), (B.10), (B.11) are Fredholm integral equations of the second kind. For the solution of a boundary value problem for elliptic equations, we cannot prescribe u and $\partial u/\partial n$ arbitrarily on S. Thus, Eq (B.6) does not allow us to construct a solution for Eq (B.4) such that u and $\partial u/\partial n$ shall have arbitrary values on S.

B.2. Dirichlet Problems

The two types of the Dirichlet boundary value problems are as follows:

Interior Dirichlet Problem :

$$\nabla^2 u_{\text{int}} = 0, \quad \mathbf{x} \in V_{\text{int}}; \quad u_{\text{int}}\big|_S = f. \qquad (B.13)$$

Exterior Dirichlet Problem :

$$\nabla^2 u_{\text{ext}} = 0, \quad \mathbf{x} \in V_{\text{ext}};$$
$$u_{\text{ext}}\big|_S = f, \; u_{\text{ext}}\big|_\infty = O\left(\frac{1}{r}\right), \; \frac{\partial u}{\partial r}\bigg|_\infty = O\left(\frac{1}{r^2}\right). \qquad (B.14)$$

Solutions: For the interior Dirichlet problem, assume that the solution u_{int} is the potential of a double–layer density τ, i.e.,

$$u_{\text{int}}(t) = \iint_S \frac{\tau(\mathbf{x}')\cos(\mathbf{x}-\mathbf{x}',n)}{r^2}\, dS. \qquad (B.15)$$

To satisfy the boundary condition, we use (B.10) and get

$$u\big|_{P_+} = f(P) = 2\pi\tau(P) - \iint_S \tau(Q)\frac{\cos(\mathbf{x}-\mathbf{x}',n)}{|\mathbf{x}-\mathbf{x}'|^2}\, dS.$$

This yields

$$\tau(P) = \frac{1}{2\pi}f(P) + \frac{1}{2\pi}\iint_S K(P,Q)\tau(Q)\, dS, \qquad (B.16)$$

which is a Fredholm integral equation of the second kind with the kernel

$$K(\mathbf{x}, \mathbf{x}') = \frac{1}{2\pi} \frac{\cos(\mathbf{x} - \mathbf{x}')}{|\mathbf{x} - \mathbf{x}'|^2}$$

with $\mathbf{x} = P, \mathbf{x}' = Q$ both on S. In order to solve the Fredholm integral equation of the second kind (B.16) for τ, we substitute this solution in (B.15). This gives the required solution for the interior Dirichlet problem.

Similarly, for the exterior Dirichlet problem we assume that u is a double-layer potential

$$u_{\text{ext}} = \iint_S \tau(\mathbf{x}') \frac{\cos(\mathbf{x} - \mathbf{x}')}{|\mathbf{x} - \mathbf{x}'|^2} \, dS.$$

Then from (B.11) we have

$$u\big|_{P_-} = f(P) = -2\pi\tau(P) - \iint_S \tau(Q) \frac{\cos(\mathbf{x} - \mathbf{x}')}{|\mathbf{x} - \mathbf{x}'|^2} \, dS,$$

or

$$\tau(P) = -\frac{1}{2\pi} f(P) + \iint_S K(P, Q)\tau(Q) \, dS$$

which is a Fredholm integral equation of the second kind.

Integral Equation Formulation in a Composite Medium. For the interior Dirichlet problem the fundamental solution $u^*(\mathbf{x}; \mathbf{x}')$ satisfies Eq (B.1). We multiply Eq (B.13) by u^*, Eq (B.1) by u_{int}, add, and integrate on V_{int}. This yields

$$\iiint_{V_{\text{int}}} \left(u^* \nabla^2 u_{\text{int}} - u_{\text{int}} \nabla^2 u^* \right) dV = \iiint_{V_{\text{int}}} u_{\text{int}} \delta(\mathbf{x}, \mathbf{x}') \, dV,$$

or using (1.8) and (1.55)

$$\iint_S \left(u^* \frac{\partial u_{\text{int}}}{\partial n} - u_{\text{int}} \frac{\partial u^*}{\partial n} \right) dS = \begin{cases} u_{\text{int}}(\mathbf{x}'), & \text{if } \mathbf{x}' \in V_{\text{int}} \\ 0, & \text{if } \mathbf{x}' \in V_{\text{ext}}. \end{cases} \tag{B.17}$$

For the exterior Dirichlet problem, multiply Eq (B.14) by u^*, Eq (B.1) by u_{ext}, add, integrate over the region bounded internally by S and externally by a sphere S_r, and apply (1.8) and (1.55). This yields (note that the contribution on S_r vanishes as $r \to \infty$)

$$\iint_{S+S_r} \left(-u^* \frac{\partial u_{\text{ext}}}{\partial n} + u_{\text{ext}} \frac{\partial u^*}{\partial n} \right) dS = \begin{cases} 0, & \text{if } \mathbf{x}' \in V_{\text{int}} \\ u_{\text{ext}}(\mathbf{x}'), & \text{if } \mathbf{x}' \in V_{\text{ext}}, \end{cases}$$

(here we have plus sign on the left side because $\hat{n}$ is outward to V_{int} on S), or

$$\iint_S \left(-u^* \frac{\partial u_{ext}}{\partial n} + u_{ext} \frac{\partial u^*}{\partial n} \right) dS = \begin{cases} 0, & \text{if } \mathbf{x}' \in V_{int} \\ u_{ext}(\mathbf{x}'), & \text{if } \mathbf{x}' \in V_{ext}. \end{cases} \quad (B.18)$$

(The outward normal to V_{ext} on S is in the $-\hat{n}$ direction). If we add (B.17) and (B.18) and note that both u_{int} and u_{ext} take the same value f on S, we get

$$\iint_S u^*(\mathbf{x}; \mathbf{x}') \left(\frac{\partial u_{int}}{\partial n} - \frac{\partial u_{ext}}{\partial n} \right) dS = \begin{cases} u_{int}(\mathbf{x}'), & \text{if } \mathbf{x}' \in V_{int} \\ u_{ext}, & \text{if } \mathbf{x}' \in V_{ext}. \end{cases}$$
$$(B.19)$$

To seek, e.g., a single–layer potential, we use (B.2) and (B.9). Then (B.19) gives

$$\iint_S \frac{\sigma(\mathbf{x})}{|\mathbf{x} - \mathbf{x}'|} dS = \begin{cases} u_{int}(\mathbf{x}'), & \text{if } \mathbf{x}' \in V_{int} \\ u_{ext}(\mathbf{x}'), & \text{if } \mathbf{x}' \in V_{ext}. \end{cases}$$

Relabeling $\mathbf{x}$ and $\mathbf{x}'$ we get

$$\iint_S \frac{\sigma(\mathbf{x}')}{|\mathbf{x} - \mathbf{x}'|} dS = \begin{cases} u_{int}(\mathbf{x}), & \text{if } \mathbf{x} \in V_{int} \\ u_{ext}(\mathbf{x}), & \text{if } \mathbf{x} \in V_{ext}. \end{cases} \quad (B.20)$$

It is a single–layer potential with unknown charge density σ. Finally, using the boundary condition $u_{int}|_S = f = u_{ext}|_S$ on the right side of (B.20), we get

$$f(\mathbf{x}) = \iint_S \frac{\sigma(\mathbf{x}')}{|\mathbf{x} - \mathbf{x}'|} dS, \quad (B.21)$$

with both $\mathbf{x}$ and $\mathbf{x}'$ on S. This is a Fredholm integral equation of the first kind.

B.3. Neumann Problems

The following boundary value problems are defined:

Interior Neumann Problem :

$$\nabla^2 u_{int} = 0, \ \mathbf{x} \in V_{int}; \quad \frac{\partial u_{int}}{\partial n} = f. \quad (B.22)$$

Exterior Neumann Problem :

$$\nabla^2 u_{ext} = 0, \ \mathbf{x} \in V_{ext}; \quad \frac{\partial u_{ext}}{\partial n} = f, \ u_{ext}|_\infty = 0. \quad (B.23)$$

For both interior and exterior Neumann problems, the function f satisfies the consistency condition

$$\iint_S f(\mathbf{x}') \, dS = 0, \quad \mathbf{x}' \in S.$$

Both interior and exterior Neumann problems can be reduced to integral equations. For the interior Neumann problem, we seek a solution in the form of, say, a single–layer potential

$$u_{\text{int}} = \iint_S \frac{\sigma(Q)}{r} \, dS, \tag{B.24}$$

which is harmonic in V_{int}. It will be the solution of (B.22) if the density σ is chosen such that

$$\left. \frac{\partial u_{\text{int}}}{\partial n} \right|_{P_-} = f(P), \quad P \in S.$$

In view of (B.8), it gives

$$2\pi\sigma(P) + \iint_S \sigma(Q) \frac{\cos(\mathbf{x}' - \mathbf{x}, n)}{|\mathbf{x} - \mathbf{x}'|^2} \, dS = f(P). \tag{B.25}$$

Thus, $\sigma(P)$ is a solution of the Fredholm integral equation of the second kind

$$\sigma(P) = \frac{1}{2\pi} f(P) - \iint_S K(P, Q)\sigma(Q) \, dS,$$

where $K(P, Q)$ is the same as (B.16).

For the exterior Neumann problem, the solution of (B.23) leads to the Fredholm integral equation of second kind

$$\sigma(P) = -\frac{1}{2\pi} f(P) - \iint_S K(P, Q)\sigma(Q) \, dS. \tag{B.26}$$

Example B.2:
(Electrostatic potential due to a thin circular disk S of radius a on which the potential V is prescribed). Choose cylindrical polar coordinates (ρ, ϕ, z) such that the center of the disk is at the origin with z–axis perpendicular to the plane of the disk. The disk occupies the region $z = 0$, $0 \leq \rho \leq a$ for all ϕ. Without loss of generality, we take $V = f(\rho) \cos n\phi$, and the charge density as $\sigma \cos n\phi$, where n is an integer. Then, since

$$f(\rho) \cos n\phi = \iint_S \frac{\sigma(\mathbf{x}')}{|\mathbf{x} - \mathbf{x}'|} \, dS$$

where $\mathbf{x} = (\rho, \phi, 0)$, $\mathbf{x}' = (t, \phi_1, 0)$, we get

$$f(\rho) \cos n\phi = \int_0^a \int_0^{2\pi} \frac{t\sigma(t) \cos n\phi_1}{\sqrt{\rho^2 + t^2 - 2\rho t \cos(\phi - \phi_1)}} \, d\phi_1 \, dt$$

$$= \int_0^a t\sigma(t) \int_{-\phi}^{2\pi-\phi} \frac{\cos n(\phi + \psi)}{\sqrt{\rho^2 + t^2 - 2\rho t \cos \psi}} \, d\psi \, dt,$$

which with $\psi = \phi_1 - \phi$ becomes

$$= \int_0^a \int_0^{2\pi} \frac{t\sigma(t) \cos n\phi \cos n\psi}{\sqrt{\rho^2 + t^2 - 2\rho t \cos \psi}} \, d\psi \, dt,$$

and gives, on cancelling $\cos n\phi$ on both sides,

$$f(\rho) = \int_0^a \int_0^{2\pi} \frac{t\sigma(t) \cos n\psi}{\sqrt{\rho^2 + t^2 - 2\rho t \cos \psi}} \, d\psi \, dt. \quad \blacksquare$$

References and Bibliography for Additional Reading

P. R. Garabedian, *Partial Differential Equations*, Wiley, New York, 1964.

R. P. Kanwal, *Linear Integral Equations: Theory and Technique*, Academic Press, New York, 1971.

O. D. Kellogg, *Foundations of Potential Theory*, Springer–Verlag, Berlin, 1929.

P. Morse and H. Feshbach, *Methods of Theoretical Physics*, vol. I, II, McGraw–Hill, New York, 1953.

G. F. Roach, *Green's Functions: Introductory Theory and Applications*, Van Nostrand Reinhold, London, 1970.

I. Stakgold, *Green's Functions and Boundary Value Problems*, Wiley, New York, 1979.

A. N. Tychonov and A. A. Samarski, *Partial Differential Equations of Mathematical Physics*, Vol. I and II, Holden–Day, San Francisco, 1964, 1967.

C

Gauss Quadrature

C.1. One–Dimensional Gauss Quadrature

Direct evaluation of integrals of the form

$$\int_a^b F(x)\, dx.$$

is time consuming, sometimes very difficult, or even impossible, depending on the form of the integrand $F(x)$. To evaluate such integrals , we shall use the numerical technique based on the Gauss–Legendre formula which is obtained by using the transformation

$$x = \frac{b+a}{2} + \frac{b-a}{2}\,\zeta, \quad dx = \frac{b-a}{2}\,d\zeta. \qquad (C.1)$$

Thus,

$$
\begin{aligned}
\int_a^b F(x)\, dx &= \frac{b-a}{2} \int_{-1}^1 F\!\left(\frac{b+a}{2} + \frac{b-a}{2}\zeta\right) d\zeta \\
&= \frac{b-a}{2}\left[W_1 F(\zeta_1) + W_2 F(\zeta_2) + \cdots + W_n F(\zeta_n)\right] + R_n \\
&\approx \frac{b-a}{2} \sum_{i=1}^n W_i F(x_i),
\end{aligned}
$$

$$(C.2)$$

where the coordinates x_i and the weights W_i are given by

$$x_i = \frac{b+a}{2} + \frac{b-a}{2}\zeta_i,$$

$$W_i = \frac{2(1-\zeta_i^2)}{(n+1)^2 P_{n+1}^2(\zeta_i)}, \quad i = 1, 2, \cdots, n; \tag{C.3}$$

$P_n(\zeta)$ are Legendre polynomials of order n:

$$P_n(\zeta) = \frac{1}{2^n n!} \frac{d^n}{d\zeta^n}\left[(\zeta^2 - 1)^n\right], \quad n = 0, 1, 2, \cdots, \tag{C.4}$$

$\zeta_1, \zeta_2, \cdots, \zeta_n$ are the zeros of $P_n(\zeta)$, and

$$R_n = \frac{(b-a)^{2n+1}(n!)^4}{(2n+1)[(2n)!]} F^{(2n)}(\eta), \quad a \le \eta \le b.$$

Note that if $a = -1$ and $b = 1$, then the factor $(b-a)/2$ in (C.2) becomes 1, and thus

$$\int_{-1}^{1} F(x)\,dx = \sum_{i=1}^{n} W_i F(x_i) + R_n. \tag{C.5}$$

The numerical values of ζ_i and W_i are given below in Table C.1 for $n = 1, 2, \cdots, 10$.

Example C.1:
Let $F(x) = x\cos x$, $a = 0$, $b = \pi/2$, $n = 3$; then $x_i = \pi(1 + \zeta_i)/4$. Note that

$$P_3(\zeta) = \frac{1}{2}\left[5\zeta^3 - 3\zeta\right].$$

Using the values of ζ_i and W_i from Table C.1, we get

$$x_1 = \pi[1 + \zeta_1]/4 \approx 0.177031, \quad x_2 = \pi[1 + \zeta_2]/4 \approx 0.785398,$$

$$x_3 = \pi[1 + \zeta_3]/4 \approx 1.393765.$$

Thus,

$$f(x_1) = x_1\cos x_1 = 0.174264, \quad f(x_2) = x_2\cos x_2 = 0.555360,$$

$$f(x_3) = x_3\cos x_3 = 0.245453,$$

and

$$\int_{0}^{\pi/2} x\cos x\,dx \approx \frac{\pi}{4}\sum_{i=1}^{3} W_i f(x_i) \approx 0.570851 - 0.000070 = 0.570781.$$

The exact value is 0.570796, which shows the accuracy of the method even when a very small number of points are used. ∎

Example C.2:

Using the interpolation functions $\phi_1(\xi)$ defined by (1.28), we shall evaluate

$$I_1 = \int_a^b \phi_1^2 \, dx, \quad \text{and} \quad I_2 = \int_a^b \left(\frac{d\phi_1}{dx}\right)^2 \, dx.$$

Let $b - a = h$. Then from (C.1)

$$dx = \frac{h}{2}\, d\xi, \quad \frac{d\phi_1}{dx} = \frac{d\phi_1}{dx}\frac{d\xi}{dx} = \frac{2}{h}\frac{d\phi_1}{d\xi} = \frac{2}{h}\left(\xi - \frac{1}{2}\right),$$

and

for $n = 2$: $\quad I_1 = \dfrac{h}{2}\displaystyle\int_{-1}^{1}(\phi_1)^2 \, d\xi$

$$= \frac{h}{2}\frac{1}{4}\left[(0.57735)^2 + (0.577735)^4\right] 2 = 0.\bar{1}\, h = \frac{h}{9},$$

for $n = 3$: $\quad I_1 = \dfrac{h}{2}\displaystyle\int_{-1}^{1}(\phi_1)^2 \, dx$

$$= \frac{h}{2}\frac{1}{4}\left[(0.77459)^2 + (0.77459)^4\right](0.\bar{5})^2 = 0.1\bar{3}\, h = \frac{2h}{15}.$$

The exact value of $I_1 = 2h/15$. Again, for I_2 we get

for $n = 1$: $\quad I_2 = \dfrac{2}{h}\displaystyle\int_{-1}^{1}\left(\frac{d\phi_1}{d\xi}\right)^2 \, d\xi = \frac{2}{h}\left(\frac{1}{4}\right)^2 = \frac{1}{h}$,

for $n = 2$: $\quad I_2 = \dfrac{2}{h}\displaystyle\int_{-1}^{1}\left(\frac{d\phi_1}{d\xi}\right)^2 \, d\xi$

$$= \frac{2}{h}\left[\frac{1}{4} + (-0.57735)^2\right]^2 2 = \frac{2.\bar{3}}{h} = \frac{7}{3h},$$

for $n = 3$: $\quad I_2 = \dfrac{2}{h}\displaystyle\int_{-1}^{1}\left(\frac{d\phi_1}{d\xi}\right)^2 \, d\xi$

$$= \frac{2}{h}\left[\frac{1}{4}(0.\bar{8}) + 2\left(\frac{1}{4} + 0.77459^2\right)(0.\bar{5})\right]$$

$$= \frac{2.\bar{3}}{h} = \frac{7}{3h}.$$

The exact value of $I_2 = 7/3h$. This again shows the accuracy of the Gauss quadrature even with a very small number of points. ∎

C.2. One–Dimensional Logarithmic Gauss Quadrature

$$\int_0^1 \left(\ln \frac{1}{r}\right) f(x)\, dx \approx \sum_{i=1}^n w_i f(\xi_i). \qquad (C.6)$$

The values of ξ_i and w_i are given below in table C.2. This is also known as the Berthod–Zaborowsky quadrature. The formula (C.6) has n integration points to compute the integral exactly if the degree of $f(x)$ is less than or equal to $(2n - 1)$.

C.3. Two–Dimensional Gauss Quadrature Formula

For triangular regions:

$$I = \int_0^1 \left[\int_0^{1-\xi_2} f(\xi_1, \xi_2, \xi_3)\, d\xi \right] d\xi_2 = \sum_{i=1}^n w_i f(\xi_1^i, \xi_2^i, \xi_3^i), \qquad (C.7)$$

where ξ_1, ξ_2, ξ_3 are coordinates of the triangle in Fig. C.1. The exact value of the integral with integral powers of the triangular coordinates is given by

$$\int_0^1 \left[\int_0^{1-\xi_2} \xi_1^i \xi_2^j \xi_3^k\, d\xi_1 \right] d\xi_2 = \frac{i!\, j!\, k!}{(i + j + k + 2)!}. \qquad (C.8)$$

In general, for an n–dimensional simplex

$$\int_0^1 \int_0^{1-x_1} \int_0^{1-x_1-x_2} \cdots \int_0^{1-x_1-x_2-\cdots-x_n} x_1^{\alpha_1} x_2^{\alpha_2} \ldots x_n^{\alpha_n}\, dx_n \ldots dx_1$$
$$= \frac{\alpha_1!\alpha_2!\ldots\alpha_n!}{(n + \alpha_1\alpha_2 + \cdots + \alpha_n)!}, \qquad (C.9)$$

where $n = 2$ is the case of a triangle in two dimensions and tetrahedron in three dimensions.

Table C.1: Gauss–Legendre Quadrature

n	ζ_i	W_i
1	0.0	2.0
2	±0.577350269189626	1.0
3	0.0	0.888888888888889
	±0.774596669241483	0.555555555555556
4	±0.861136311594053	0.347854845137454
	±0.339981043584856	0.652145154862546
5	0.0	0.568888888888889
	±0.906179845938664	0.236926885056189
	±0.538469310105683	0.478628670499366
6	±0.932469514203152	0.171324492379170
	±0.661209386466265	0.360761573048139
	±0.238619186083197	0.467913934572691
7	0.0	0.417959183673469
	±0.949107912342759	0.129484966168870
	±0.741531185599394	0.279705391489277
	±0.405845151377397	0.381830050505119
8	±0.960289856497536	0.101228536290376
	±0.796666477413627	0.222381034453374
	±0.525532409916329	0.313707645877887
	±0.183434642495650	0.362683783378362
9	0.0	0.330239355001260
	±0.968160239507626	0.081274388361574
	±0.836031107326636	0.180648160694857
	±0.613371432700590	0.260610696402935
	±0.324253423403809	0.312347077040003
10	±0.973906528517172	0.066671344308688
	±0.865063366688985	0.149451349150581
	±0.679409568299024	0.219086362515982
	±0.433395394129247	0.269266719309996
	±0.148874338981631	0.295524224714753

Table C.2: Logarithmic Gauss Quadrature

n	ξ_i	w_i
2	0.11200820	0.71853931
	0.60227690	0.28146068
3	0.6389079	0.51340455
	0.36899706	0.39128004
	0.76688030	0.09461540
4	0.04144848	0.38346406
	0.24527491	0.38687531
	0.55616545	0.19043512
	0.84898239	0.03922548
5	0.02913447	0.29789347
	0.17397721	0.34977622
	0.41170252	0.23448829
	0.67731417	0.09893045
	0.89477136	0.01891155
6	0.02163400	0.23876366
	0.12958339	0.30828657
	0.31402044	0.24531742
	0.53865721	0.14200875
	0.75691533	0.05545462
	0.92266885	0.01016895
7	0.01671935	0.19616938
	0.10018567	0.27930264
	0.24629424	0.23968187
	0.43346349	0.16577577
	0.63235098	0.08894322
	0.81111862	0.03319430
	0.94084816	0.00593278
8	0.01332024	0.16441660
	0.07750427	0.23752560
	0.19787102	0.22684198
	0.35415398	0.17575408
	075181452	0.05787221
	0.84937932	0.02097907
	0.95332645	0.00368640

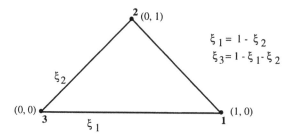

Fig. C.1. Triangular Coordinates.

References and Bibliography for Additional Reading

M. Abramowitz and I. A. Stegun, *Handbook of Mathematical Functions*, Dover, New York, 1965.

M. Cristescu and G. Loubignac, *Gaussian Quadrature Formulas for Functions with Singularities in $1/R$ over Triangles and Quadrangles*, in Recent Advances in Boundary Element Methods (C. A. Brebbia, Ed.), Pentech Press, London, 1978.

A. H. Stroud and D. Secrest, *Gaussian Quadrature Formulas*, Prentice–Hall, New York, 1966.

J. J. Tuma, *Handbook of Numerical Calculations in Engineering*, McGraw–Hill, New York, 1989.

D

Software Development

D.1. Computer Software

The computer software required for this book is provided on the enclosed disk which can be used on a PC. The programs are written in C which is a case–sensitive structural language. It combines subtlety and elegance with raw power and flexibility. However, care should be taken about the global quantities which may be mishandled when the programs are combined with others. The modules for each of the programs Be1.c, Be2.c, Be5.c, and Be11.c have the header cbox1.h, cbox2.h, cbox5.h, cbox11.h respectively, together with the C library headers ⟨stdio.h⟩ and ⟨math.h⟩. These modules can be compiled and linked separately for each program.

The diskette contains two directories, one for all versions of UNIX C, and the other for C programs in DOS. The programs are already compiled and linked. They are run, e.g., for Be1 as follows:

```
> cd unix/Be1          or        > cd dos\Be1
> a.out                          > Be1
```

```
NOTE: FIRST LINE IN THE INPUT FILE SHOULD BE EITHER BLANK OR THE TITLE
NOT DATA
    Enter the name of the input file:ex5_2.in
    Enter the name of the output file:ex5_2.out
    > open ex5_2.out
```

Other programs can be similarly run. UNIX C has been used to test run these programs on some benchmark problems in Chapters 5 and 6. Since there are different C programs, the .exe files in the DOS directory may sometimes fail to run. In such a situation, depending on the C compiler being used, the files Be1.c, Be2.c, Be5.c, or Be11.c must be first compiled and linked with an appropriate command which is bcc for C++, ttc for Turbo C, and cl for Microsoft C; some compilers, like Turbo C, will compile and run from within its integrated environment.

The file a.out in the unix directory runs for all versions of UNIX C; it is also the recommended directory since the results obtained even in single precision are more accurate and closer to exact solutions in this directory.

The user is advised to make a copy of the diskette immediately, and use this copy; also, if possible, copy the diskette on a hard drive.

D.2. DRM and MRM

The following schematic representation of the matrix Eq (9.17) is helpful in writing the computer code for the DRM:

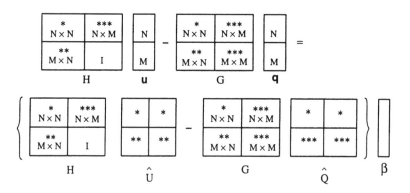

where each of the matrices $H, G, \hat{U}, \hat{Q}$ is $(N \times M) \times (N \times M)$, i is an $(M \times M)$ identity matrix, and $\hat{U}$ is an $(N \times M) \times (N \times M)$ matrix with zeros on the leading diagonal. We have identified the submatrices by $*$, $**$, and $* * *$ which mean as follows:

$*$ These submatrices are used in the boundary solution.

$**$ These are submatrices of H and G and are used to obtain the results at the interior points once all the boundary values of u and q are known.

$***$ For an interior point j, these submatrices of H, G, and q have zero elements.

The above matrix partition scheme contains parts for both the boundary and the interior nodes. The interior part is schematically like this:

There are some good Fortran subroutines available to solve potential and elastostatic problems. Noteworthy are those contained in the work of Banerjee and Butterfield (1981), Becker (1992), Brebbia and Dominguez (1992), Gipson (1987), and Wrobel (1986). These are cited in the bibliography given below.

References and Bibliography for Additional Reading

M. H. Aliabadi and C. A. Brebbia (Eds.), *Advanced Formulations in Boundary Element Methods*, Computational Mechanics Publications, Southampton, and Elsevier Applied Science, London, 1993.

P. K. Banerjee, *Boundary Element Methods in Engineering*, 2nd Ed., McGraw–Hill, New York, 1993.

_____ and R. Butterfield, *Boundary Element Methods in Engineering Science*, McGraw–Hill, New York, 1981.

A. A. Becker, *The Boundary Element Method in Engineering*, McGraw–Hill, London, 1992.

C. A. Brebbia and J. Dominguez, *Boundary Elements*, Computational Mechanics Publications, Southampton and McGraw–Hill, New york, 1992.

C. A. Brebbia and Ferrante, *Computational Methods for the solution of Engineering Problems*, Pentech Press, London, 1978.

P. C. Das, *A disc bases block elimination technique used for the solution of non-symmetrical fully populated matrix systems encountered in the boundary element method*, in Recent Advances in Boundary Element Methods (C. A. Brebbia, Ed.), Pentech Press, London, 1978, pp. 391–404.

G. S. Gipson, *Boundary Element Fundamentals*, Computational Mechanics Publications, Southampton, 1987.

J. Mackerle and C. A. Brebbia (Eds.), *The Boundary Element Reference Book*, Computational Mechanics Publications, Southampton, and Springer–Verlag, Berlin, 1988.

P.W. Partridge and C. A. Brebbia, *Computer implementation of the BEM dual reciprocity method for the solution of the Poisson type equations*, Software for Engg. Workstations **5** (1989), 199–206.

L. C. Wrobel, *Two–Dimensional Potential Analysis Using Boundary Elements*, Software Series, Computational Mechanics Publications, Southampton, 1986.

Selected Answers

Note: In the case when a problem has an exact solution, compare it with the numerical solution.

1.1. Using $\nabla \equiv \hat{i}\,\partial/\partial x + \hat{j}\,\partial/\partial y + \hat{k}\,\partial/\partial z$, the right side

$$
= u\left[\frac{\partial}{\partial x}\left(h\frac{\partial v}{\partial x}\right) + \frac{\partial}{\partial y}\left(h\frac{\partial v}{\partial y}\right) + \frac{\partial}{\partial z}\left(h\frac{\partial v}{\partial z}\right)\right]
$$

$$
= u\left(\hat{i}\frac{\partial}{\partial x} + \hat{j}\frac{\partial}{\partial y} + \hat{k}\frac{\partial}{\partial z}\right) \cdot \left(\hat{i}h\frac{\partial v}{\partial x} + \hat{j}h\frac{\partial v}{\partial y} + \hat{k}h\frac{\partial v}{\partial z}\right)
$$

$$
= u\nabla \cdot (h\nabla v) = \text{the left side.}
$$

1.2. Take $h = 1$ in **1.1.**

1.3.

$$
\iint_R \nabla^2 u\,\nabla^2 w\,dx\,dy =
$$

$$
= \iint_R \left[\frac{\partial}{\partial x}\left(\nabla^2 u\frac{\partial w}{\partial x}\right) + \frac{\partial}{\partial y}\left(\nabla^2 u\frac{\partial w}{\partial y}\right)\right]dx\,dy-
$$

$$
- \iint_R \left[\frac{\partial}{\partial x}\left(w\frac{\partial}{\partial x}\nabla^2 u\right) + \frac{\partial}{\partial y}\left(w\frac{\partial}{\partial y}\nabla^2 u\right)\right] +
$$

$$
+ \iint_R w\left(\frac{\partial^2}{\partial x^2}\nabla^2 u + \frac{\partial^2}{\partial y^2}\nabla^2 u\right)dx\,dy
$$

$$
= \iint_R (\nabla^4 u)w\,dx\,dy + \int_C \nabla^2 u\left(\frac{\partial w}{\partial x}\,dy - \frac{\partial w}{\partial y}\,dx\right) -
$$

$$
- \int_C w\left(\frac{\partial(\nabla^2 u)}{\partial x}\,dy - \frac{\partial(\nabla^2 u)}{\partial y}\,dx\right)
$$

$$
= \iint_R \nabla^4 u\,w\,dx\,dy + \int_C \nabla^2 u\frac{\partial w}{\partial n}\,ds - \int_C \frac{\partial}{\partial n}\nabla^2 u\,w\,ds.
$$

1.4.

$$\iint_R \{uLv - vLu\}\, dx\, dy =$$

$$= \iint_R \{v\nabla \cdot (h\nabla u) - u\nabla \cdot (h\nabla v)\}\, dx\, dy$$

$$= \int_C h(u\frac{\partial v}{\partial n} - v\frac{\partial u}{\partial n})\, ds,$$

by divergence theorem; thus L is self–adjoint. Also

$$\iint_R uLu\, dx\, dy = \iint_R h|\nabla u|^2\, dx\, dy - \int_C hu\frac{\partial u}{\partial n}\, ds.$$

If u satisfies (1.50a) or (1.50b), then $\int_C hu\partial u/\partial n\, ds = 0$; so L is positive definite if $h > 0$. If u satisfies (1.50c), then $\int_C hu\partial u/\partial n\, ds = -\int_C huku\, ds < 0$
if $k > 0$; so L is positive definite if $h > 0$.

1.5.

$$\iint_R \{vLu - uLv\}\, dx\, dy =$$

$$= \iint_R \{v(\nabla^2 + h^2)u - u(\nabla^2 + h^2)v\}\, dx\, dy$$

$$= \iint_R (v\nabla^2 u - u\nabla^2 v)\, dx\, dy$$

$$= \int_C \left(v\frac{\partial u}{\partial n} - u\frac{\partial v}{\partial n}\right) ds,$$

by using the second form of the Green theorem. Thus L is self–adjoint.

1.11. (a) $\dfrac{4}{\pi}\left(\cos x - \dfrac{1}{3}\cos 3x - +\dfrac{1}{5}\cos 5x + \cdots\right)$

(b) $2\left(\sin x - \dfrac{1}{2}\sin 2x + \dfrac{1}{3}\sin 3x - \dfrac{1}{4}\sin 4x + - \cdots\right)$

(c) $\dfrac{\pi^2}{3} - 4\left(\cos x - \dfrac{1}{4}\cos 2x + \dfrac{1}{9}\cos 3x - \dfrac{1}{16}\cos 4x + - \cdots\right)$

1.12. (a)$\dfrac{4}{\pi}\left(\sin \pi x + \dfrac{1}{3}\sin 3\pi x + \dfrac{1}{5}\sin 5\pi x + \cdots\right)$

(b) $\dfrac{2}{3} + \dfrac{4}{\pi^2}\left(\cos \pi x - \dfrac{1}{4}\cos 2\pi x + \dfrac{1}{9}\cos 3\pi x - + \cdots\right)$

2.1.

$$b(w,u) = \int_0^1 a\frac{dw}{dx}\frac{du}{dx}\,dx,$$

$$l(w) = \int_0^1 wf\,dx - w(0)\left[a\frac{du}{dx}\right]_{x=0} + q_0 w(l).$$

2.3.

$$b(w,u) = \int_0^1 \left(a\frac{dw}{dx}\frac{du}{dx} - cwu\right)dx, \quad l(w) = -\int_0^1 wx^2\,dx + w(1).$$

2.6.

$$b(w,u) = \int_0^l a\frac{d^2w}{dx^2}\frac{d^2u}{dx^2}\,dx,$$

$$l(w) = -\int_0^l wf\,dx + w(0)\left[\frac{d}{dx}\left(a\frac{dw}{dx}\frac{du}{dx}\right)\right]_{x=0} -$$

$$-\left[\frac{dw}{dx}\right]_{x=0}\left[a\frac{du}{dx}\right]_{x=0} - f_0 w(l) + m_0\left[\frac{dw}{dx}\right]_{x=l}.$$

2.7.

$$B(w,u) = \iint_R \left[\frac{dw}{dx}\left(c_{11}\frac{du}{dx} + c_{12}\frac{du}{dy}\right) + \frac{dw}{dx}\left(c_{21}\frac{du}{dx} + c_{22}\frac{du}{dy}\right)\right.$$

$$\left. + +wf\right]dx\,dy,$$

$$l(w) = \int_C wq_n\,ds,$$

where $q_n = \left(c_{11}\frac{du}{dx} + c_{12}\frac{du}{dy}\right)n_x + \left(c_{21}\frac{du}{dx} + c_{22}\frac{du}{dy}\right)n_y.$

2.8.

$$0 = -\iint_R k\left(\frac{d^2T}{dx^2} + \frac{d^2T}{dy^2}\right)w\,dx\,dy,$$

$$= \iint_R k\left(\frac{dw}{dx}\frac{dT}{dx} + \frac{dw}{dy}\frac{dT}{dy}\right)dx\,dy - \int_C kw\left(\frac{dT}{dx}n_x + \frac{dT}{dy}n_y\right)ds.$$

The boundary conditions on $C_1 = AB$ (prescribed temperature T_0): $n_x = 0$, $n_y = -1$; on $C_2 = BC$ (convective boundary, T_∞): $n_x = 1$, $n_y = 0$; on

$C_3 = CDEFGH$ (insulated boundary): $q = \partial T/\partial n = 0$; and on $C_4 = HA$
(prescribed conduction $q_0(y)$): $n_x = -1$, $n_y = 0$. Thus,

$$b(w, u) = \iint_R k \left(\frac{dw}{dx}\frac{dT}{dx} + \frac{dw}{dy}\frac{dT}{dy} \right) dx\, dy + h \int_0^b w(a, y)T(a, y)\, dy,$$

$$l(w) = - \int_0^b w(0, y)q_0(y)\, dy + hT_\infty \int_0^b w(a, y)\, dy.$$

2.9. Let w_1 and w_2 be the two test functions, one for each equation, such that
they satisfy the essential boundary conditions on u and v. Then

$$0 = \int_0^l \left[a\frac{dw_1}{dx}\left\{ \frac{du}{dx} + \frac{1}{2}\left(\frac{dv}{dx}\right)^2 \right\} + w_1 g \right] dx,$$

$$0 = \int_0^l \left[b\frac{d^2 w_2}{dx^2}\frac{d^2 v}{dx^2} + a\frac{dw_2}{dx}\frac{dv}{dx}\left\{ \frac{du}{dx} + \frac{1}{2}\left(\frac{dv}{dx}\right)^2 \right\} + w_2 f \right] dx -$$

$$- m_0 \frac{dw_2}{dx}(l).$$

Thus,

$$b((w_1, w_2), (u, v)) = \int_0^l \left[a\frac{dw_1}{dx}\left\{ a\frac{du}{dx} + \frac{1}{2}\left(\frac{dv}{dx}\right)^2 \right\} + b\frac{d^2 w_2}{dx^2}\frac{d^2 v}{dx^2} + \right.$$

$$\left. + a\frac{dw_2}{dx}\frac{dv}{dx}\left\{ \frac{du}{dx} + \frac{1}{2}\left(\frac{dv}{dx}\right)^2 \right\} \right] dx,$$

$$l((w_1, w_2)) = - \int_0^l \left(w_1 g + w_2 f \right) dx + m_0 \frac{dw_2}{dx}(l),$$

$$I[(u, v)] = \int_0^l \left[\frac{1}{2}\left\{ a\left(\frac{du}{dx}\right)^2 + b\left(\frac{d^2 v}{dx^2}\right)^2 + a\frac{du}{dx}\left(\frac{dv}{dx}\right)^2 + \frac{a}{4}\left(\frac{dv}{dx}\right)^4 \right\} + \right.$$

$$\left. + w_1 g + w_2 f \right] dx - m_0 \frac{dw_2}{dx}(l).$$

2.10. The variation of the total work done by the force f/T is

$$\delta \iint_R \frac{fu}{T}\, dx\, dy =$$

$$= \iint_R \frac{f\delta u}{T}\, dx\, dy$$

$$= - \iint_R \nabla^2 u\, \delta u\, dx\, dy$$

$$= - \iint_R [\nabla \cdot (\nabla u \delta u) - \nabla(\delta u) \cdot \nabla u]\, dx\, dy$$

$$= \frac{1}{2}\iint_R \delta|\nabla u|^2\, dx\, dy - \int_C \frac{\partial u}{\partial n}\delta u\, ds,$$

which leads to

$$I[u] = \frac{1}{2} \iint_R \{|\nabla u|^2 - 2fu\} \, dx \, dy.$$

Note that **1.2** and Example A.2 are used. Thus,

$$\nabla \cdot (\nabla u \delta u) = \left(\hat{i} \frac{\partial}{\partial x} + \hat{j} \frac{\partial}{\partial y} \right) \cdot \left(\hat{i} \frac{\partial u}{\partial x} \delta u + \hat{j} \frac{\partial u}{\partial y} \delta u \right)$$

$$= \nabla^2 u \delta u = -\frac{f}{T} \delta u.$$

2.11.

$$b(w, u) = \iint_R \left(\frac{\partial w}{\partial x} \frac{\partial u}{\partial x} + \frac{\partial w}{\partial y} \frac{\partial u}{\partial y} \right) \, dx \, dy,$$

$$l(w) = \iint_R fw \, dx \, dy - \int_C w \frac{\partial u}{\partial n} \, ds.$$

2.13. Take $u_1 = x(1 - x)(a + bx)$. Then $a = 0.70529$, $b = 0.27973$. The exact solution is $u = 1 + (e - 1)x - e^x$.

2.15. Take $u_1(x) = x(1 - x)(a + bx)$.

2.17. Same as **2.6.** with $a = EI$. Exact solution is obtained by direct integration.

3.1. $T(4) = 82.88°\,\mathrm{C}, T(10) = 62.54°\,\mathrm{C}, q(0) = -325.28\,\mathrm{W/cm}^2, q(10) = 0$ (with $f = 0$).

3.2. $\phi(0) = 46.05$. The exact solution is $\phi = 2l \ln \left(l - \frac{x}{l} \right)$.

3.3. $u(30) = -0.361\,\mathrm{in}, \theta(30) = 0.022\,\mathrm{rad}, u(60) = -0.915\,\mathrm{in}, \theta(60) = 0.025\,\mathrm{rad}, u(90) = -2.303\,\mathrm{in}, \theta(90) = 0.038\,\mathrm{rad}, s(0) = 7000\,\mathrm{lb}, s(90) = -562500\,\mathrm{lb}, m(0) = 0.0002625\,\mathrm{lb\,in}.$

3.4. $u(75) = -0.97\,\mathrm{in}, \theta(75) = 0.021\,\mathrm{rad}, u(150) = -2.64\,\mathrm{in}, \theta(150) = 0.025\,\mathrm{rad}, s(0) = 23000\,\mathrm{lb}, m(0) = 2.3 \times 10^6\,\mathrm{lb\,in}.$

3.5. $u(10) = -0.716\,\mathrm{m}, \theta(5) = 0.065\,\mathrm{rad}, \theta(10) = 0.169\,\mathrm{rad}, m(7.5) = -20832\,\mathrm{N\cdot m}.$

3.6. $s_1 = s_2 = -6, m_2 = -60, \theta_1 = -300.$

5.1. (a) By symmetry, consider the half–region $\{0 \le x \le 1/2, 0 \le y \le 1\}$. Then, with $u(0, 1) = 1$, we get $u(0.25, 0.25) = 0.0714, u(0.25, 0.5) = 0.1875, u(0.25, 0.75) = 0.4285, u(0.5, 0.25) = 0.0982, u(0.5, 0.5) = 0.25, u(0.5, 0.75) = 0.5268.$

(b) $u(0.25, 0.25) = 0.5834$, $u(0.25, 0.5) = 0.1508$, $u(0.25, 0.75) = 0.3318$,
$u(0.5, 0.25) = 0.0825$, $u(0.5, 0.5) = 0.2134$, $u(0.5, 0.75) = 0.4693$.

5.7. Take N=12, L=5; write X, Y, Xi, Yi;

    Code: 0 0 0 1 1 1 0 0 0 1 1 1
    Bc: −250.0 − 183.2 − 95.4 21.0 48.3 48.3 21.0
    95.4 183.2 250.0 − 21.0 − 48.3 − 48.3 − 21.0

Let the side of the square be unity. Then
$u(0.25, 0.25) = 170.3 = u(0.25, 0.75)$, $u(0.5, 0.5) = 125.0$,
$u(0.75, 0.25) = 91.8 = u(0.75, 0.75)$.

5.8. Input:

    Problem 5.8
    12 5
    0. 0.   3. 0.   6. 0.   9. 0.   9. 2.   9. 4.
    9. 6.   6. 6.   3. 6.   0. 6.   0. 4.   0. 2.
    0 0.   0 0.   0 0.   0 50.   0 50.   0 50.
    0 0.   0 0.   0 0.   0 50.   0 50.   0 50.
    2.25 1.5   6.75 1.5   7.75 4.5   2.25 4.5   4.5 3.

The exact solutions are: $u(2.25, 1.5) = 19.758 = u(6.75, 1.5) = u(6.75, 4.5) = u(2.25, 4.5)$, and $u(4.5, 3.0) = 18.562$.

5.11. $T(A) = 327$, $T(B) = 307$, $T(C) = 302$, $T(D) = 300$.

5.17. (a) Take $a = 1$, $v_0 = 5$. Then $u(-1/\sqrt{2}, 0) = -5 = u(0, -1/\sqrt{2})$, $u(1/\sqrt{2}, 0) = 5 = u(0, 1/\sqrt{2})$, $u(0, 0) = 0.08$. The exact solution is

$$u(x, y) = \frac{\sqrt{2}v_0}{a}(x + y) - $$
$$- \frac{4v_0}{\pi} \sum_{n=1}^{\infty} \left((-1)^n \sinh \frac{n\pi[a - \sqrt{2}(y - x)]}{2a} + \right.$$
$$\left. + \sin \frac{n\pi[a + \sqrt{2}(y - x)]}{2a} \right) \cdot \frac{\sin \frac{n\pi[a - \sqrt{2}(y - x)]}{2a}}{n \sinh n\pi}.$$

(b) Exact solution:

$$u(x, y) = v_0 \left[\frac{x}{a} + \frac{2}{\pi} \sum_{n=1}^{\infty} \frac{\cosh(n\pi y/b)}{\cosh(n\pi a/b)} \frac{\sin(n\pi x/a)}{n} \right].$$

5.18. Data: $J = 1.198 \times 10^3$A (total current); $\sigma = 0.599 \times 10^8 \Omega^{-1} \cdot m^{-1}$ (conductivity); $h = 0.1$ cm $= 1 \times 10^{-3}$ m (sheet thickness); $\varepsilon = 0.1$ cm; $a = 5$ cm; $b = 3$ cm. Thus $A = -10$A $\cdot \Omega \cdot m^{-1}$. Consider the upper half region. Then $u(0, 0) = 71.64$, $u(2.5, 0) = 71.49$, $u(5, 0) = 71.97$,

$u(0,0.1) = 71.71$, $u(2.5,0.1) = 71.35$, $u(5,0.1) = 72.65$, $u(0,3) = 71.49$, $u(2.5,3) = 71.87$, $u(5,3) = 71.73$.

5.19. Take the top side as 4 units and the origin of the coordinates at the left bottom corner. Then $u(A) = 0.99975$, $u(B) = 2.98199$, $u(C) = 1.96537$, $u(D) = 0.99362$. The exact solution is $u = x$.

5.23. With origin of coordinates at the lower left corner of Fig. 5.9(b), $T(1.5,0) = 50.32°$ C; $T(3.5,1.5) = 64.19°$ C.

6.2. The input file is:

```
12  9  1  12  0  0  0  94500  0.1
3.0  0.0
2.1213203 1.0606601 −3.0  0.0
−2.1213203  − 1.0606601  2.8284271  1.4142135  − 2.8284271
−1.4142135
5.0  0.0  10.0  0.0  20.0  0.0
1.7320508  −0.5  1.0  −0.8660254  0.0  −1.0  −1.0  −0.8660254
−1.7320508  − 0.5  − 2.0  0.0  − 1.7320508  0.5  − 1.0
0.8660254  0.0  1.0  1.0  0.8660254  1.7320508  0.5  1.0  0.0
1  86.60  1  − 50.00
1  50.00  1  − 86.60
0  0.0  1  − 100.00
1  − 50.00  1  − 86.60
1  − 86.60  1  − 50.00
1  − 100.00  1  0.0
1  − 86.60  1  50.00
1  − 50.00  1  86.00
0  0.0  1  100.00
1  50.00  1  86.00
1  86.00  1  50.00
1  100.00  0  0.0
```

6.6. $u(120,0) = 11.2 \times 10^{-4}$, $v(120,0) = 1.9 \times 10^{-4}$, $u(120,160) = 10.1 \times 10^{-4}$, $v(120,160) = 1.1 \times 10^{-4}$.

6.7. The plate has dimensions 8×2 in^2. Because of symmetry, consider the upper right rectangle. The boundary conditions are: $u = 0$, $t_y = 0$ on the side $x = 0$, $0 \le y \le 1$; $u = -1$, $t_y = 0$ on the side $x = 1$, $0 \le y \le 1$; $v = 0$, $t_x = 0$ on the sides $y = 0, y = 1$ and $0 \le x \le 4$. Then $u(1,0) = u(1,0.5) = u(1,1) = -0.25$, $u(2,0) = u(2,0.5) = u(2,1) = -0.25$, $u(3,0) = u(3,0.5) = u(3,1) = -0.075$.

6.9. Because of symmetry, consider the right top quarter region. Dispacements: $u(1,0) = u(1,0.5) = u(1,1) = -0.025$, $u(2,0) = u(2,0.5) = u(2,1) = -0.05$, $u(3,0) = u(3,0.5) = u(3,1) = -0.075$, $u(4,0) = u(4,0.5) = u(4,1) = -0.1$, and $v = 0$ at all points. Exact solution;

$u = x\delta/l$, $\delta = 0.2$, $l = 8$. Stresses (with $\nu = 0.3$): $\sigma_x = -0.0274E$, $\sigma_y = -0.008E$, $\tau_{xy} = 0$. Exact stresses are given by

$$\sigma_x = -\frac{E\delta}{(1 - \nu^2)l}, \quad \sigma_y = -\frac{\nu E\delta}{(1 - \nu^2)l}, \quad \tau_{xy} = 0.$$

6.15. Timoshenk and Goodier (page 80) give the stress concentration factor $\max \sigma_\theta = 3.0$ for an infinite plate, i.e., the maximum tensile stress is three times the uniform stress. This value increases to 3.024 for the finite plate in this problem.

6.18. See Timoshenko and Goodier (page 391, twist of a circular ring sector). The stress function ϕ is defined in the polar cylindrical coordinate system by

$$\frac{\partial^2 \phi}{\partial r^2} + \frac{\partial^2 \phi}{\partial z^2} - \frac{3}{r}\frac{\partial \phi}{\partial r} + 2c = 0.$$

Introducing the transformation $\xi = R - r$, $\zeta = z$, and assuming $\phi = \phi_0 + \phi_1 + \cdots$, the first order approximation is governed by

$$\frac{\partial^2 \phi}{\partial \xi^2} + \frac{\partial^2 \phi}{\partial \zeta^2} + 2c = 0,$$

which can be solved by the subroutine Bell. The analytical solution gives

$$(\tau_{r\theta})_0 = G\frac{\partial \phi_0}{\partial \zeta}, \quad (\tau_{\theta z})_0 = G\frac{\partial \phi_0}{\partial \xi}.$$

The equation of the boundary is $\xi^2 + \zeta^2 = a^2$, thus giving $\phi_0 = -c(\xi^2 + \zeta^2 - a^2)/2$, hence $(\tau_{r\theta})_0 = -cG\zeta$, and $(\tau_{\theta z})_0 = -cG\xi$.

6.19.

Point (x, y)	u_r	u_z	σ_{rr}	σ_{zz}	σ_{rz}
$(1.48, 0.26)$	-0.953	-0.168	-3.36	-2.41	-0.17
$(1.41, 0.51)$	-0.911	-0.331	-3.27	-2.49	-0.32
$(1.3, 0.75)$	-0.838	-0.484	-3.13	-2.63	-0.44
$(1.15, 0.96)$	-0.742	-0.622	-2.97	-2.79	-0.5

7.19. The characteristic wavenumbers for a box–like structure, defined by $|x| \le a, |y| \le b, |z| \le c$, and subject to the homogeneous boundary conditions $p(\pm a, y, z) = 0$, $p(x, \pm b, z) = 0$, $p(x, y, \pm c) = 0$, are given by

$$k_{lmn} = \frac{\pi}{2}\sqrt{\left(\frac{l}{a}\right)^2 + \left(\frac{m}{b}\right)^2 + \left(\frac{n}{c}\right)^2},$$

where l, m, n are positive integers. Fir a cube of side $s = 2a = 2b = 2c$, $k_{lmn} = \dfrac{\pi}{s}\sqrt{l^2 + m^2 + n^2}$. If $s = \lambda/4 = \pi/2k$, then the wavenumber k is less than the first characteristic wavenumber $k_{111} = 2\sqrt{3}k$. The computation can be repeated for $s = \sqrt{3}\lambda/2 = \sqrt{3}\pi/k$, which yields $k_{111} = k$. The box–like structure can be partitioned into cubes of side s and the results superposed. Schenck's work describes the arrays of cubes and the spectrum of characteristic wavenumbers for each box array.

8.16. $\psi(1.5, 1) = 0.939$, $\psi(3, 1) = 0.604$, $\psi(3.5, 1.5) = 1.086$, $\psi(4, 1.5) = 1.043$

$\phi(0,0) = 4.865$, $\phi(0,1) = 4.894$, $\phi(0,2) = 4.837$, $\phi(1.5,0) = 3.402$, $\phi(1.5,1) = 3.345$, $\phi(1.5,2) = 3.309$, $\phi(3,0) = 2.194$, $\phi(3,1) = 1.792$, $\phi(3,2) = 1.622$, $\phi(3.5,1) = 1.206$, $\phi(3.5,1.5) = 0.931$.

9.7. (a) With $f = 1 + r$

Node	u	u_{exact}
17 (1.5, 0)	0.188	0.187
18 (1.2, .35)	0.181	0.177
19 (.6, .45)	0.124	0.121
20 (0, .45)	0	0
29 (.9, 0)	0.208	0.205
30 (.3, 0)	0.088	0.083
31 (0, 0)	0	0

Node	q	q_{exact}
1 (2, 0)	−0.497	−0.571
2 (1.706, .522)	−0.635	−0.620
3 (1.179, .808)	−0.586	−0.556
4 (.598, .954)	−0.334	−0.326
5 (0, 1)	0	0

$$\text{Exact } u = -\frac{2x}{7}\left(\frac{x^2}{4} + y^2 - 1\right),$$

$$\text{Exact } q = -\frac{x}{14}\left(\frac{3x^2}{2} + 2y^2 - 2\right) - \frac{4xy^2}{7}.$$

(b)

$$\text{Exact } u = -\frac{1}{123}\left(25x^2 - 4y^2 + \frac{84}{5}\right)\left(\frac{x^2}{4} + y^2 - 1\right),$$

$$\text{Exact } q = -\frac{1}{246}\left(-25x^4 + 2x^3y - 50x^2y^2 + \frac{208}{5}x^2\right) -$$

$$-\frac{1}{123}\left(50x^2y^2 - 2xy^3 - 16y^4 + \frac{208}{5}x^2\right).$$

9.10. The results, using (9.59), with $N = 5$ are as folloes:

Shear stress	A	B	C	D	E
τ_{zy}	104.8	204.8	156.7	63.6	91.8
τ_{zx}	0	0	−271.4	−90.8	−53.0

10.1. Part (a) is direct, by substitution. For part (b), note that

$$\frac{\partial}{\partial t'}[uu^*] = u^*\frac{\partial u}{\partial t'} + u\frac{\partial u^*}{\partial t'} = k\left(u^*\nabla^2 u - u\nabla^2 u^*\right).$$

If we integrate both sides of this equation over the cylinder $C \times (t_0, t)$, we get

$$\int_{t_0}^t \iint_R \frac{\partial}{\partial t'}[uu^*]\, dA(\mathbf{x'})\, dt' = k\int_{t_0}^t \iint_R \left(u^*\nabla^2 u - u\nabla^2 u^*\right)\, dA(\mathbf{x'})\, dt',$$

$$(*)$$

where $dA(\mathbf{x'}) = dx'\, dy'$. Now, using (1.8),

$$\iint_R \left(u^*\nabla^2 u - u\nabla^2 u^*\right)\, dA(\mathbf{x'}) = \int_C (u^*q - uq^*)\, ds(\mathbf{x'}),$$

where $ds(\mathbf{x'})$ is the differential on the line integral along C from the source point $\mathbf{x'}$. Thus,

right side of Eq (*) $=$

$$= k\int_{t_0}^t \iint_R \left(u^*q - uq^*\right)\, ds(\mathbf{x'})\, dt' = k\int_C \int_{t_0}^t \left(u^*q - uq^*\right)\, dt'\, ds(\mathbf{x'}),$$

left side of Eq (*) $=$

$$= \iint_R u(\mathbf{x}, t')u^*(\mathbf{x}, t; \mathbf{x'}, t')\, dA(\mathbf{x'}) - \iint_R u(\mathbf{x}, t_0)u^*(\mathbf{x}, t; \mathbf{x'}, t_0)\, dA(\mathbf{x'})$$

$$= \iint_R u(\mathbf{x}, t)\delta(\mathbf{x} - \mathbf{x'})\, dA(\mathbf{x'}) - \iint_R u(\mathbf{x}, t_0)u^*(\mathbf{x}, t; \mathbf{x'}, t_0)\, dA(\mathbf{x'}),$$

where we have used (10.7) in the first integral over R. If we use the notation

$$c(\mathbf{x}) = \iint_R \delta(\mathbf{x}, \mathbf{x}')\, dA(\mathbf{x}'),$$

$$u_0^* \equiv u_0^*(\mathbf{x}, t; \mathbf{x}') = u_0^*(\mathbf{x}, t; \mathbf{x}'' t_0), \quad u_0 = u(\mathbf{x}, t_0),$$

then equating both sides of Eq (*), we obtain (10.8).

10.4. Use (10.53), which leads to

$$\left\{\begin{matrix} p(0,t) \\ p(1^-, t) \end{matrix}\right\} = \left\{\begin{matrix} p_1 \\ p_2 \end{matrix}\right\} = \begin{bmatrix} -u_1^* & u_2^* \\ -u_2^* & u_1^* \end{bmatrix} \star \left\{\begin{matrix} v_1 \\ v_2 \end{matrix}\right\} +$$

$$+ \begin{bmatrix} 1/2 & (1/2)\,\mathrm{erfc}(1/2\sqrt{t}) \\ \mathrm{sym} & 1/2 \end{bmatrix} \left\{\begin{matrix} p_1 \\ p_2 \end{matrix}\right\} + 2b \left\{\begin{matrix} 1 \\ 1 \end{matrix}\right\} \mathrm{erf}\left(\frac{1}{2\sqrt{t}}\right),$$

with $u_1^* = u^*(0, t; 1^-, \tau)$ and $u_2^* = u^*(1, t; 1^-, \tau)$, where u^* is defined by (10.46) (see the last result in Example 10.1 also).

10.6. With $\Delta t = 0.5$, and four linear boundary elements on the circular boundary, take the ambient temperature as zero, and assume all material constants unity, except the heat transfer coefficient $h = 0.2$. Then

t	u
1	8.782
2	8.567
3	8.235
4	7.976
5	7.815

A.1. From (A.16), integrating by parts twice and using $\phi(x)\dfrac{\partial F}{\partial U'}\Big|_a^b = 0$, $\phi'(x)\dfrac{\partial F}{\partial U'} - \phi(x)\dfrac{d}{dx}\left(\dfrac{\partial F}{\partial U''}\right)\Big|_a^b = 0$, we get

$$0 = \frac{\partial I}{\partial \alpha} = \int_a^b \phi(x)\left[\frac{\partial F}{\partial u} - \frac{d}{dx}\left(\frac{\partial F}{\partial u'}\right) + \frac{d^2}{dx^2}\left(\frac{\partial F}{\partial u''}\right)\right] dx.$$

A.2. Introduce two functions $\phi(x)$ and $\psi(x)$, and two parameters α and β respectively, such that $U = u + \alpha\phi(x)$, $V = v + \beta\psi(x)$. Then $\partial I/\partial\alpha = 0$ and $\partial I/\partial\beta = 0$ lead to (A.20).

A.3. (a) Straight lines $y = c_1 x + c_2$; (b) Straight lines $r\cos(\theta - c_1) = c_2$; (c) $z = c_1 t + c_2$.

A.5. (a) $y = c_1 e^x + c_2 e^{-x}$; (b) $y = \left(c_1 e^{x/\sqrt{2}} + c_2 e^{-x/\sqrt{2}} \right) \cos(y/\sqrt{2}) + \left(c_3 e^{x/\sqrt{2}} + c_4 e^{-x/\sqrt{2}} \right) \sin(y/\sqrt{2})$.

A.8. $(x - c_1)^2 + (y - c_2)^2 = k^2$, where c_1, c_2 and k make the arc C pass through A and B and have length l.

Index